The Sociology of Health, Illness, and Health Care

A Critical Approach

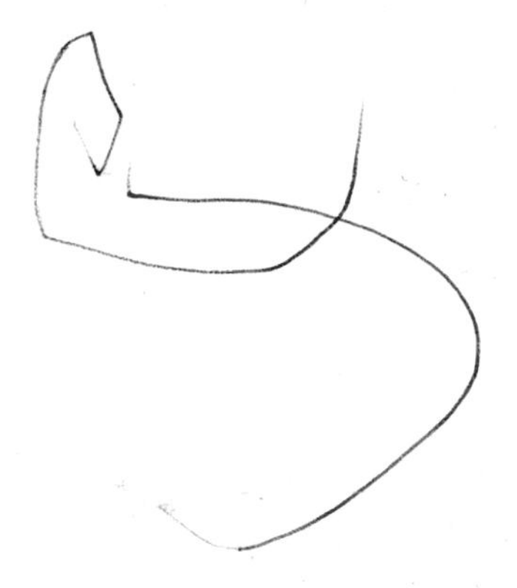

The Sociology of Health, Illness, and Health Care

A Critical Approach

SEVENTH EDITION

ROSE WEITZ
Arizona State University

Australia • Brazil • Mexico • Singapore • United Kingdom • United States

***The Sociology of Health, Illness, and Health Care: A Critical Approach*, Seventh Edition**

Rose Weitz

Product Director: Marta Lee-Perriard

Product Manager: Elizabeth A. Beiting-Lipps

Content Developer and Project Manager: Cheri Palmer

Product Assistant: Chelsea Meredith

Marketing Manager: Kara Kindstrom

Art Director: Vernon Boes

Manufacturing Planner: Judy Inouye

Production Service: MPS Limited

Photo Researcher: Lumina Datamatics

Text Researcher: Lumina Datamatics

Copy Editor: Jill Pellarin

Illustrator: MPS Limited

Cover Designer: Paula Goldstein

Cover Images: Doctors pushing patient on gurney: Ryan McVay/ Getty Images. Seniors doing stretching exercises: vm/ iStockphoto.com. Heart rate monitor in hospital: fivepointsix/ Fotolia LLC. Woman in wheelchair with yellow Labrador: Don Farrall/ Photodisc/Getty Images.

Compositor: MPS Limited

WCN: 01-100-101

For product information and technology assistance, contact us at **Cengage Learning Customer & Sales Support, 1-800-354-9706**.

For permission to use material from this text or product, submit all requests online at **www.cengage.com/permissions**.

Further permissions questions can be e-mailed to **permissionrequest@cengage.com**.

Library of Congress Control Number: 2015937109

ISBN: 978-1-305-58370-2

Cengage Learning
20 Channel Center Street
Boston, MA 02210
USA

Cengage Learning is a leading provider of customized learning solutions with employees residing in nearly 40 different countries and sales in more than 125 countries around the world. Find your local representative at **www.cengage.com**.

Cengage Learning products are represented in Canada by Nelson Education, Ltd.

To learn more about Cengage Learning Solutions, visit **www.cengage.com**.

Purchase any of our products at your local college store or at our preferred online store **www.cengagebrain.com**.

Printed in the United States of America
Print Number: 01 Print Year: 2015

In memory of my mother, Lilly Weitz, with love

Brief Contents

Contents

Preface

The sociology of health, illness, and health care has changed dramatically over the past few decades. Although the field was started primarily by sociologists who worked closely with doctors, taking for granted doctors' assumptions about health and health care, and primarily asking questions that doctors deemed important, by the 1970s the field had begun shifting toward a very different set of questions. Some of these new questions challenged doctors' assumptions, whereas others focused on issues that lay outside most doctors' areas of interest or expertise, such as how poverty affects health or how individuals develop meaningful lives despite chronic illness.

I entered graduate school during this shift, drawn by the opportunity to study how health and illness are socially created and defined and how gender; ethnicity; social class; and, more broadly, power affect both the health care system and individual experiences of health and illness. As a result, over the years I have researched such topics as how medical values affect doctors' use of genetic testing, how midwives and doctors have battled for control over childbirth, and how social ideas about AIDS affect the lives of those who have this disease.

Although I had no trouble incorporating this new vision of the sociology of health, illness, and health care into my research, I consistently found myself frustrated by the lack of a textbook that would help me incorporate it into my teaching. Instead, most textbooks still seemed to reflect older ideas about the field and to take for granted medical definitions of the situation. Most basically, the books assumed that doctors define illness according to objective biological criteria, and therefore failed to question whether political and social forces underlie the process of defining illnesses. Similarly, most textbooks ignored existing power relationships rather than investigating the sources, nature, and health consequences of those relationships. For example, the textbooks gave relatively little attention to how doctors gained control over health care or how the power of the more developed nations has affected health in less developed nations. As a result, the available textbooks used sociology primarily to answer questions posed by those working in the health care field, such as what social factors lead to heart disease and why patients might ignore their doctors' orders. Consequently, these textbooks often seemed to offer a surprisingly unsociological perspective, with their coverage of some topics differing little from coverage of those topics in health education textbooks.

Because the textbooks available when I first began working on this book often avoided critical questions about health, illness, and health care, those textbooks seemed unlikely to encourage students to engage with the materials and to question either the presented materials or their own assumptions, such as the assumptions that the United States has the world's best health care system; that medical advances explain the modern rise in life expectancy; or that all Americans receive the same quality of health care regardless of their ethnicity, gender, or social class. Instead, the textbooks primarily gave students already-processed information to memorize.

My purpose in writing this textbook was to fill these gaps by presenting a critical approach to the sociology of health, illness, and health care. This did not mean presenting research findings in a biased fashion or presenting only research that supported my preexisting assumptions, but it did mean using critical skills to interpret the available research and to pull it together into a coherent "story" in each chapter. In addition, I hoped to tell these stories in a manner that would engage students—whether in sociology classes, medical schools, or nursing schools—and encourage them to learn actively and think independently. These remain the primary goals of this seventh edition. Both of these goals led me to decide against trying to please all sides or cover all topics, as I believe such a strategy leads to a grab-bag approach that makes textbooks hard to follow and to an intellectual homogenization that makes them seem lifeless.

THE CRITICAL APPROACH

The critical approach, as I have defined it, means using the "sociological imagination" to question previously taken-for-granted aspects of social life. For example, most of the available textbooks in the sociology of health, illness, and health care still view patients who do not comply with prescribed medical regimens essentially through doctors' eyes, starting from the assumption that patients should do so. More broadly, previous textbooks have highlighted the concept of a sick role—a concept that embodies medical and social assumptions regarding "proper" illnesses and "proper" patients and that downplays all aspects of individuals' lives other than the time they spend as patients.

In contrast, I emphasize recent research that questions such assumptions. For example, I discuss patient compliance by examining how patients view medical regimens and compliance, why doctors sometimes have promoted medical treatments (such as hormone therapy for menopausal women) that later proved dangerous and how doctors' tendency to cut short patients' questions can foster patient noncompliance. Similarly, this textbook explains the concept of a sick role but pays more attention to the broader experience of illness—a topic that has generated far more sociological research than has the sick role over the past 20 years.

CHAPTER ORGANIZATION

This textbook demonstrates the breadth of both micro- and macro-level topics that constitute the sociology of health, illness, and health care. Part I discusses the role social factors play in fostering illness and in determining which social groups experience which illnesses. Chapter 1 offers an introduction to the field, the sociological approach, and the history of disease. Chapter 2 describes the major causes of preventable deaths in the United States, demonstrating how social as well as biological factors affect health and illness. Building on this basis, Chapter 3 describes how age, sex, gender, social class, race, and ethnicity affect the likelihood, nature, and consequences of illness in the United States. Finally, Chapter 4 explores the nature and sources of illness in the poorer countries of Asia, Africa, and Latin America.

Part II analyzes the meaning and experience of illness and disability in the United States. Chapter 5 explores the social meanings of illness and social explanations for illness, as well as the social consequences of defining behaviors and conditions as illnesses. With this as a basis, Chapter 6 first explores the meaning of disability and then offers a sociological overview of the experience of living with chronic pain, chronic illness, or disability, including the experience of seeking care from both medical doctors and alternative health care providers. Finally, Chapter 7 provides a parallel assessment of mental illness.

Part III moves the analysis to the macro-level. Chapter 8 describes the US health care system, the likely impact of the 2010 Patient Protection and Affordable Care Act, and the continuing crises in health care costs and accessibility. Chapter 9 begins by suggesting some basic measures for evaluating health care systems and then uses these measures to evaluate the systems found in Canada, Great Britain, Germany, the People's Republic of China, Mexico, and the Democratic Republic of Congo. Finally, Chapter 10 examines four common health care settings—hospitals, hospices, nursing homes, and family homes—and provides a social analysis of the technologies used in those settings.

Part IV shifts the focus from the health care system to health care providers. Chapter 11 analyzes the nature and source of doctors' professional status as well as the threats to that status. The chapter also describes the process of becoming a doctor, the values embedded in medical culture, and the impact of those values on doctor–patient relationships. Chapter 12 describes the history and social position of various health care occupations, including dentistry, nursing, osteopaths, and acupuncturists. Finally, Chapter 13 presents a sociological overview of bioethics.

COVERAGE

Although I have tried in this book to present a coherent critical view, I have not sacrificed coverage of topics professors have come to expect. Consequently, this book covers essentially all the topics—both micro level and macro level—that have become standard over the years, including doctor–patient relationships, the nature of the US health care system, and the social distribution of illness. In addition, I include several topics that until recently received relatively little coverage in other textbooks in the field, including bioethics, mental illness, the medical value system, the experience of illness and disability, and the social sources of illness in both more and less developed nations. As a result, this text includes more materials than most teachers can cover effectively in a semester. To assist those who choose to skip some chapters, each important term is printed in bold the first time it appears in *each* chapter, alerting students that they can find a definition in the book's Glossary. (Each term is both printed in bold and defined the first time it appears in the book.)

In addition, reflecting my belief that sociology neither can nor should exist in isolation, but must be informed by and in turn inform other related fields, several chapters begin with historical overviews. For example, the chapter on health care institutions discusses the political and social forces that led to the development of the modern hospital, and the chapter on medicine as a profession discusses how and why the status of medicine grew so dramatically after 1850. These discussions provide a context to help students better understand the current situation.

CHANGES IN THE SEVENTH EDITION

Throughout the textbook, I have thoroughly updated not only statistics and discussions of topical issues (such as health care reform or the health consequences of US military engagements overseas) but also all reviews of the theoretical and empirical literature. As a result, two-thirds of references in this new edition are from 2000 or later, and fewer than 10 percent are from books or articles written before 1990—a level of timeliness that significantly surpasses that of most textbooks. The reader can thus safely assume that, wherever possible, the statistics, policy summaries, and legal information are the latest available.

New and Updated Chapter Topics

Chapter 1

- "Big Data"

Chapter 3

- Cumulative inequality

Chapter 4

- Structural violence
- Global health

Chapter 5

- Epigenetic effects

Chapter 6

- Self-diagnosis

Chapter 7

- Social capital

Chapter 8

- Chapter reorganized in wake of Affordable Care Act
- Health care costs and the Affordable Care Act
- Health care access and the Affordable Care Act
- Neoliberalism
- Updated discussion of state-level health care reform

Chapter 9

- Updated statistics and descriptions of health care in five nations

Chapter 10

- Updated statistics on hospice care, hospitals, nursing homes, and home care
- The shift away from hospitals

Chapter 11

- Cultural health capital
- Updated discussion on gender, class, and ethnicity in medical education

Chapter 12

- Profession of dentistry
- Updated statistics on each health care occupation

Chapter 13

- Chapter reorganized to highlight contemporary issues

New or Revised Tables and Figures

- Figure 2.1 Substance Use in Past 30 Days Among US High School Seniors
- Figure 2.2 Obesity Among US Children and Adults
- Table 2.1 Main Causes of Deaths, 1900 and 2011
- Table 2.2 Underlying Causes of Premature Death in the United States
- Figure 3.1 Life Expectancy by Race/Ethnicity and Sex
- Table 3.1 Infant Mortality Rates in Different Nations and US Ethnic Groups
- Table 3.2 Top Causes of Death by Ethnicity

- Map 4.1 Lifetime Prevalence of Intimate Partner Violence against Women
- Table 4.1 Life Expectancy and Infant Mortality by Development Level
- Table 4.2 Leading Causes of Death around the World
- Table 6.1 Percentage of Americans with Basic Activity Limitations
- Table 7.1 Sex, Ethnicity, and Social Class Groups with the *Highest* Lifetime Risks of Specific Mental Illnesses
- Figure 7.1 Antidepressant Use in the Past 30 Days, United States
- Table 8.1 Citizens' Views on and Experiences with Health Care
- Figure 8.1 Health Expenses and Inpatient Days in Acute Care Hospitals in 30 Nations
- Figure 8.2 Health Expenses and Number of Doctor Visits in 30 Nations
- Figure 8.3 Health Expenses and Life Expectancy in 30 Nations
- Table 9.1 Characteristics of Health Care Systems in Seven Countries
- Figure 11.1 Median Salaries by Percentage Women in Specialty
- Figure 11.2 Geographic Variations in Use of Spinal Fusion for Back Pain

New Feature Content

Chapter 2

- Key Concepts: Understanding Rates
- Contemporary Issues: Poverty and Ebola Virus Disease

Chapter 4

- Key Concepts: Structural Violence
- Contemporary Issues: Toilets and Malnutrition
- Chapter opener: Insomnia as Disease

Chapter 5

- Contemporary Issues: Citizenship and Biomedicine

Chapter 6

- Contemporary Issues: Mobile Digital Health Devices

Chapter 7

- Ethical Debate: Mental Illness and Gun Control

Chapter 10

- Ethical Debate: "Savior Siblings"
- Contemporary Issues: Newborn Screening

Chapter 11

- Ethical Debate: A Duty to Care?

Chapter 13

- Ethical Debate: The Right to Refuse to Treat

PEDAGOGICAL FEATURES FOR STUDENTS

Learning Objectives

Each chapter opens with a list of learning objectives matched to the chapter's main sections. These objectives help students to focus their studying by alerting them to the chapter's main themes. The objectives also can help students demonstrate their ability to apply what they have learned and can help both students and faculty to assess students' understanding.

Chapter Openings

Unfortunately, many students take courses only to fill a requirement. As a result, the first problem professors face is interesting students in the topic. For this reason, the main text of each chapter begins with a vignette taken from a sociological or literary source and chosen to spark students' interest by demonstrating that the topic has real consequences for real people—that, for example, stigma is not simply an abstract concept, but something that can cost ill persons their friends, jobs, and social standing.

Chapter "Roadmaps"

To help orient students to the chapters, each chapter's introductory section ends with a brief overview of what is to come.

Contemporary Issues

To further raise student interest and add to their knowledge, most chapters include a boxed discussion of a relevant topic taken from recent news reports. Topics include the debate over full-body computed tomography scans and the decline of primary care. These boxes should spark student interest while helping them to make connections between textbook topics and the world around them.

Ethical Debates

To teach students that ethical dilemmas pervade health care, most chapters include a discussion of a relevant ethical debate. The debates are complex enough that students must use critical thinking skills to assess them; teachers can use these debates for classroom discussions, group exercises, or written assignments.

Key Concepts

To help students understand particularly important and complex topics, such as the difference between the sociological and medical models of illness or the strengths and weaknesses of the sick role model, I have included Key Concepts tables or boxes in several chapters.

Implications Essays

Each chapter ends with a brief essay that discusses the implications of the chapter and points the reader toward new questions and issues. These essays should stimulate critical thinking and can serve as the basis for class discussions.

Chapter Summaries

Each chapter ends with a detailed, bulleted summary that will help students to review the material and identify key points.

Review Questions and Critical Thinking Questions

Each chapter includes both Review Questions that take students through the main points of the chapter and Critical Thinking Questions that push students to extrapolate from the chapter to other issues or to think more deeply about issues discussed within the chapter.

Glossary

The book includes an extensive Glossary that defines all important terms used in the book. Each Glossary term is printed in bold and defined the first time it appears in the text. In addition, each term is also printed in bold the first time it appears in *each* chapter, so students will know that they can find a definition in the Glossary.

SUPPLEMENTAL AND PEDAGOGICAL FEATURES FOR FACULTY

Instructor's Manual with Test Bank

For each chapter, the *Instructor's Manual* contains a detailed summary; a set of multiple-choice questions; and list of relevant books, narrative films, and documentaries. In addition, the *Instructor's Manual* includes several questions for each chapter that require critical-thinking skills to answer and that teachers can use for essay exams, written assignments, in-class discussions, or group projects. The manual also includes for each chapter a set of Internet exercises designed both to familiarize students with materials available on the Web and to facilitate critical reading and use of those materials. Finally, the manual lists for each chapter a few relevant nonprofit organizations. Organizations listed in the manual can serve as sources for more information or as sites for out-of-class assignments.

The Test Bank contains 25 to 30 multiple choice and three to five essay questions per chapter, fully updated according to match the seventh edition's content. Each question includes page references to the textbook.

To guarantee the quality of the *Instructor's Manual with Test Bank*, I wrote everything in it rather than relying on student assistants. The manual is available for downloading at http://login.cengage.com.

PowerPoint Lectures

PowerPoint lectures for each chapter, including all tables and figures, can be downloaded from this textbook's website. These lectures should prove useful both for new adopters and for past users who would like to incorporate more visual materials into their classrooms. As with the *Instructor's Manual*, I put these lectures together myself to ensure their quality.

Critical Thinking

In this textbook, I have aimed not only to present a large body of data in a coherent fashion but also to create an intellectually rigorous textbook that will stimulate students to think critically. I have tried to keep this purpose in mind in writing each chapter. Debates discussed within the chapters, as well as the various chapter features, all encourage students to use critical thinking, and all serve as resources that teachers can use in building their class sessions.

ACKNOWLEDGMENTS

In writing this textbook, I have benefited enormously from the generous assistance of my colleagues. I am very fortunate to have had several exceptional scholars as colleagues over the years—Victor Agadjanian, Jill Fisher, Verna Keith, Bradford Kirkman-Liff, Jennie Jacobs Kronenfeld, and Deborah Sullivan—who shared my interest in health issues and helped me to improve various chapters. I am also exceptionally fortunate to have had the assistance of several research assistants—Ashley Fenzl, Allison Hickey, Ann Jensby, Melinda Konicke, Christopher Lisowski, Stephanie Mayer, Leslie Padrnos, Zina Schwartz, Diane Sicotte, Lisa Tichavsky, Caroleena Von Trapp, and especially Karl Bryant, Lisa Comer, and Amy Weinberg, who worked on the first edition.

Because, of necessity, this textbook covers a wealth of topics that range far beyond my own areas of expertise, I have had to rely heavily on the kindness of strangers in writing it. One of the most rewarding aspects of writing this book has been the pleasure of receiving information, ideas, critiques, and references from individuals I did not previously know. This edition was undoubtedly improved by suggestions from Astrid Eich-Krohm (University Hospital Magdeburg),

Krista Hodges (University of Hawaii), Michael Polgar (Penn State University), Jennifer Schumann, Ian Shaw (University of Nottingham), Lisa Strohschein (University of Alberta), Julia Stumkat, Diane Kholos Wysocki (University of Nebraska–Kearney), and Wei Zhang (University of Hawaii).

In addition; I would like to take this opportunity to once again thank those who gave me the benefit of their expert advice on previous editions: Emily Abel (University of California–Los Angeles); James Akré (World Health Organization); Ellen Annandale (University of Leicester); Ofra Anson (Ben Gurion University of the Negev); Judy Aulette (University of North Carolina; Charlotte); Miriam Axelrod; James Bachman (Valparaiso University); Kristin Barker (University of New Mexico); Paul Basch (Stanford University); Phil Brown (Brown University); Peter Conrad (Brandeis University); Timothy Diamond (California State University–Los Angeles); Luis Durán (Mexican Institute of Social Security); Elizabeth Ettorre (University of Liverpool); Michael Farrall (Creighton University); Kitty Felker; Arthur Frank (University of Alberta); María Hilda García-Pérez (Arizona State University); Alya Guseva (Boston University); Frederic W. Hafferty (University of Minnesota–Duluth); Harlan Hahn (University of Southern California); Ida Hellander (Physicians for a National Health Program); Paul Higgins (University of South Carolina); Allan Horwitz (Rutgers University); David J. Hunter (University of Durham); Joseph Inungu (Central Michigan University); Michael Johnston (University of California–Los Angeles); Stephen J. Kunitz (University of Rochester); Donald W. Light (University of Medicine and Dentistry of New Jersey); Judith Lorber (City University of New York); William Magee (University of Toronto); Judy Mayo; Peggy McDonough (University of Toronto); Jack Meyer (Economic & Social Research Institute); Cindy Miller; Jeanine Mount (University of Wisconsin); Marilynn M. Rosenthal (University of Michigan); Beth Rushing (Kent State University); C. J. Schumaker (Walden University); Wendy Simonds (Georgia State University); Teresa Scheid (University of North Carolina at Charlotte); Clemencia Vargas (Centers for Disease Control and Prevention); Olaf von dem Knesebeck (University of Hamburg); Robert Weaver and his students, especially Cheryl Kratzer (Youngstown State University); Daniel Whitaker; David R. Williams (University of Michigan); Irving Kenneth Zola (Brandeis University); and Robert Zussman (University of Massachusetts–Amherst). This book undoubtedly would have been better if I had paid closer attention to their comments. I apologize sincerely if I have left anyone off this list.

Similarly, I am very grateful for the advice received from reviewers of this edition: Jennifer Bulanda (Miami University), Benjamin Drury (Indiana University at Columbus), Jamie Gusrang (Community College of Philadelphia), David Mullins (University of Saint Francis), Claire Norris (Xavier University of Louisiana), Michael Polgar (Penn State University), Richard Scotch (University of Texas at Dallas), Nicole Vadino (Community College of Philadelphia). And I remain grateful for the suggestions from reviewers on previous editions: Thomas Allen (University of South Dakota), Karen Bettez (Boston College), Linda Liska Belgrave (University of Miami), Pamela Cooper-Porter (Santa Monica College), Karen Frederick (St. Anselm College), Stephen Glazier (University of Nebraska),

Linda Grant (University of Georgia), Janet Hankin (Wayne State University), Heather Hartley (Portland State University), Alan Henderson (California State University–Long Beach), Simona Hill (Susquehanna University), Frances Hoffman (University of Missouri), Joseph Kotarba (University of Houston), Lilly M. Langer (Florida International University), Christine Malcom (Roosevelt University), Keith Mann (Cardinal Stritch University), Phylis Martinelli (St. Mary's College of California), Dan Morgan (Hawaii Pacific University), Larry R. Ridener (Pfeiffer University), Susan Smith (Walla Walla University), Kathy Stolley (Virginia Wesleyan College), Deborah Sullivan (Arizona State University), Gary Tiedman (Oregon State University), Diana Torrez (University of North Texas), Linda Treiber (Kennesaw State University), Robert Weaver (Youngstown State University), and Diane Zablotsky (University of North Carolina–Charlotte).

Finally, I would like to express my appreciation to the current and former Cengage staff who made the process of revising this book for its seventh edition as smooth as possible: Marta Lee-Perriard, Erik Fortier, Cheri Palmer, Ruchika Abrol, and Jennifer Harrison.

About the Author

Rose Weitz received her doctoral degree from Yale University in 1978. Since then, she has carved an exceptional record as both a scholar and a teacher. She is the author of numerous scholarly articles, the book *Life with AIDS*, and the book *Rapunzel's Daughters: What Women's Hair Tells Us About Women's Lives*. She also is co-author of *Labor Pains: Modern Midwives and Home Birth* and co-editor of *The Politics of Women's Bodies: Appearance, Sexuality, and Behavior.*

Professor Weitz has won several teaching awards (including the Pacific Sociological Association's Distinguished Contributions to Teaching Award, the ASU Last Lecture Award, and the ASU College of Liberal Arts and Sciences Outstanding Teaching Award) and has served as Director of ASU's graduate and undergraduate sociology programs. In addition, she has served as president of Sociologists for Women in Society, as chair of the Sociologists AIDS Network, and as chair of the Medical Sociology Section of the American Sociological Association.

PART I

Social Factors and Illness

Illness is a fact of life. Everyone experiences illness sooner or later, and everyone eventually must cope with illness among close friends and relatives.

To the ill individual, illness can seem a purely internal and personal experience. But illness is also a social phenomenon with social roots and social consequences. In this first part, we look at the role that social factors play in fostering illness within societies and in determining which groups in a given society will experience which illnesses with which consequences.

Chapter 1 introduces the sociological perspective and illustrates how sociology can help us understand issues related to health, illness, and health care. The chapter also provides a brief history of disease in the Western world, which highlights how social factors can foster disease. In the subsequent chapters, we explore the role social forces play in causing disease and in determining who gets ill in the modern world. In Chapter 2, we review the basic concepts needed to discuss diseases and look at modern patterns of disease. After that, we look at the social sources of illness in the contemporary United States and at some social factors that help predict individual health and illness. In Chapter 3, we investigate how four social factors—age, sex and gender, social class, and race or ethnicity—affect the distribution of illness in the United States and explore why some social groups bear a greater burden of illness than others. Finally, in Chapter 4, we analyze the very different pattern of illnesses found in poorer countries and explain how social forces—from the low status of women to the rise of migrant labor—can foster illness in these countries.

CHAPTER

1

The Sociology of Health, Illness, and Health Care

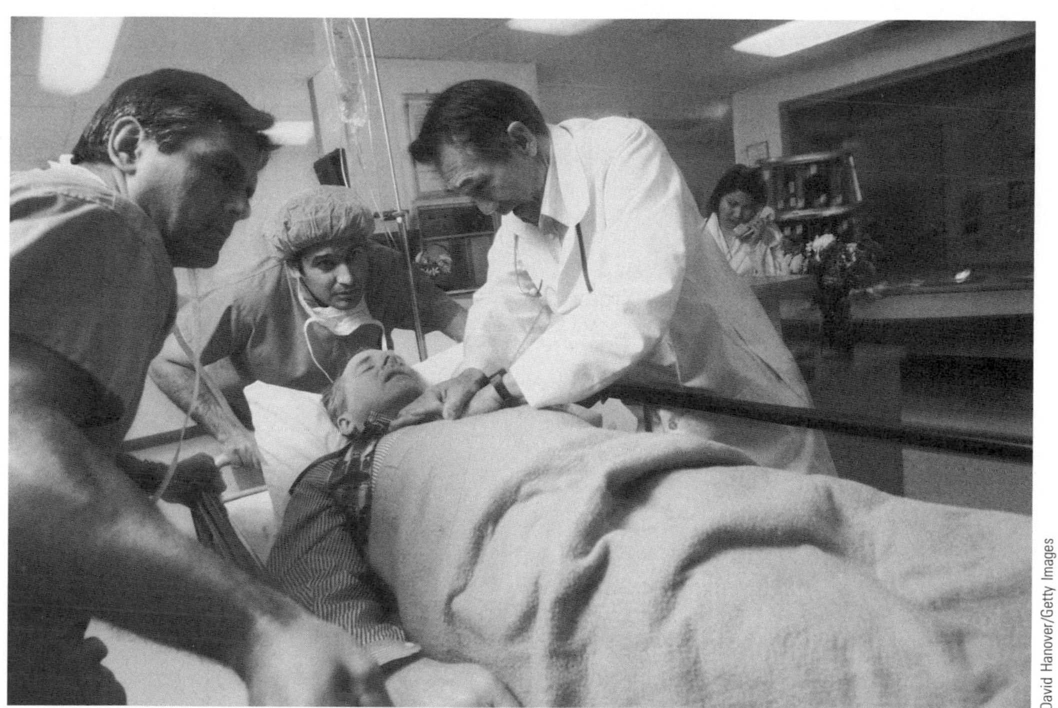

David Hanover/Getty Images

LEARNING OBJECTIVES

After reading this chapter, students should be able to:

- Describe the sociological perspective.
- Identify the difference between sociology *in* medicine and sociology *of* medicine.
- Understand how social changes have affected the health of populations historically.
- Evaluate research methods and sources.

Shortly before her 46th birthday, my friend Lara found a lump in her breast. A mammogram (a type of X-ray) soon identified the lump as potentially cancerous, and so a surgeon removed part of the lump for further testing. A few days later, Lara learned that she did indeed have breast cancer. That week, she got her affairs in order and signed a "living will," specifying the circumstances in which she would want all treatment stopped, and a "medical power of attorney" giving me legal authority to make medical decisions for her if she could not do so herself. These two documents, she hoped, would protect her from aggressive medical treatments that might prolong her suffering without improving her quality of life or chances of survival.

Two weeks after the initial tests, her surgeon removed the rest of the lump as well as the lymph nodes under her arm (where breast cancer most often spreads). The surgery went well, but the subsequent laboratory tests showed that the cancer had spread to her nodes.

Yet in many ways, Lara was fortunate. Her breast cancer was detected at an early stage, improving her odds of surviving. Although she had neither husband nor children to turn to, her friends proved uniformly supportive. She received health insurance through her employer and had no fears of losing either her job or her insurance.

Nevertheless, cancer changed Lara's life irrevocably, making it, at times, a nightmare. Having breast cancer shook Lara's faith in her body and changed her sense of her physical self. At the same time, her illness threatened her relationships with others. Despite the supportive responses she received from friends and coworkers, she feared that they would drift away as her illness continued or that she would chase them away with her all-too-reasonable complaints, worries, and needs.

Although she had far better health insurance than many Americans have, her debts for items not covered by insurance mounted. In addition, she had to spend hours fighting her insurance company to obtain relaxation training and expensive but effective anti-nausea drugs to cope with chemotherapy's side effects. Without the drugs, chemotherapy made her so ill that she could barely function, let alone fight her insurance company. In addition, chemotherapy proved so toxic that it damaged her veins with each painful intravenous treatment. As a result, her doctors suggested inserting a semipermanent plastic tube into her chest wall so they could instead administer the chemotherapy through the tube. Although doing so would have reduced her pain, Lara rejected the suggestion because she felt that, with this sign of her illness physically attached to her body, cancer would become integral to her very self, rather than merely one part of her life.

After a year of surgery, chemotherapy, and radiation, Lara's physical traumas ended, although it took another year before she regained her former energy.

Lara's story demonstrates the diverse ways that illness affects individuals' lives, as well as the diverse range of topics that sociologists of health, illness, and health care can study. This chapter opens with an overview of those topics, the sociological perspective, and the critical approach within sociology. We then look briefly at the history of disease, which helps put sociological research on health into context, before exploring the research sources used by sociologists.

THE SOCIOLOGY OF HEALTH, ILLNESS, AND HEALTH CARE: AN OVERVIEW

Sociologists' research into health, illness, and health care falls into three main categories. First, some sociologists study how social forces promote health and illness and why some social groups suffer more illness than others do. For example, researchers have explored whether working conditions in US factories help explain why poorer Americans get certain cancers more often than do wealthier Americans. Similarly, sociologists can study how historical changes in social life can explain changes in patterns of illness. To understand why rates of breast cancer have increased, for example, researchers have studied the possible impact of environmental pollution, increased meat consumption, and women's changing work lives.

Second, instead of studying broad patterns of illness, sociologists can study the experiences of those, like Lara, who live with illness each day—exploring, for example, how illness affects individuals' sense of identity, relationships with family, or ideas about what causes illness. Similarly, sociologists can study the experiences of health care providers. Some sociologists have analyzed how doctors' status and power have shifted over time, and others have investigated how power affects interactions among doctors, nurses, and other health care workers. Still others have examined interactions between health care workers and patients, asking, for example, how doctors maintain control over patients or whether doctors treat male and female patients differently.

Finally, sociologists can analyze the health care system as a whole. Sociologists have examined how health care systems have developed; compared the strengths and weaknesses of different systems; and explored how systems can be improved. For example, some have studied how US health insurance companies can make it difficult for people like Lara to get needed care, explored why European countries do better than the United States at providing health care to all who need it, and examined whether European health care policies could work in the United States.

The topics researched by sociologists of health, illness, and health care overlap in many ways with those studied by health psychologists, medical anthropologists, public health workers, and others. What most clearly differentiates sociologists from these other researchers is the *sociological perspective*. The next section describes that perspective.

THE SOCIOLOGICAL PERSPECTIVE

The **sociological perspective** is a view of the world that focuses on *social patterns* rather than on *individual behaviors*. Whereas a psychologist might help a battered wife develop a greater sense of her own self-worth so she might eventually leave her abusive husband, a sociologist likely would consider therapy a useful but inefficient means of addressing the root causes of wife abuse. Most battered wives, after all, don't have the time, money, or freedom to get help from psychologists. Moreover, even when therapy helps, it takes place only after the women have experienced physical and emotional damage. The sociologist would not deny that individual personalities play a role in wife battering, but would find it more useful to explore whether social forces can explain why wife battering is much more common than husband battering, or why battered wives so often remain with abusive husbands. Consequently, whereas the psychologist hopes to enable the individual battered wife eventually to leave her husband, the sociologist hopes to uncover the knowledge needed by legislators, social workers, activists, and others to prevent wife abuse in the first place.

As this example demonstrates, using the sociological perspective means framing problems as *public issues*, rather than simply *personal troubles*. According to C. Wright Mills (1959: 8–9), the sociologist who first drew attention to this dichotomy:

> [*Personal*] *troubles* occur within the character of the individual and within the range of his immediate relations with others; they have to do with his self and with those limited areas of social life of which he is directly and personally aware. Accordingly, the statements and the resolutions of troubles properly lie within the individual as a biographical entity and within the scope of his immediate milieu.... [In contrast, *public*] *issues* have to do with matters that transcend these local environments of the

> individual and the range of his inner life. They have to do with the organization of many such milieus into the institutions of an historical society as a whole.

For example, whenever a child dies from leukemia, it is a tragedy and a personal trouble for the child's family. If, on the other hand, several children in a neighborhood die of leukemia during the same year, it could suggest a broader public issue such as a contaminated water system. A sociologist would be likely to look for such a pattern and to explore why, for example, polluting industries are more likely to build factories in poor, minority neighborhoods than in affluent, white neighborhoods. The sociological perspective, then, departs radically from the popular American belief that individuals create their own fates and that anyone can succeed if he or she tries hard enough.

The sociological perspective can help us identify critical research questions that might otherwise go unasked. For example, in the book *Forgive and Remember: Managing Medical Failure*, sociologist Charles Bosk (2003: 62–63) described a situation he observed one day on "rounds," the time each day when recently graduated doctors (known as residents) and more senior doctors jointly examine the patients on a service, or ward:

> Dr. Arthur [the senior doctor] was examining the incision [surgical cut] of Mrs. Anders, a young woman who had just received her second mastectomy. After reassuring her that everything was fine, everyone left her room.
>
> We walked a bit down the hall and Arthur exploded: "That wound looks like a walking piece of dogshit. We don't close wounds with continuous suture on this service. We worked for hours giving this lady the best possible operation and then you screw it up on the closure. That's not how we close wounds on this service, do you understand? These are the fine points that separate good surgeons from butchers, and that's what you are here to learn. I never want to see another wound closed like that. Never!"
>
> Arthur then was silent, he walked a few feet, and then he began speaking again: "I don't give a shit how Dr. Henry [another senior doctor] does it on the Charlie Service or how Dr. Gray does it on Dogface; when you're on my service, you'll do it the way I want."

Dr. Arthur and the residents he supervised undoubtedly viewed this situation as a personal trouble, requiring a personal solution—the residents seeking to appease Dr. Arthur, and Dr. Arthur seeking to intimidate and shame the residents into doing things the way he considered best. Similarly, depending on their viewpoint, most observers probably would view this as a story about either careless residents or an autocratic senior doctor. Sociologists, however, would first ask whether residents and senior doctors *typically* interact like this. If they do, sociologists then would look for the social patterns underlying such interactions, rather than focusing on the personalities of these particular individuals. So, for example, based on his observations in this and other cases, Bosk discovered that cultural expectations within the medical world regarding authority, medical errors, and the importance of personal, surgical experience gave Dr. Arthur and the other supervising doctors power and allowed them to humiliate residents publicly and to set policies based more on personal preferences than on scientific data.

Whereas Charles Bosk studied relations among doctors, sociologist Kristin Barker (2008) looked at interactions among individuals who believe they have fibromyalgia. Fibromyalgia is a relatively new disease label given to individuals who experience a wide variety of disabling symptoms. Because there are no biological tests for fibromyalgia, many doctors doubt whether it should be considered a disease.

To explore what it means to live with fibromyalgia, Barker looked at posts to an online fibromyalgia support group. In a typical post, a woman named Sarah wrote:

> My new doctor appointment was today. Was not good!! First of all she is 4 months out of medical school. She looked over my chart and immediately wanted to change all medications that I am taking.... [Then she said,] "Now about your fibromyalgia, I will not prescribe pain killers for fibro." I sat there with my mouth open. She went on to tell me the fresh-out-of-med-school approach to fibro is exercise, diet. I said what about the pain? She proceeded to tell me the pain was "ALL IN MY HEAD, THERE IS NO PAIN, YOU JUST IMAGINE THERE IS." My first thought was [to] jump up out of this chair and slap the B——!! Instead I said "You are an idiot"!! Then I walked out.

Depending on one's perspective, Sarah's post either suggests an ignorant and insensitive doctor or a rude patient with delusions of grandeur. To a sociologist, however, this post raises several questions that go beyond these individuals, to look at the surrounding culture and social structure. Barker, for example, explored how the online support group increased patients' power to negotiate with their doctors, how the broader social structure nevertheless allowed doctors to control most interactions with patients, and how these struggles between doctors and patients reflected wider social questions regarding what constitutes an illness.

In sum, the sociological perspective shifts our focus from individuals to social groups and institutions. One effect of this shift is to highlight the role of power. **Power** refers to the ability to get others to do what one wants, whether willingly or unwillingly. Power is what allowed Dr. Arthur to treat his residents so rudely and what allowed Sarah to reject her doctor's advice. Because sociologists study groups rather than individuals, the sociological analysis of power focuses on why some social groups have more power than others, how groups use their power, and the consequences of **differential** (i.e., unequal) access to power, rather than on how specific individuals get or use power. For example, sociologists have examined how doctors use their power in negotiations with hospitals over working conditions and how *lack* of power exposes poor persons to unhealthy living conditions.

A CRITICAL APPROACH

Although the concept of power underlies the sociological perspective, sociologists don't necessarily emphasize power in their research and writing. For example, some sociologists have researched unhealthy eating patterns among poor people without exploring how lack of power may force individuals to work two jobs and, in turn, leave them insufficient time to prepare healthy meals.

Those sociologists, on the other hand, who focus on the sources, nature, and consequences of power relationships can be said to use a critical approach. Critical sociologists recognize that, regardless of how power is measured, men typically have more power than do women, adults more power than do children, whites more power than do African Americans, heterosexuals more power than do gays and lesbians, and so on. Critical sociologists who study health, illness, and health care have raised questions such as how differing levels of power affects individuals' access to health care and healthy living conditions.

Critical sociologists also emphasize how social institutions and popular beliefs can reflect or reinforce the existing distribution of power. For example, many researchers who study the US health care system have looked simply for ways to improve access to care within that system, such as by providing subsidies to doctors who practice in low-income neighborhoods. Those who use a critical approach have asked instead whether we could provide better care to more people if we changed the basic structure of the system, such as by removing the profit motive from health care to reduce the costs of care for everyone.

Similarly, critical sociologists have drawn attention to how doctors' power enables them to shape our ideas about health, illness, and health care. Most basically, these sociologists have questioned the very terms *health, illness*, and *disability* and have explored how these terms can reflect social values as well as physical characteristics.

In any sociological field, therefore, those who adopt a critical approach will ask quite different research questions than will others. Within the sociology of health, illness, and health care, this approach translates largely to whether sociologists limit their research to questions about social life that doctors consider useful—a strategy referred to as **sociology *in* medicine**—or design their research to answer questions of interest to sociologists in general—a strategy referred to as the **sociology *of* medicine** (Straus, 1957). Research using the latter strategy often challenges both medical views of the world and existing power relationships within health care.

To understand the difference between sociology *in* medicine and sociology *of* medicine, consider the sociological literature on patients who don't follow their doctors' advice. Reflecting doctors' view of such patients as problematic, many sociologists (practicing sociology *in* medicine) have explored ways to encourage patients to comply with medical advice. In contrast, sociologists *of* medicine have looked at the issue of compliance through patients' eyes. As a result, they have learned that patients sometimes ignore medical advice not out of foolishness, but because their doctors did not clearly explain the prescribed regimens or because the emotional or financial costs of following that advice seem to outweigh the potential benefits. Similarly, whereas those practicing sociology in medicine have studied the experience of *patienthood*, those practicing sociology *of* medicine instead have studied the broader experience of *illness*, which includes but is not limited to the experience of patienthood. The growing emphasis on sociology of medicine and on the critical approach has led to a proliferation of research on the many ways illness affects everyday life and on how ill individuals, their families, and their friends respond to illness.

A BRIEF HISTORY OF DISEASE

One of the most important questions raised by critical sociologists is how social conditions cause or amplify disease. Their research (along with research conducted by historians and others) demonstrates that across history, social factors such as poverty, urbanization, and living conditions have fostered illness. This section provides a brief overview of disease throughout Western history, highlighting the role played by social forces.

The European Background

To understand health in the modern world, it helps to begin with the Middle Ages (approximately A.D. 800 to 1300), when commerce, trade, and cities began to swell (Kiple, 1993). These shifts sparked a devastating series of epidemics. The term **epidemic** refers to any significant increase in the numbers affected by a disease *or* to the first appearance of a new disease. In the fledgling European cities, people lived in close and filthy quarters, along with rats, fleas, and lice—perfect conditions for transmitting infectious diseases such as bubonic plague and smallpox. In addition, because cities lacked sewer systems, families would dump human waste into the streets, where it eventually would be washed into local rivers. As a result, typhoid, cholera, and other waterborne diseases that live in human waste flourished. Simultaneously, the growth of long-distance trade helped epidemics spread to Europe from the Middle East, where cities had long existed and many diseases were **endemic** (i.e., established within a population at a fairly stable level). In addition, religious pilgrimages and crusades to Jerusalem helped spread diseases to Europe.

The resulting epidemics ravaged Europe. Waves of disease, including bubonic plague, leprosy, and smallpox, swept the continent. The worst of these was bubonic plague, popularly known as the "Black Death," which killed at least 25 million Europeans—up to half the population—between 1347 and 1351 (Gottfried, 1983; J. Kelly, 2005).

Although the great **pandemics** (worldwide epidemics) began diminishing during the fifteenth and sixteenth centuries, average life expectancy increased only slightly, for malnutrition continued to threaten health (Kiple, 1993). By the early 1700s, however, life expectancy began to increase. This change can't be attributed to any developments in health care, for folk healers had nothing new to offer, and medical doctors and surgeons (as will be described in more detail in Chapter 11) harmed at least as often as they helped. For example, US President George Washington died after his doctors, following contemporary medical procedures, "treated" his sore throat by cutting into a vein and draining much of his blood.

If advances in medicine did not cause the eighteenth-century increase in life expectancy, what did? Historians commonly trace this change to a combination of social factors (Kiple, 1993). First, changes in warfare moved battles and soldiers away from cities, protecting citizens from both violence and the diseases that followed in soldiers' wakes. Second, the development of new crops and new lands improved people's diets and their ability to resist disease. Third, women began to have fewer children at later ages, increasing both women's and children's chances of survival. Fourth, women less often spent long hours in exhausting fieldwork and so more often were strong enough to survive childbirth. Infants, too, more often survived because mothers could more easily keep their children with them and breastfeed. (This would change soon, however, for the many women who would become workers in emerging factories.)

Disease in the New World

In the New World, meanwhile, colonization by Europeans was decimating Native Americans (Kiple, 1993). The colonizers brought with them about 14 new diseases—including influenza, measles, and smallpox—for which the Native Americans had no natural immunities. These diseases ravaged the Native American population, in some cases wiping out entire tribes (Crosby, 1986). Conversely, life expectancy *increased* for those who emigrated from Europe to the colonies, for the New World's vast lands and agricultural resources protected them against the malnutrition and overcrowding common in Europe.

The Epidemiological Transition

As industrialization and urbanization increased, many—especially the urban poor—began dying at younger ages. The main killer was tuberculosis, followed by influenza, pneumonia, typhus, and other infectious diseases. Except for tuberculosis, these diseases are all considered **acute diseases**, that is, diseases that strike suddenly and disappear quickly—sometimes killing their victims, sometimes causing only a mild illness. By the late nineteenth century, however, deaths during infancy and childhood and deaths at all ages from infectious disease began to decline rapidly. Between 1900 and 1930, **life expectancy**—the average number of years individuals can expect to live—increased from 47 to 60 years for whites and from 33 to 48 years for African Americans (US Bureau of the Census, 1975).

As infectious diseases declined in importance, chronic diseases gained importance. **Chronic diseases** are those that typically last several years or more, such as muscular dystrophy and asthma. During this period, heart disease, stroke, arthritis, and diabetes became major causes of illness and disability. Increasingly, too, these conditions shifted from primarily diseases of the affluent (formerly

KEY CONCEPTS

The Epidemiological Transition

The *epidemiological transition* refers to the point in a society's history when deaths from infectious and parasitic diseases fall significantly; life expectancies increase significantly; and, consequently, degenerative and chronic diseases become more common. For example:

	Transition Stage		
Stage Indicators	**Pre-Transition: United States, 1850**	**In Transition: India, 2013**	**Post-Transition: United States, 2013**
Life Expectancy	40 years for whites, 23 for African Americans	66 years	79 years for whites, 75 for African Americans
Deaths from parasitic and infectious diseases	Frequent	Less frequent than in the past, rare among the growing middle and upper classes	Rare
Deaths from chronic diseases	Rare	Increasingly frequent	Frequent

the only people likely to enjoy long lives) to disproportionately diseases of the poor.

The shift from a society characterized by infectious and parasitic diseases and low life expectancy to one characterized by degenerative and chronic diseases and high life expectancy is referred to as the **epidemiological transition** (Omran, 1971). This transition seems to occur around the world once a nation's mean per capita income reaches a threshold level (in 2015 dollars) of about $11,400 (Wilkinson, 1996). (See *Key Concepts: The Epidemiological Transition.*)

Contrary to conventional wisdom, medical interventions such as vaccinations, new drugs, and new surgical techniques played little role in the epidemiological transition (Leavitt and Numbers, 1985; McKeown, 1979; McKinlay and McKinlay, 1977). For example, deaths from tuberculosis, scarlet fever, and typhoid were all declining steadily by the early 1900s, even though doctors had no effective treatments for these diseases until the 1940s (McKinlay and McKinlay, 1977). The same was true for most other infectious diseases of that era. Similarly, historians and other scholars suggest that medical care explains no more than one-sixth of the overall increase in life expectancy during the twentieth century and no more than 10 percent of the risk of dying in any given year (Bunker, Frazier, and Mosteller, 1994; McGinnis, Williams-Russo, and Knickman, 2002; Schroeder, 2007).

How, then, can we explain the epidemiological transition? The answer appears to lie in changing social conditions. Even though doctors at the time misunderstood the causes of various diseases, public health measures such as the development of clean water supplies and sanitary sewage systems virtually eliminated waterborne diseases like typhoid. These measures also reduced the number of minor infections individuals acquired, increasing their overall health and reducing the risks of dying from diseases such as pneumonia and tuberculosis. Cleaner water systems accounted for almost half of the overall rise in life expectancy and two-thirds of the decrease in deaths among infants between 1900 and 1940 (Cutler and Miller, 2005). Similarly, as living conditions and access to healthy foods improved, so did individuals' ability to resist infection and to survive if they became infected.

All these forces reduced the likelihood that babies would die young. As a result, parents no longer needed to have many children to ensure that one or two would survive long enough to support them in old age. At the same time, as adults increasingly shifted to working in factories—some of which offered pensions—their need to have numerous children to help on the farm or to care for them in old age declined. As a result, adults had fewer children and could devote more resources to each child, further increasing their children's chances of survival.

UNDERSTANDING RESEARCH SOURCES

The preceding discussion on the history of disease, as well as numerous other discussions you will read in this book, may well challenge your previous ideas about health and illness. To respond intelligently to these challenges, you need

CONTEMPORARY ISSUES

"Scienciness"

The rise of the Internet has made it particularly easy for individuals and corporations to use "scienciness"—the aura of scientific research—to sell the public on highly dubious measures for treating or preventing illness. For example, the first web page identified in a November 3, 2010, Google search for "cure cancer" declared, "Cancer can be cured and prevented naturally and scientifically.... [These are] facts that have been proven in scientific labs, by doctors" (www.1cure4cancer.com). The main purpose of the site is to convince viewers to buy laetrile, an extract typically made from apricot or almond pits. To do so, the website uses "scientific" language ("Laetrile is a decomposition product resulting from the hydrolysis of amygdalin") and "scientific credentials" (mentioning the names and university affiliations of doctors who, it claims, have proven laetrile's effectiveness).

In reality, no true scientific research has ever found evidence that laetrile works, and no reputable scientific journals have ever published articles in favor of laetrile. Studies have, however, found that cyanide found naturally in apricot and almond pits can cause liver damage, coma, or even death (National Cancer Institute, 2010). Meanwhile, a vast number of other websites use scientific-looking charts, "doctors" with degrees in art history, and self-published "journal articles" to sell useless or potentially dangerous treatments for everything, from acne to acquired immunodeficiency syndrome (AIDS), to individuals who don't understand how to evaluate research sources and data (Goldacre, 2010).

to understand how researchers evaluate research sources and data. *Contemporary Issues: "Scienciness"* explores some of the problems that may arise when the public lacks the tools to evaluate research data.

Evaluating Research Sources

In any field, scholars give the most credibility to research that is published in the most prestigious journals. In sociology of health, these journals include the *Journal of the American Medical Association*, the *New England Journal of Medicine*, the *American Journal of Sociology,* and the *Journal of Health and Social Behavior.* These journals are held in high esteem because each has existed for a long time, has a large readership, publishes only a small percentage of the manuscripts it receives, and uses peer review (i.e., has several scholars review each manuscript before the journal decides which to publish).

Similarly, scholars are most likely to trust data that come from reliable, nonprofit and nonpartisan sources. Thus this book draws heavily on statistics collected by the US government and by nonprofit groups such as the Mayo Clinic, the Kaiser Foundation, and the **World Health Organization (WHO)**, a United Nations organization charged with documenting health problems and improving world health. Because these statistics are collected by bureaucrats whose employment typically continues regardless of shifts in the political climate, rather than by groups with a particular political agenda, they are generally regarded as the most objective data available. This holds true whether the data are found in print sources or on the Internet.

Box 1.1 suggests some long-standing Internet sites that provide particularly useful information about health issues.

Box 1.1 Useful Internet Sources

- www.healthfinder.gov: Run by the United States Department of Health and Human Services, this site offers a wide range of health information as well as an extensive set of links to other government and nongovernmental health-related sites.
- www.nlm.nih.gov: This site provides access to both published and unpublished materials available at the National Library of Medicine, the largest medical library in the world.
- www.mayoclinic.org: Run by the nonprofit Mayo Clinic, this site offers consumer health information plus the opportunity to email questions to physicians.
- www.gapminder.org: Gapminder is a Swedish nonprofit organization that provides free access to international health-related data and to software for vividly displaying that data.
- www.who.int: WHO's website provides a vast array of information about health, illness, and health care around the world.
- Scholar.google.com: This branch of the Google® search engine takes viewers only to scholarly journal articles, on health as well as other topics. This is an excellent starting point for finding reputable information on any topic.

Evaluating Research Data

Sociological studies can be broadly divided into qualitative research and quantitative research. In qualitative studies, researchers may spend months conducting in-depth interviews or observing a community. Qualitative research is evaluated most highly if the researcher actively sought data that might support *or* challenge his or her preconceived notions *and* if the data the researcher offers seem to logically support the conclusions he or she draws from that data. Qualitative research is most useful for helping us understand how people understand their lives, why people behave or think in specific ways, and how social interaction works.

In contrast, quantitative research aims to understand social life by finding ways to turn observations into numbers and statistics. It is most useful for assessing how often something occurs in social life and how one thing may cause another. Quantitative research is especially useful when based on large samples because such data is more likely to reflect trends among the population as a whole. Research based on **random samples** is held in particularly high regard. In a random sample, each member of a population has an equal chance of being selected (such as when names are drawn out of a hat), and so researchers can be fairly certain that selected individuals will be representative of the population as a whole.

Quantitative research is regarded most highly when the researchers **control** for the potential influence of extraneous factors (i.e., use statistical techniques to eliminate the possibility that something else caused whatever the researchers are studying). For example, compared to non-smokers, Americans who smoke cigarettes are more likely to get lung cancer *and* to be white. To determine whether smoking or race causes lung cancer, researchers could divide their sample into four groups—white smokers, white *non*-smokers, nonwhite smokers, and nonwhite *non*-smokers—and then see what percentage of each group gets cancer. (In fact, among both whites and nonwhites, those who smoke are more likely to get lung cancer.)

One important recent trend in quantitative research is the rise of **big data**: huge studies that pull together multiple sets of data from entire populations. For example, researchers were able to combine two national Danish registries—one of every case of brain cancer in Denmark and one of every person who owned a cell phone, beginning when cell phones were first introduced to the country—to convincingly argue that cell phone use does not cause brain cancers (Frei et al., 2011).

Research based on big data is particularly convincing because it reflects virtually the entire population, and so research results are likely to apply to everyone in that population. In addition, because big data were typically collected initially for some other purpose, they are less likely to be biased (as might happen, for example, if a pharmaceutical company collects data on the effects of a drug it produces). Big data are likely to play a growing role in sociological research in coming years, especially as doctors and health care systems increasingly record patient data electronically (Bates et al., 2014).

SUMMARY

1. Topics in the sociology of health, illness, and health care include the nature of the health care system; how social forces promote health and illness; the experience of living with illness or disability; and the status, power, training, and values of health care providers.
2. The sociological perspective sets sociologists apart from other health and social researchers. This perspective focuses on explaining social patterns rather than individual behavior and on identifying and resolving public issues rather than personal problems.
3. Sociology *in* medicine refers to sociological research that focuses on answering questions that doctors consider useful.
4. Sociology *of* medicine focuses on how power affects health, illness, and health care. Sociology of medicine is a branch of critical sociology, which focuses more generally on the sources, nature, and consequences of power.
5. Although disease has always accompanied human life, patterns of diseases have varied over time and place, for both social and biological reasons.
6. Epidemics refer to any significant increase in the numbers affected by a disease or to the first appearance of a new disease. Pandemic diseases are worldwide epidemics, such as HIV/AIDS currently. Endemic diseases are those that continue to appear in a population at a relatively stable rate.
7. Devastating epidemics accompanied the rise of cities in medieval Europe. By the early eighteenth century, however, life expectancy began to increase, primarily due to improved living and working conditions, later and less frequent childbirth, and changes in military strategies that separated soldiers and civilians.
8. Life expectancy increased further in the late nineteenth century, primarily due to improvements in nutrition, living conditions, and public sanitation, rather than to medical advances. Still, as in previous eras, the main causes of death were infectious and parasitic diseases.
9. Currently, the main causes of death in the Western nations are chronic illnesses of middle and old age. The shift from a society characterized by low life expectancy and infectious and parasitic diseases to one characterized by high life expectancy and chronic diseases is known as the epidemiological transition.
10. To evaluate printed or online sources, readers must ask whether their data come from a reputable source, were peer reviewed or otherwise checked for quality and bias, were based on a representative sample of reasonable size, and were controlled statistically for possible confounding factors.

REVIEW QUESTIONS

1. What is the sociological perspective?
2. How do the questions sociologists ask differ from those asked by psychologists or by health care workers?

3. What does this textbook mean by a critical approach?
4. What is the epidemiological transition?
5. What is the difference between acute and chronic diseases?
6. What factors caused the decline in mortality between the nineteenth and early twentieth centuries?
7. What are some ways a reader can tell whether a journal article or Internet website is a reliable source of data?

CRITICAL THINKING QUESTIONS

1. How can knowing the history of disease help us to understand (a) current health problems and (b) how health patterns might change in the future?
2. Write three research questions about the causes of cancer. The first should be a question that a doctor might ask; the second, a question a psychologist might ask; and the third, a question that a sociologist might ask.
3. Assume you have found a website that argues for vitamin C as a cure for the common cold. List three questions you would want to ask before deciding whether to believe the website.

CHAPTER

2

The Social Sources of Modern Illness

Janine Wiedel Photolibrary/Alamy

LEARNING OBJECTIVES

After reading this chapter, students should be able to:

- Use key epidemiological terms to describe the health of a population.
- Describe the modern disease profile in Western nations.
- Understand how social factors can lead to preventable deaths.
- Analyze the social processes that encourage or discourage individuals from adopting healthy behaviors.
- Identify the impact of social networks and social stress on health.

Judy Cude and her daughter Jenny live in the small city of Dickson, Tennessee. When Jenny was about to give birth to her first child, Judy went with her to the hospital. At first, everything seemed fine:

> *The attending physician kept up a cheerful, reassuring stream of talk as he assisted Jenny with her labor. "Peyton came out face down. When Dr. Booker turned [the baby] over, he stopped talking," Judy recalled. "He had his back to us, but when the nurse gave him a shove, he turned around, and he had tears running down his cheeks."*
>
> *The baby's face was badly disfigured with a cleft lip and a bilateral cleft palate. And though they did not know it immediately, Peyton also had a damaged heart, a valve that failed to close properly....*
>
> *Two weeks after Peyton was born, Jenny was given the name of another mother in Dickson whose child, born a couple of months earlier, also had a cleft lip and palate. Then a woman called Judy at her day care center and asked if she could accommodate children with special needs because ultrasound tests found that her child was about to be born with a cleft palate. "That made three," Judy said. She and the other mothers kept a tally. Soon they had counted six. Judy placed a newspaper advertisement asking families with similar defects to contact her. And, as it turned out, nineteen children had been born in Dickson with a cleft lip and palate in a little over two years. The odds against such a series of identical birth defects were almost certainly too high to be coincidental. [Moreover,] within a brief period, four babies were born with a rare brain malformation, where the two hemispheres of the brain are not connected. There have been a large number of cases of hypospadias, a condition in male children where the urethra is inverted. Dickson families also reported a high incidence of heart defects and leukemia among their babies.*
>
> *Eventually Judy would learn that most of the affected children lived near the Dickson County landfill and that toxic wastes, including chemicals known to cause cancers and birth defects, had been dumped there for years (Shabecoff and Shabecoff, 2010:4–5).*

At one level, disease is a biological process existing within an individual body. Yet as Judy's story suggests, disease can stem from social conditions as well as from individual biology. In this chapter, we explore the *social* sources of disease. After reviewing the basic concepts needed to discuss diseases, we review modern disease patterns. Then we focus on the major social sources of premature death in the United States today (including environmental toxins). Finally, we explore social factors that affect whether people will adopt healthier behaviors and whether individuals will remain healthy regardless of their behaviors.

AN INTRODUCTION TO EPIDEMIOLOGY

The first essential concepts that students of health and illness need to understand are disease and illness. To researchers and clinicians working with physical health problems, **disease** refers to *biological* problems within organisms. In contrast, **illness** refers to the *social* experience and consequences of living with a disease. Using these terms, we might say that a man who is infected with the poliomyelitis virus must adapt to physical disability caused by the *disease* called polio and to changes in how others view him because he has the *illness* called polio. (Chapter 5 will discuss the meaning of *illness* in more detail, and Chapter 7 will discuss how these concepts are applied a bit differently to mental health problems.)

The study of the distribution of disease within a population is known as **epidemiology**. This chapter and the next focus more specifically on **social epidemiology**, or how *social* behaviors and factors (such as social class or use of tobacco) affect the distribution of disease within a population. For example, whereas biologists might investigate whether heart disease is more common among those with high cholesterol levels, social epidemiologists might investigate whether it is more common among those with high incomes.

But what do we mean when we say a disease is "more common" among one group than another? Data on infection with **human immunodeficiency virus (HIV)**, the virus that in its later stages causes **acquired immunodeficiency syndrome (AIDS)**, provide a useful example. (This book uses the term **HIV/AIDS** to refer to the full range of illness from the earliest stages of infection with HIV to full-blown AIDS.)

Currently, more than twice as many Brazilians have HIV/AIDS than do Botswanans (Population Reference Bureau, 2014). At first glance, this might suggest that HIV/AIDS is a much greater problem in Brazil than in Botswana. However, Brazil's population is much larger than that of Botswana. To take this difference into account, epidemiologists typically look at the *rate* rather than the *number* of HIV/AIDS cases in a population. **Rate** refers to the *proportion* of a specified population that experiences a given circumstance. For example, to learn the rate of a HIV/AIDS in Brazil, we would divide the number of people in Brazil who have the disease into the total number of people in the country. In 2013, the rate of HIV/AIDS among adults was 3 per 1000 adults in Brazil

KEY CONCEPTS

Understanding Rates

Concept	Example	Advantages and Disadvantages
Raw number	About 10,000 Americans get tuberculosis each year.	Hard to interpret on its own: 10,000 out of 100,000 would be much scarier than 10,000 out of 300 million (the size of the US population).
Incidence rate	The incidence of tuberculosis in the United States is currently 3/100,000: there were 3 *new* cases last year for every 100,000 Americans. In contrast, 10 years ago the incidence of tuberculosis was 6/100,000.	Provides overview of the *increase* or *decrease* of a disease, behavior, or condition: how many *new* cases have been identified.
Prevalence rate	The prevalence of arthritis in the United States is 230/1000: out of every 1,000 US residents, 230 have arthritis (whether diagnosed recently or years ago.)	Provides overview of the *magnitude* of a disease, behavior, or condition: how many people have it *now*, regardless of when they got it.

compared with *234* per 1,000 adults in Botswana (Population Reference Bureau, 2014). This tells us that HIV/AIDS affects a greater *proportion* of the population in Botswana than in Brazil. It also demonstrates the advantage of using rates rather than raw numbers.

Two particularly useful types of rates are incidence and prevalence rates. **Incidence** refers to the number of *new* occurrences of an event (disease, births, deaths, etc.) within a population during a specified period. **Prevalence** refers to the *total* number of cases within a population at a specified time—both those newly diagnosed (incidence) and those diagnosed in previous years but still living with the condition under study. *Key Concepts: Understanding Rates* provides a helpful comparison.

In general, incidence better measures the spread of **acute disease**. The term *incidence* also better measures rapidly spreading diseases such as HIV/AIDS. For example, to see how HIV/AIDS has spread, we might compare its incidence in 1981 with its incidence today. Prevalence, on the other hand, better measures the frequency of **chronic diseases**.

Two final terms often used in epidemiology are *morbidity* and *mortality*. **Morbidity** refers to symptoms, illnesses, and impairments; **mortality** refers to deaths. To assess the overall health of a population, epidemiologists typically calculate the rate of serious morbidity in a population (i.e., the proportion with serious illness) and the rates of infant mortality and maternal mortality (i.e., the proportion of infants and childbearing women who die during or soon after childbirth). In addition, they typically calculate **life expectancy** (the average number of years individuals born in a certain year are likely to live).

But what if one population is much older than another? Because younger people have very different health risks than do older people, it would be misleading to compare these populations without taking this into account. For example, Arizona's population is younger on average than is North Dakota's, so we would expect Arizona to have more deaths from drunk driving and fewer from heart disease than would North Dakota. To deal with this issue, epidemiologists use **age-adjusted rates**. These rates are calculated using standard statistical procedures that, as Chapter 1 described, **control** for the effect of age differences among populations.

The next section uses epidemiological concepts and data to describe current disease patterns.

THE MODERN DISEASE PROFILE

As the previous chapter discussed, by the beginning of the twentieth century, patterns of disease had shifted markedly. No longer did most people in Western nations die young of infectious or parasitic diseases (even though such diseases still ravaged poorer nations). Instead, most children survived childhood without major illnesses, and most adults died in old age of chronic disease. Yet infectious disease once again is on the rise.

The New Rise in Infectious Disease

The rising threat from infectious diseases first made medical news in 1981, when the first cases of what would become known as HIV/AIDS were identified. Since then, other new infectious diseases (such as Ebola virus disease) have been identified, previously known diseases (such as cholera and streptococcus) have become deadlier, and previously harmless microorganisms (such as the virus that causes avian influenza, or "bird flu") have caused important disease outbreaks (Armelagos and Harper, 2010; Oldstone, 2010). *Contemporary Issues: Poverty and Ebola Virus Disease* explores one such case.

The renewed dangers posed by infectious disease partly reflect basic principles of natural selection. Just as natural selection favors animals whose camouflaging coloration hides them from predators long enough to reproduce, natural selection favors microorganisms that can resist drug treatments. As doctors prescribed antibiotics more widely, often under pressure from patients who feel cheated if they don't receive a prescription at each visit, the drugs killed all susceptible variants of disease-causing microorganisms while allowing variants resistant to the drugs to flourish. Similarly, drug-resistant tuberculosis is increasing in nations where both HIV/AIDS and poverty leave individuals both more susceptible to infection and less able to afford consistent, effective treatment. Meanwhile, the growing use of antibiotics in everything from cutting boards to kitty litter, chicken feed, and soaps also encourages the rise of drug-resistant bacteria.

CONTEMPORARY ISSUES

Poverty and Ebola Virus Disease

In 2014, Ebola virus disease began spreading rapidly in west Africa. Ebola is a classic example of the links between poverty and disease. The disease probably originated in fruit bats and spread to humans after poverty and environmental decline pressed people to seek wood for fuel and land for small farms in ever-deeper rainforests where infected bats (or other creatures) lived (Quammen, 2013). Ebola disease is highly deadly, causing intense fevers, vomiting, diarrhea, bleeding, kidney damage, liver damage, and eventual death in up to 90 percent of those infected with it, at least in poor nations (WHO, Global Alert and Response, 2014). But many of those deaths can be prevented. The disease is typically spread when body fluids (blood, urine, semen, etc.) from an infected individual come into contact with another person's open skin or mucus membranes, whether directly or via used needles, bed linens, or the like. Tragically, many of those killed by Ebola in Africa have been health care workers whose deaths could have been prevented if the hospitals and clinics they worked at could have afforded sufficient gloves, hospital gowns, and face masks. Many other deaths have occurred among individuals who were vulnerable to infection when forced to care for ill family members at home because hospitals were overwhelmed. Once individuals are infected, their best hope is to receive intensive care to keep them hydrated and stable—care that is largely unavailable in impoverished regions.

Sadly, Ebola disease not only reflects poverty but also can increase it (Nossiter, 2014). In the worse-hit areas of Africa, the disease is placing a heavy toll on local economies, as both skilled and unskilled workers die, markets collapse, international investors and travelers flee, and neighboring countries impose quarantines that keep goods from going into or out of the hardest hit nations. Moreover, because the disease kills young as well as old, the economies of these nations will find it particularly difficult to rebuild.

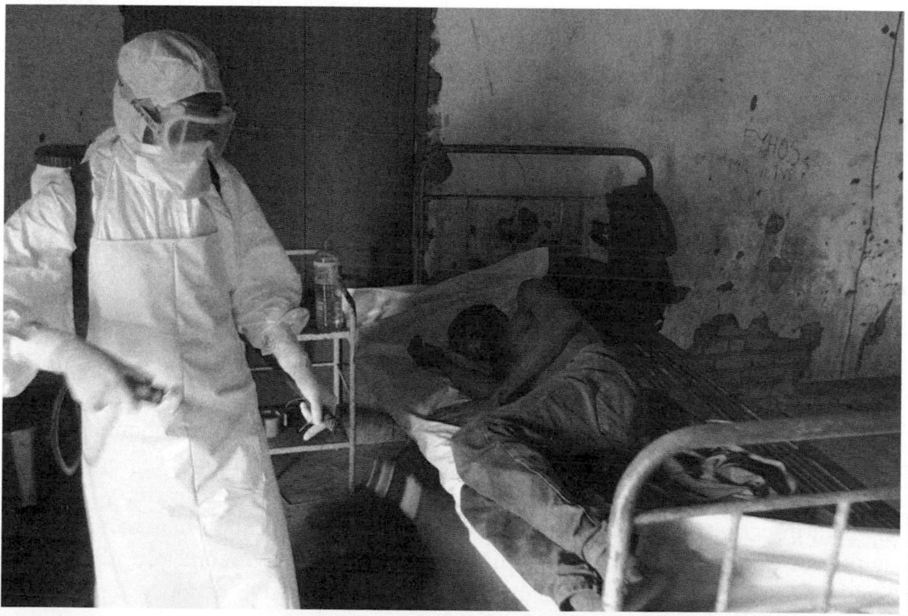

DESIREY MINKOH/AFP/Getty Images

Other forces also promoted the rise in infectious diseases (Oldstone, 2010). In the same way that population growth and the rise of cities once fostered the spread of infectious diseases in Europe, they now are causing new epidemics in the rapidly growing cities of Africa, Asia, and Latin America (Armelagos and Harper, 2010). Meanwhile, older cultural traditions often erode among those who move to these cities, making health-endangering activities such as tobacco smoking and sexual experimentation more likely. At the same time, industrial plans and cities are growing into former forests and farmlands, bringing wild animals increasingly in contact with humans. As a result, microorganisms that previously had infected only animals now have the opportunity to infect humans as well (Quammen, 2013).

All of these factors have been heightened by **globalization**, the process through which ideas, resources, people, and trade increasingly operate in a worldwide rather than local framework. The erosion of cultural traditions in Asia, Africa, and Latin America reflects, among other things, the increasingly global spread of Western ideas by tourists, the mass media, businesspeople, and nongovernmental organizations such as the United Nations and the International Monetary Fund. Similarly, environmental changes that encourage disease partly stem from actions taken by Western-based industries and corporations, which now find it easy to operate internationally because of trade agreements such as NAFTA (the North American Free Trade Agreement). In addition, the globalization of business investment and tourism has globalized disease simply by increasing the number of people traveling from one region to another (Oldstone, 2010).

Finally, the rise in infectious disease reflects political decisions as well as biological realities. For example, providing clean needles to those who use illegal drugs is a proven way of controlling the spread of HIV/AIDS and does not seem to lead to greater use of drugs (Epstein, 1996; Holtgrave and Curran, 2006). Yet in most US cities and states, it remains illegal to provide needles to drug users. Similarly, the Russian government's policy of imprisoning large numbers of individuals in miserable conditions has led to a rapid increase in tuberculosis, both in prisons and (as prisoners are eventually released) in the society at large (Goozner, 2008).

Today's Top Killers

Despite the recent reemergence of infectious diseases, however, such diseases still play a relatively small role in US mortality rates. Table 2.1 shows the top ten causes of death in the United States in 2011 (the latest data available as of 2014) and illustrates how these causes have changed since 1900.

As the table demonstrates, whereas the top killers in 1900—influenza, pneumonia, and tuberculosis—were infectious diseases that could strike at any age, most of today's top killers—including heart disease, cancer, Alzheimer's, and diabetes—are chronic diseases primarily associated with older populations. These diseases now far outpace infectious diseases as causes of death.

But infectious diseases have not disappeared. Influenza and pneumonia remain significant causes of death, although mostly they kill elderly people already in poor health. More importantly, although deaths from HIV/AIDS dropped significantly

TABLE 2.1 Main Causes of Deaths, 1900 and 2011

1900	Rate per 100,000	2011	Rate per 100,000
Influenza and pneumonia	202	Heart disease	192
Tuberculosis	194	Cancer	185
Diarrhea, enteritis, intestinal ulcers	143	Chronic respiratory disease	46
Disease of the heart	137	Cerebrovascular disease (strokes)	41
Cerebrovascular diseases (strokes)	107	Accidents	41
Chronic kidney disease	89	Alzheimer's disease	27
Accidents	72	Diabetes	24
Cancer	64	Influenza and pneumonia	17
Senility	50	Kidney disease	15
Diphtheria	40	Suicide	13

SOURCE: National Center for Health Statistics (2014b).

after new treatments became available in the late 1990s, the disease remains a significant cause of death among African American men (National Center for Health Statistics, 2014a). Those treatments can increase life expectancy dramatically, but only for those who can tolerate significant drug side effects, manage the required regimen of as many as 20 pills per day taken at strictly regulated times, and afford the cost of about $24,000 to $60,000 per year (Aguirre, 2012).

Finally, Table 2.1 illustrates the role that social factors play in mortality rates. Accidental deaths mostly stem from motor vehicle accidents (many of them linked to alcohol use), while tobacco use is the main cause of chronic respiratory disease and is a common contributor to heart disease, cancer, and cerebrovascular disease (strokes). Similarly, diabetes (which is a main cause of kidney disease) largely reflects diet and exercise patterns. Each of these causes of death reflects social behaviors rooted in social conditions. The remainder of this chapter discusses the role social forces play in mortality and morbidity.

THE SOCIAL SOURCES OF PREMATURE DEATHS

In a widely cited article titled "A Case for Refocusing Upstream," sociologist John McKinlay (1994) offered the following oft-told tale as a metaphor for the modern doctor's dilemma:

> Sometimes it feels like this. There I am standing by the shore of a swiftly flowing river and I hear the cry of a drowning man. So I jump into the river, put my arms around him, pull him to shore and apply artificial respiration. Just when he begins to breathe, there is another cry for help.

> So I jump into the river, reach him, pull him to shore, apply artificial respiration, and then just as he begins to breathe, another cry for help. So back in the river again, reaching, pulling, applying, breathing, and then another yell. Again and again, without end, goes the sequence. You know, I am so busy jumping in, pulling them to shore, applying artificial respiration, that I have no time to see who the hell is upstream pushing them all in (McKinlay, 1994 509–10).

This story illustrates the traditional emphasis within medicine on diagnosing and treating illness and disability rather than preventing it. Moreover, even when doctors, researchers, and others do focus on preventing illness, they typically look only far enough upstream to see how individual psychological characteristics (such as poor impulse control) or biological characteristics (such as a gene) may make some people more susceptible than others to disease or unhealthy behaviors. In contrast, although sociologists agree that biological and psychological factors affect health, they also recognize that these factors don't operate in a vacuum. For example, adolescents are most likely to drink alcohol dangerously if their friends and family do so. Similarly, the high rates of diabetes found among contemporary Native Americans partially reflect individual decisions regarding exercise and diet. But they also reflect the effects of living on reservations with ready access to fatty and sugary foods, limited access to fresh fruits and vegetables, and high rates of poverty, which can lead to poor nutrition and in the long run to diabetes (Benyshek, Martin, and Johnston, 2001). In both cases, to blame unhealthy behavior patterns on individual choices seems simplistic.

As these examples suggest, truly refocusing upstream requires us to look beyond individual behavior or characteristics to what McKinlay refers to as the **manufacturers of illness**: those groups that promote illness-causing behaviors and social conditions. These groups include alcohol distributors, auto manufacturers that fight against vehicle safety standards, and politicians who vote to subsidize tobacco production.

An article by Ali Mokdad and his colleagues (2004), published in the *Journal of the American Medical Association*, provides a useful starting point for refocusing upstream. The article synthesizes the available literature on the major underlying causes of premature deaths—that is, deaths caused neither by old age nor by genetic disease. Table 2.2 shows these causes and their prevalences (listed not by disease but by the factors that cause disease), as well as a tenth cause (medical errors) that other researchers have identified. The next section looks at these ten causes of illness.

Tobacco

As Table 2.2 shows, tobacco is the number one source of preventable deaths in the United States. Whether smoked, chewed, or used as snuff, tobacco can cause an enormous range of disabling and fatal diseases, including heart disease, strokes, emphysema, and numerous cancers (WHO, 2014a). Up to half of all smokers will die because of their tobacco use, losing an average of 15 years from their normal life expectancy. Tobacco use also increases morbidity and mortality

TABLE 2.2 Underlying Causes of Premature Death in the United States

Cause	Number	Percentage of All Deaths
Tobacco	435,000	18
Medical errors	400,000	>13
Diet or activity patterns[a]	100,000–400,000	17
Alcohol	85,000	4
Bacteria and viruses[b]	75,000	3
Toxic agents	55,000	2
Motor vehicles[c]	43,000	2
Firearms	29,000	1
Sexual behavior	20,000	1
Illicit use of drugs	17,000	1

[a]Estimates vary.

[b]Does not include deaths related to HIV, tobacco, alcohol, illicit drugs, or infections caused by nonmicrobial diseases.

[c]Includes motor vehicle accidents linked to drug use, but *not* to alcohol use.

SOURCES: Mokdad et al., 2004; HealthGrades, 2004; James, 2013.

among "passive smokers," those who live and work around smokers (WHO, 2014a). Similarly, both active and passive smoking can cause birth defects and infant mortality. Unfortunately, quitting smoking is difficult because nicotine (the active ingredient in tobacco) is more addictive than heroin (Weil and Rosen, 1998).

Given nicotine's addictiveness, it's easy to understand why individuals continue smoking once they have started. But why do individuals begin smoking in the first place, especially when many initially find tobacco vile tasting and even nauseating? To answer this question, we need to look at the role played by tobacco manufacturers.

Since the 1960s, when research first proved the link between smoking and lung cancer, tobacco manufacturers have labored to convince the public—especially youths, women, and minorities—to associate tobacco with positive attributes rather than with death and disability (Luke, Esmundo, and Bloom, 2000). To target youths and minorities, manufacturers have advertised in movie theatres and at sports events. To target women, manufacturers have played on women's desire for equality, excitement, personal fulfillment, and weight loss. This strategy was exemplified by the campaign for Virginia Slims—the name was not accidental—and its slogan, "You've come a long way, baby."

Over the past decade, successful legal attacks on tobacco manufacturers and advertisers have begun to erode their ability to attract new customers. For example, tobacco companies can no longer use cartoon characters in advertisements and now must limit their sponsorship of sports and entertainment events. Public health campaigns have also had an impact. Partly because of these campaigns, Americans

FIGURE 2.1 Substance Use in Past 30 Days Among US High School Seniors

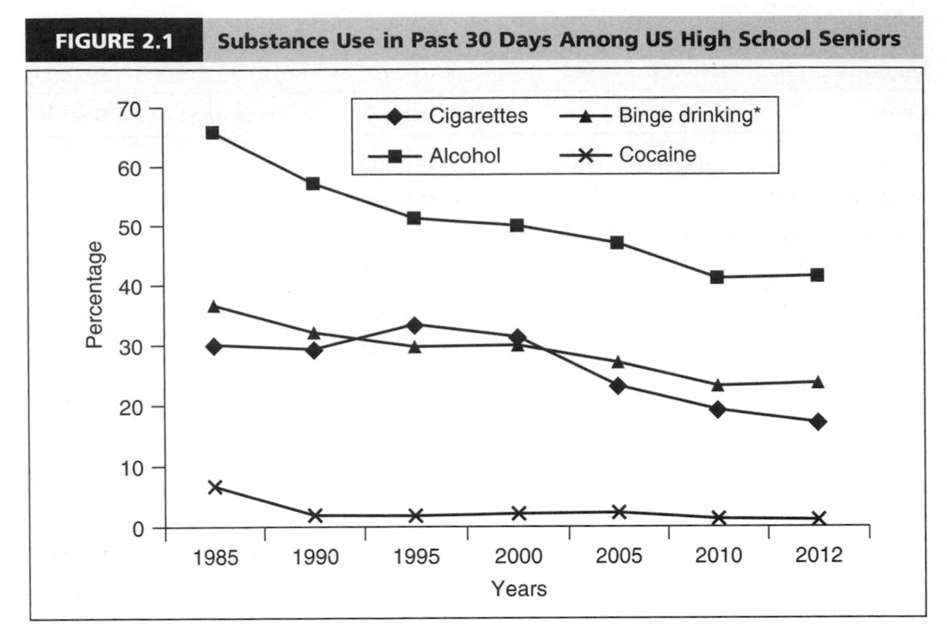

*Five or more alcoholic drinks in a row at least once in last two weeks.

SOURCE: National Center for Health Statistics (2014a).

increasingly support the idea of "smoke-free" areas and a smoke-free culture (Givel and Glantz, 2004; National Center for Health Statistics, 2014a). Figure 2.1 illustrates the decline of cigarette use among high school seniors.

Medical Errors

Surprisingly, recent research suggests that medical errors cause more preventable deaths than any factor other than tobacco (James, 2013; HealthGrades, 2004; Leape and Berwick, 2005). Medical errors include conducting surgery on the wrong patient, giving a patient two drugs that interact dangerously, or misdiagnosing and hence mistreating a patient.

When errors occur, it is natural to focus on identifying who is to blame. Yet most errors occur despite the best efforts of health care workers (Kohn, Corrigan, and Donaldson, 1999). Consequently, most researchers argue that we need to shift our focus from individual errors to problematic systems. For example, many hospitals stock certain drugs only at full strength, even though the drugs must be diluted to use safely. Stocking these drugs in diluted form would eliminate this source of death much more effectively than trying to identify every doctor or nurse who might administer the wrong dosage. Similarly, fatal errors can easily occur when different drugs have similar names: Someone with epilepsy, for example, who receives the antifungal drug Lamisil instead of the antiepileptic Lamictal can die if his seizures continue unabated. As this suggests, most fatalities result from the combination of human error with systems that facilitate errors.

The lack of a system for identifying deaths caused by medical errors has hampered efforts to prevent such deaths. During the 1950s and 1960s, hospitals routinely autopsied about half of the patients who died in their care. Now, because of a combination of economic costs and fear that identifying errors might lead to malpractice claims, hospitals autopsy only about 5 percent, thus virtually eliminating one of medicine's most basic tools for identifying medical errors (Burton and Collins, 2014).

Medical culture, too, makes it difficult to control medical errors. Research consistently finds that doctors rarely focus on identifying such errors (Bosk, 2003; Orlander and Fincke, 2003; Pierluissi et al., 2003). Instead, because of professional etiquette, the need to maintain good relations with colleagues, and a medical culture that values individual doctors' right to make their own decisions, most errors are ignored, labeled unavoidable, blamed on nonmedical staff, or blamed on doctors in other divisions (Pierluissi et al., 2003).

Despite all of these problems, the situation has improved (Leape and Berwick, 2005). Most importantly, there is now widespread agreement among doctors, insurers, researchers, the public, and the government that medical errors are a problem, and there is growing agreement that systemic changes are needed. For example, Veterans Administration hospitals now use a computerized record system that gives nurses and doctors access to comprehensive information on their patients. In addition, the record system generates bar-coded strips that are attached to each nurse, patient, and medication. Before administering medications, nurses must scan their own bar code, their patients' bar codes, and the medications' bar codes into a computer. The computer then checks that the nurse has the right drug for the right patient and that the drug won't interact dangerously with any other drug taken by that patient. Since adopting this system, medication errors have dropped 70 percent (Leape and Berwick, 2005). The federal government now provides this record system for free to all US doctors who treat patients under **Medicare**, the federally funded insurance program for elderly and permanently disabled individuals. In addition, the Affordable Care Act (sometimes referred to as "Obamacare") now offers various incentives aimed at encouraging health care providers to adopt computerized record systems. (The Act and its effects are discussed further in Chapter 8.)

Alcohol

Like tobacco, alcohol kills far more people than do all illegal drugs combined. Heavy alcohol use can cause irreversible brain damage, hepatitis, heart disease, cirrhosis of the liver, and cancers of the digestive system while reducing the body's ability to fight infections such as tuberculosis and pneumonia. In addition, by diminishing individuals' ability to make rational choices, alcohol use contributes to deaths from drownings, fires, violence, and accidents, and increases the odds of engaging in unsafe sexual behavior. Yet the US government's "War on Drugs" targets only illegal drugs.

ETHICAL DEBATE

Drug Testing in Schools and Workplaces

Currently, the federal government requires all federal job applicants, as well as randomly selected federal employees who hold "safety sensitive" positions, to take urine or blood tests to detect illegal drug use. Similarly, many businesses use tests to identify employees or job applicants who use either illegal drugs *or* legally prescribed drugs (such as hydrocodone) that can impair performance (Zezima and Goodnough, 2010). Many schools, too, require students to test negative for illegal drugs, and sometimes for alcohol and tobacco, before they can participate in extracurricular activities such as sports, chess clubs, and language clubs (Steinberg, 1999).

US courts generally have ruled that use of drug tests by government agencies breaches the Fourth Amendment right to privacy unless the tests are necessary to protect public safety or unless other evidence suggests that an individual uses drugs. Courts generally have placed no restrictions on private employers' use of drug tests. Nor have they restricted schools from requiring drug tests for extracurricular activities, although schools may not use drug tests to determine eligibility for academic courses.

At first glance, the benefits of drug testing seem obvious. Students, employees, and potential employees who know they will be tested may refrain from using drugs, potentially reducing rates of accidents and violence. Moreover, reducing drug use may reduce absenteeism, tardiness, and insurance costs while improving student and worker performance.

But drug testing comes with a price. Those opposed to drug testing argue that testing inherently invades privacy because it involves taking urine or blood from an individual's body. Moreover, the only way to ensure a urine sample comes from a specific individual is to watch that individual urinate—an obvious invasion of Western norms of privacy. In addition, drug testing constitutes an invasion of privacy because it can reveal much more than just illegal drug use. For example, the same tests that identify use of illegal drugs can identify legal use of drugs to control epilepsy, manic depression, or schizophrenia. Individuals identified in this way may experience not only social embarrassment but also discrimination and even loss of employment.

To ensure that the government continues to treat alcohol as a beverage rather than a drug, alcohol manufacturers contribute heavily to political campaigns (Center for Responsive Politics, 2011). Manufacturers also have worked to define the individual drinker rather than alcohol itself as the problem by promoting the idea that alcoholism only affects susceptible individuals, funding research on presumed biological roots of alcoholism, supporting laws that make it illegal for minors to drink, and opposing laws that would make it illegal to sell alcohol to minors (Mosher, 1995).

At the same time, alcohol manufacturers have endeavored to sell drinking to the public as a pleasurable "lifestyle." Much of this marketing either directly or indirectly targets youths. For example, manufacturers are most likely to advertise in magazines, on television and radio shows, during athletic events that attract large youth audiences, and at popular spring break destinations (Centers for Disease Control and Prevention [CDC], 2006; Garfield, Chung, and Rathouz, 2003; Kwate, Jernigan, and Lee, 2007; Zwarun, 2006). In addition, alcohol manufacturers have increased sales to youths by developing "alcopops": extra-sweet, fruit-flavored

Finally, drug testing invades privacy because it measures not only what a person does in school or on the job but also what he or she does during his or her free time, given that some drugs can linger in the body for months.

In addition, those who oppose drug testing in the workplace also question why, if the purpose of testing is to identify workers whose performance is impaired, we measure drug use rather than job performance. After all, some individuals who use drugs nevertheless perform adequately, and some who *abstain* from drugs perform poorly; many doctors believe, for example, that the vast majority of those who use legally prescribed narcotics under medical supervision experience no impairment. At any rate, most drug-related impairment in the workplace stems from alcohol use, which goes untested.

Finally, opponents of drug testing argue that the potential benefits of testing are far outweighed by the potential for harm when individuals are falsely labeled as drug users. In fact, as many as 40 percent of those identified as drug users by urine tests have not actually used such drugs.

As these problems suggest, developing a responsible policy regarding drug testing will require us to find a balance between public safety and protection of individual rights.

Sociological Questions

1. What social views and values about medicine, society, and the body are reflected in current drug testing policies? Whose views are these?
2. Which social groups are in conflict over this issue? Whose interests are served by strict drug testing policies? Whose interests are served by more lenient policies?
3. Which of these groups has more power to enforce its view? What kinds of power do they have?
4. What are the intended consequences of this policy? What are the unintended social, economic, political, and health consequences of this policy?

alcoholic beverages such as Hard Lemonade, Blast, and Skyy Blue. Advertisements for these and other alcoholic beverages typically associate alcohol with adulthood, sexual adventure, status, freedom, excitement, and pleasure. On the other hand, a combination of stricter laws, stricter enforcement of laws, and public health campaigns about the dangers of alcohol use have led to significant declines in alcohol use and binge drinking among high school seniors, as Figure 2.1 shows.

Illegal Drugs

Although far less deadly than tobacco or alcohol, illegal drugs nevertheless rank among the top ten causes of premature death. Illegal drugs can kill users through overdose, suicide, motor vehicle injury, HIV infection, pneumonia, hepatitis, and endocarditis (heart infections). In addition, they can kill nonusers by contributing to homicide and birth defects (Mokdad et al., 2004). Similarly, illegal drug use can contribute to dangerous behaviors. *Ethical Debate: Drug Testing in Schools and Workplaces* discusses some policies that have emerged in response to concerns about illegal drug use.

Common Illegal Drugs The two illegal drugs that most often cause mortality and morbidity are heroin and cocaine (including "crack" cocaine). Both drugs can cause physical addiction, although cocaine is usually used in quantities too small to do so (Weil and Rosen, 1998). Cocaine provides such great pleasure so briefly, however, that some individuals use it as often as possible, creating the appearance that they are addicted. As a result, both heroin and cocaine can cause people's lives to spin out of control.

The main health risk linked to cocaine use is severe sleep disturbance, which can lead to paranoia and occasionally to violence (Liska, 1997; Weil and Rosen, 1998). Cocaine also may increase the risk of death, although evidence is limited (Degenhardt et al., 2011). The good news is that its use among high school seniors has dropped considerably, as Figure 2.1 shows. (No comparable data is available on use of heroin or methamphetamines.)

Although heroin causes no direct damage to the human body, an overdose can lead to death, usually by suppressing the natural inclination to breathe. However, new, easily administered nasal sprays are now available that can quickly counteract overdoses of heroin (and other similar drugs). These sprays have been used very successfully by police forces and emergency medical providers around the country, and will soon be available by prescription to individuals as well.

The Impact of Illegality Added to the inherent dangers of heroin, cocaine, methamphetamine, and other illegal drugs are the dangers caused by their illegality. As mentioned earlier, when drug users can't obtain clean needles legally, they are likely to share needles and thus increase their risks of HIV/AIDS, hepatitis, and other infections. Similarly, users who buy drugs on the street can't know how powerful the drugs are. For example, individuals who typically inject heroin that is 30 percent pure can die if they accidentally buy heroin that is 60 percent pure, thus doubling their usual dosage.

Similarly, individuals who must purchase drugs at the extraordinarily high prices charged by illegal sellers are quickly ground into poverty. Once they no longer can afford proper food, clothing, or shelter, their vulnerability to all sorts of illnesses increases. Moreover, the high prices of illegal drugs can pressure users to engage in crime, including violent crimes that can damage the health of others.

Research consistently shows that prevention and treatment programs are both cheaper and more effective than criminal sanctions in reducing the use and social costs of illegal drugs (Amaro, 1999). Unfortunately, most government funding for drug control goes to the criminal justice system rather than to prevention or treatment.

Diet, Exercise, and Obesity

According to Mokdad and his colleagues (2004), a high-fat diet and sedentary modern lifestyles have led to soaring rates of obesity across all age groups, as Figure 2.2 shows. That obesity, they argue, has led to premature deaths from cardiovascular disease, strokes, certain cancers, and diabetes. Moreover, even when

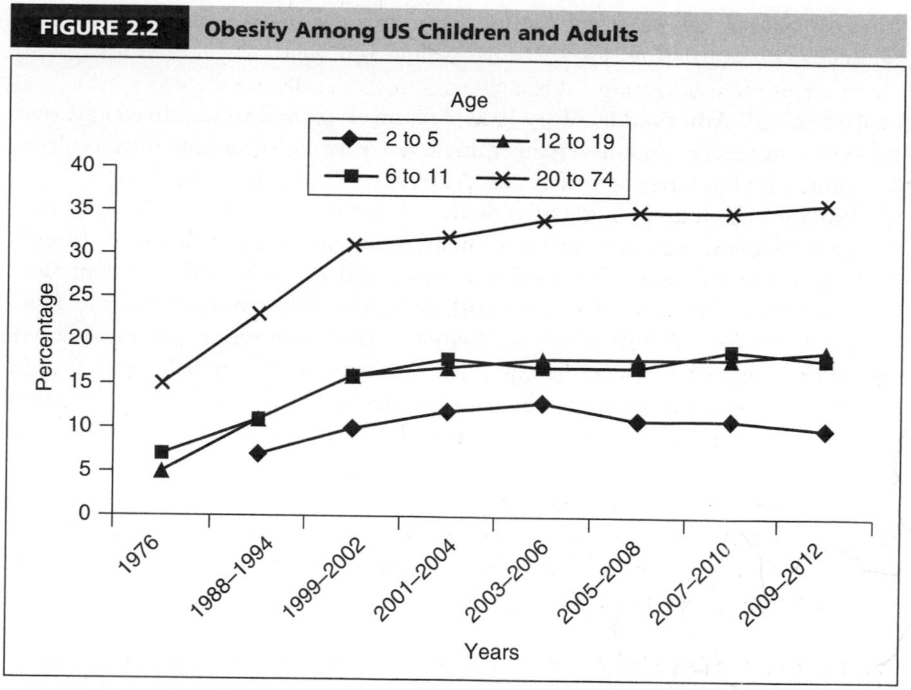

FIGURE 2.2 **Obesity Among US Children and Adults**

SOURCE: National Center for Health Statistics (2014a).

obesity doesn't kill, it can lead to heart conditions, diabetes, sleep difficulties, and other problems that diminish individuals' quality of life (CDC, 2012).

The Obesity Myth? At the same time, however, research has accumulated suggesting that the dangers of excess weight have been overstated (Campos, 2004; Gibbs, 2005). The best current research (Flegal et al., 2005, 2013) strongly suggests that death rates (after controlling for smoking, illness, and other factors) are indeed highest among those who are obese (e.g., 5′ 6″ and more than 180 pounds). But death rates are *second highest* among those who are *underweight* (e.g., 5′ 6″ and less than 112 pounds). Moreover, death rates are *lowest* in those who are overweight, but not obese; those considered normal weight have the second lowest death rates.

Why have studies disagreed so dramatically about the impact of weight on mortality? Earlier studies were based on narrow populations (such as middle-aged nurses), relied on self-reported weights and heights, controlled statistically for few variables, and may not reflect current conditions. In addition, many of the studies that stressed the health risks of obesity were funded by the diet industry, which may have colored their findings. In contrast, recent studies that have questioned the dangers of overweight and obesity were conducted by government researchers whose research studies were primarily federally funded. Finally, considerable research suggests that the real danger is not obesity, but physical activity and fitness (Blair and Church, 2004): Those who are obese but physically fit have half the death rate of those who are normal weight but *unfit*.

"Supersizing" Americans Why have Americans gained so much weight over the past generation? To answer this question, we need to look at how biology, economics, and politics interact (Critser, 2003).

Biologically, humans naturally desire sweet and fatty foods. In past eras, when food was scarce, these cravings helped humans stay alive. Now, though, most Americans have plentiful access to food and eat more calories than their bodies can use, leading in the long run to weight gain. In addition, the rapid adoption by food manufacturers of high-fructose corn syrup (an inexpensive sweetener) and palm oil (an inexpensive fat) may have spurred rapid weight gains because the former is metabolized by the body differently than are other sugars, and the latter is an especially saturated fat (Critser, 2003).

To these changes in *what* Americans eat were added changes in *how much* Americans eat. Longer work weeks and the increase in two-earner and single-parent families have led Americans to eat out more than ever before: Between 1970 and 2009, the percentage of Americans' food expenditures devoted to eating out increased from 25 to 41 percent (Bureau of Labor Statistics, 2010). Yet restaurant foods typically include far more fat and sugar than do home-made meals. Moreover, since the 1970s, restaurants increasingly have offered "supersized" portions, buffet tables, and packaged "value meals." Unfortunately, few individuals can restrict their calorie intake when offered large, varied meals, so this trend has increased calorie consumption. For these reasons, the rise in eating away from home has increased Americans' calorie consumption by an average of 1400 calories per person per week (Critser, 2003:33). For example, teenage boys who eat at fast-food restaurants three or more times a week consume 800 more calories *per day* than do those who avoid fast food (French et al., 2001).

Food manufacturers and the fast-food industry have used advertising to further encourage Americans to eat a sweet, fatty, high-calorie diet. Because manufacturers earn far less money selling healthy foods (such as fruits and vegetables) than by selling highly refined products loaded with fat, sugar, and salt (such as soft drinks and convenience foods), they spend 30 times more money on advertising the latter foods (Nestlé, 2002:22). Such advertising has grown increasingly insidious and now pervades every sphere of our society—especially those where children can be found. Soft-drink companies, for example, encourage sales of their products to children not only by advertising on television, in magazines, and on the Internet but also through such tactics as placing their products in movies, sponsoring school sports teams, and offering cash bonuses to schools that place soda machines (provided by manufacturers) at cafeteria entrances. Consumption of sugar-sweetened soft drinks is directly and substantially related to obesity and diabetes among both adults and children (Apovian, 2004).

Meanwhile, as caloric consumption has increased, physical exercise has decreased. Because of budget cuts, almost half of all US high school students no longer take any physical education classes (Gerberding and Marks, 2004). And at home, few children these days are allowed to spend their afternoons running free or playing non-organized sports. Instead, poor children are admonished to stay indoors to stay safe, and more-affluent children are shepherded from tutors to classes, to the occasional sports activity. Finally, few children or adults nowadays commute by foot or bicycle to work, play, or shopping, so physical activity is no longer built into most Americans' daily lives.

Motor Vehicles

Mokdad and his colleagues (2004) attribute 2 percent of all premature deaths to motor vehicle accidents (including accidents involving drug but not alcohol use). These deaths are not a necessary byproduct of modern life. Rather, they reflect in part a series of decisions regarding the design of automobiles and transportation systems.

The good news is that deaths from motor vehicle accidents have declined substantially since 1966, when Congress established the National Highway Traffic Safety Administration to regulate motor vehicle design and oversee highway safety programs. Changes in street and highway design, greater enforcement against drunk driving, and public education campaigns against drunk driving have reduced the number of accidents. Meanwhile, required changes in car design, coupled with use of seat belts and child car seats, have dramatically reduce the chances that accidents can cause death or serious injury.

That said, automobile manufacturers have continued to oppose inexpensive improvements that could save thousands of lives yearly, such as strengthening bumpers or covering instrument panels with softer materials to prevent head injuries. Equally important, legislators and government regulators have continued to exempt vans, multipurpose vehicles, and light trucks—which now account for more than 50 percent of all noncommercial vehicle sales—from passenger car safety regulations, even though most consumers use these vehicles as family cars.

It's also important to recognize that Americans die so often in motor vehicle accidents simply because they drive so much. Partly this is because driving distances here can be so long. But it's also because e.g., Americans have far fewer options than do citizens of other wealthy nations. Through a series of local and federal decisions, public transportation in this country has declined significantly since its apex in the 1920s (Hayden, 2003). Trains and railroad tracks have decayed while federal dollars have subsidized highway construction and motor vehicle production. Long-distance bus systems run for profit have eliminated money-losing connections to many smaller communities. Meanwhile, cities spend billions for parking facilities, road construction, and road maintenance but offer bus service only to limited locations, during limited hours, and on a limited schedule. Consequently, whereas a French citizen can use publicly subsidized trains or buses to go to any town or city in France on any given day and probably at several different times, an American citizen often has no way to go by public transportation from one town to the next. For example, Phoenix, Arizona, is the sixth largest city in the United States but has no passenger rail service.

Firearms

According to Mokdad and his colleagues (2004), firearms account for 1 percent of all premature deaths in the United States: 16,586 suicides, 10,801 homicides, 776 accidental deaths, and 270 deaths by police. Importantly, firearms ownership and death from firearms are exceptionally common in the United States (Krug, Powell, and Dahlberg, 1998). Studies have found that having a gun in the home significantly increases the odds of suicide, homicide, and unintentional shooting deaths of children (Kellerman et al., 1993).

Those who support firearm ownership typically argue that guns protect honest citizens from attacks by criminals. Yet guns are used far more often against noncriminals than against criminals. Furthermore, owning a gun increases the chances of being killed even when a home is forcibly entered or a victim attempts to resist (Kellerman et al., 1993).

Although interest in gun control rises sharply after each mass murder (such as the 2014 shooting in which 6 were killed and 13 injured near the University of California–San Diego campus), this interest has not translated into widespread legislative changes. Those favoring gun control face heavy financial odds, for the "gun rights" lobby routinely donates about ten times more to federal candidates than does the "gun control" lobby (Center for Responsive Politics, 2010). Moreover, in June 2008, the US Supreme Court affirmed for the first time that individuals have a right to own guns (with certain exceptions). Nevertheless, the battle against gun violence still continues.

Sexual Behavior

Mokdad and his colleagues (2004) attribute 1 percent of premature deaths to sexual behavior, primarily via hepatitis B, HIV/AIDS, and cervical cancer. The first two are directly transmitted through sex and the last is most often caused by human papillomavirus (HPV), a sexually transmitted virus. Mokdad and his colleagues also include in this category infant mortality after unplanned and unwanted pregnancies.

No "manufacturer of illness" benefits from convincing people to engage in sexual activity without protecting themselves against disease or pregnancy, but social conditions can encourage such behavior. First, those forced by economic necessity to turn to prostitution to support themselves, whether male or female, often find that they can't suggest safer sex to clients without losing business or risking violence. Similarly, those whose intimate relationships are not based on mutual respect and equality sometimes find that suggesting safer sex to their romantic partners results in violence or abandonment (Wingood and DiClemente, 1997). Finally, those who have learned to have little hope for the future—a sentiment particularly common among youths in communities wracked by racism and poverty—sometimes believe they have little to lose by engaging in unsafe sexual activity (Plotnick, 1992).

Other sexually active individuals, however, do fear sexually transmitted diseases (STDs) and pregnancy but lack knowledge about safer sexual practices or access to birth control. Most US schools now solely provide "abstinence-only" education, even though research overwhelmingly suggests it doesn't work (Kohler, Manhart, and Lafferty, 2007). Such education delays individuals' first sexual intercourse by only about three months, while significantly reducing the odds that condoms will be used.

Meanwhile, access to birth control and abortion has declined. Cuts in public funding for contraceptive services have reduced options for teenagers and low-income women, the groups most at risk for unplanned pregnancies and infant mortality. Similarly, the federal government will pay for abortions for women on **Medicaid** (the government-funded health insurance program for poor persons) only if the woman's life is endangered. Meanwhile, cutbacks in government funding for abortions, strict new laws and regulations, and harassment or even violence against abortion providers, have reduced the number and geographic distribution of abortion providers. Moreover, few medical schools now teach how to perform abortions, making it the only medical procedure that doctors can refuse to learn. Currently, 38 percent of US women live in counties without any abortion provider (Allan Guttmacher Institute, 2014). Other restrictions, such as requiring waiting periods or parental consent before abortions, also limit access, especially for poor and young women. Yet despite these restrictions, abortion remains common: An estimated one-third of all US women will have an abortion at some time during their lives (Allan Guttmacher Institute, 2010). As a result, preserving the safety of abortion services is an important health issue.

Bacteria and Viruses

Bacteria and viruses surround us all the time. Yet only rarely do individuals become infected, and even more rarely do these infections lead to deaths. Under what conditions do these deaths occur?

First, individuals won't develop fatal diseases if they are vaccinated against them. Virtually all US children are vaccinated before they begin school, but about one-quarter don't receive all the required vaccinations by the recommended ages (National Center for Health Statistics, 2014a).

Second, even in the absence of vaccinations, individuals exposed to microorganisms may not become infected unless they already are physically weakened. For example, a significant percentage of all persons admitted to hospitals—a population that obviously is already physically vulnerable—develop infections while in the hospitals, some of them life threatening. Similarly, individuals are far more susceptible to infection if age, malnutrition, poor housing, insufficient clothing, or other difficulties weaken their bodies. This explains why American tourists rarely contract tropical diseases when they travel to countries where disease is endemic, even if they don't get vaccinated and don't take drugs to prevent infection.

Third, the same factors that leave some susceptible to infection help explain why, among those who do become infected with a given disease, some die whereas others experience only minor health problems. Measles, for example, is a minor childhood disease in the United States but a major killer in poorer countries (as described in Chapter 4).

Fourth, among those who become ill, death or long-term disability may not occur if individuals have ready access to good health care. For example, doctors can cure most bacterial infections in otherwise healthy individuals, and simply providing intravenous nutrition and fluids can save the lives of many infants who experience life-threatening diarrhea.

Toxic Agents and Risk Societies

Mokdad and his colleagues (2004) trace 2 percent of premature deaths to **toxic agents**: substances that can harm or kill people or other organisms. These agents can be divided into occupational hazards and environmental pollutants. In "light" industries such as electronics, workers are often exposed to a wide variety of potentially toxic solvents, such as trichloroethylene (TCE); in traditional industries such as mining and construction, welders often face substantially increased risks of lung cancer caused by toxic levels of chromium and nickel. Similarly, agricultural workers are often exposed to dangerous pesticides (as described in Chapter 3).

Unlike occupational hazards, environmental pollution most threatens children because of their still-growing bodies and immune systems, the time they spend playing outdoors, and their tendency to play on the ground and put things in their mouths (US Environmental Protection Agency, 2008). Many forms of environmental pollution threaten children. Flaking lead paint in old houses or apartments, which can be tempting to very young children, can cause mental retardation, learning disabilities, hearing deficiencies, hyperactivity, and other problems. Pesticides on fruits and vegetables or in the air near farm fields can cause cancers and other disabilities. Air pollution can lead to asthma, bronchitis, and other respiratory problems. And hazardous waste sites and contaminated water can cause birth defects, bacterial infections, and other health problems.

In the long run, the greatest environmental health threat may be climate change. During the past quarter century, carbon dioxide and synthetic gases, especially chlorofluorocarbons (CFCs) such as Freon, have mushroomed. According to the Intergovernmental Panel on Climate Change (2007), a joint venture of the World Meteorological Organization and the United Nations Environment Programme, these chemical byproducts of industrial manufacturing have damaged the ozone level surrounding the planet and have caused increased smog, dangerously erratic rainfall patterns, and extreme temperature changes—both up and down—around the globe. Debate continues about the consequences of climate change, but many scientists suspect that damage to the ozone layer will lead to more cancers (especially skin cancer), increased smog will lead to more cases of bronchitis and emphysema, and rising temperatures will lead to more mosquitoes and thus to more people infected with mosquito-borne diseases such as malaria and dengue fever (Armelagos and Harper, 2010).

All of these hazards posed by toxic agents result directly from the modern risk society. Sociologists use the term **risk society** to refer to any society that depends so heavily on potentially dangerous modern technologies that the risks from such technologies become commonplace and accepted (Beck, 1992, 2006).

Individuals who become ill due to these risks face an uphill battle in gaining recognition for their illnesses (Brown et al., 2002; Brown, Kroll-Smith, and Gunter, 2000). By definition, these risks stem from technologies deemed crucial to a society, such as the production of oil, chemical fertilizers, and biological weapons. Because these technologies bring considerable wealth and power to governments and corporations, those institutions have a vested interest in maintaining the status quo. As a result, individuals can find it exceptionally difficult to win acknowledgment for their health problems, especially when they have access only to doctors who work for the government or corporations.

THE HEALTH BELIEF MODEL, HEALTH LIFESTYLES, AND HEALTH "PROJECTS"

It is no secret that tobacco, guns, and sex without condoms can kill. So why do some people engage in behaviors that place their health at risk? Conversely, why do others make maintaining their health an all-consuming project? To address these questions, sociologists turn to the concepts of the health belief model, health lifestyles, and health projects.

The Health Belief Model

Within the health care world, **compliance** refers to individuals' willingness to follow medical advice. The most commonly used framework for studying compliance to medical advice is the **health belief model** (Becker, 1974, 1993; Rosenstock, 1966). The model was developed to explain why healthy individuals adopt healthy behaviors. According to the model, four factors affect these decisions: Individuals must believe (a) that they are susceptible to a particular health problem, (b) that the problem is serious, (c) that adopting preventive measures will reduce their risks significantly, and (d) that no significant barriers make it difficult for them to adopt those measures. For example, people are most likely to adopt a low-fat diet if they believe that otherwise they will face high risks of heart disease; that heart disease will substantially decrease their life expectancy; that a low-fat diet will substantially reduce their risk of heart disease; and that adopting such a diet won't be too costly, inconvenient, or unpleasant. In turn, according to the health belief model, these four factors are affected by demographic variables (such as the individual's gender and age), psychosocial variables (such as personality characteristics and peer group pressures), structural factors (such as access to knowledge about the problem and contact with those who experience the problem), and external cues to action (such as media campaigns about the problem or doctors' advice). *Key Concepts: The Health Belief Model* outlines how this model works.

Although this model recognizes that social factors as well as individual psychological factors affect health decision making, in practice, it is most often used to

KEY CONCEPTS

The Health Belief Model

People Are Most Likely to Adopt Healthy Behaviors When They ...	Example: Adopting Healthy Behaviors Likely	Example: Adopting Healthy Behaviors Unlikely
Believe they are susceptible	40-year-old smoker with chronic bronchitis who believes he is at risk for lung cancer	16-year-old boy who believes he is too healthy and strong to contract an STD
Believe risk is serious	Believes lung cancer would be painful and fatal and does not want to leave his young children fatherless	Believes that STDs can all be easily treated
Believe compliance will reduce risk	Believes he can reduce risk by stopping smoking	Doesn't believe that condoms really prevent sexual diseases
Have no significant barriers to compliance	Friends and family urge him to quit smoking, and he can save money by so doing	Enjoys sexual intercourse more without condoms

explain individual choices. In other words, researchers who use this model tend to emphasize **agency**—individual free will to make choices—over **structure**—social forces that limit the choices individuals realistically can make (Cockerham, 2005). As a result, such researchers, along with most policy makers, more often promote policies such as educating consumers about the dangers of smoking than policies such as banning smoking in public places. The debate over the relative importance of agency and structure—sometimes referred to as "life choices" versus "life chances"—is at the center of many theoretical discussions within sociology and, even more so, between sociology and other fields such as psychology and medicine.

Health Lifestyles

All human behavior is affected by both agency and structure. No one blindly follows every social rule and expectation. Nor is anyone fully free of socialization, cultural expectations, and social limitations on what options are truly available. Nevertheless, knowing the social groups that individuals belong to helps us predict their odds of adopting various health behaviors: Lower-class citizens are far more likely than upper-class citizens to smoke, men are far more likely than women to drink heavily, and so on. Consistent patterns such as these led sociologist William Cockerham to propose the **health lifestyle theory**. This theory acknowledges both agency and structure but emphasizes group rather than individual behaviors. Compared with the health belief model, this new theory offers a more comprehensive analysis of why healthy behaviors are or are not adopted (see *Key Concepts: Health Lifestyle Theory*).

Cockerham (2005:55) defines health lifestyles as "collective patterns of health-related behavior based on [life] *choices* from options available to people according to their life *chances*" (emphasis mine). According to this theory, decisions about healthy and unhealthy behavior begin with demographic circumstances, cultural memberships, and living conditions. These factors *directly* affect

KEY CONCEPTS

Health Lifestyle Theory

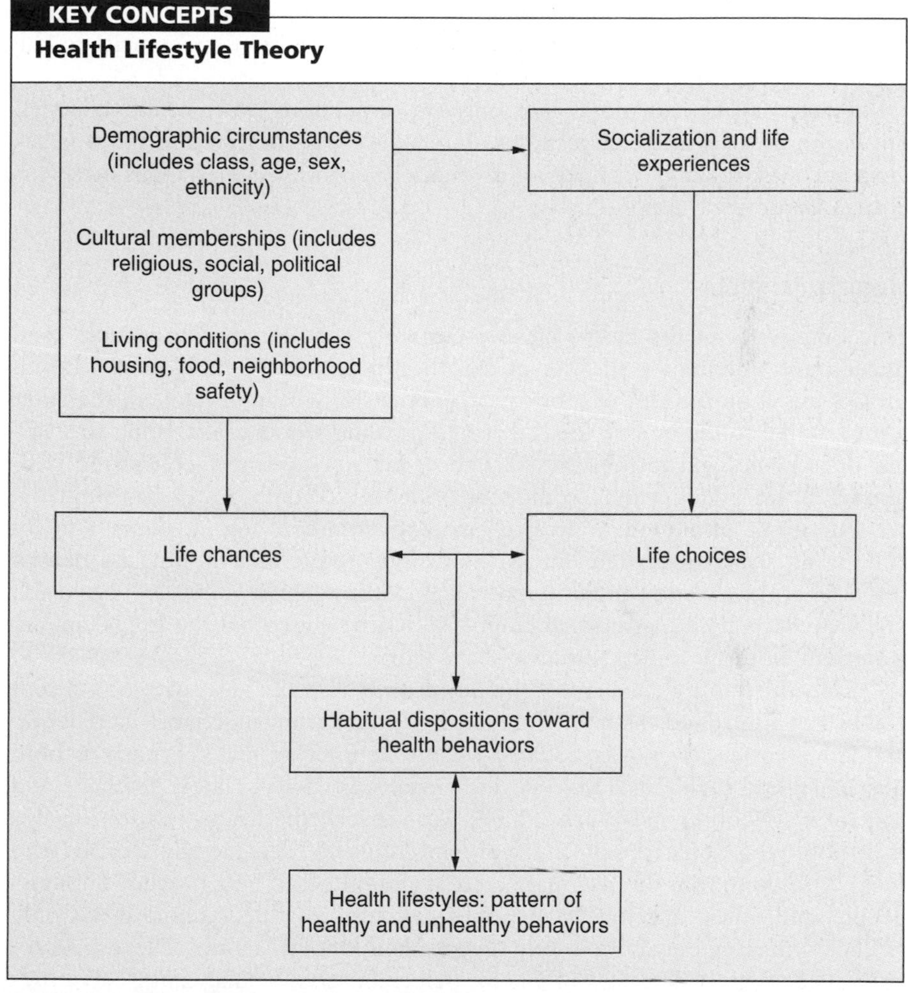

SOURCE: Cockerham (2005).

individuals' life *chances*, such as whether they have the education needed to avoid physically dangerous jobs. In addition, demographic circumstances, cultural memberships, and living conditions *indirectly* affect life *choices* through their effect on socialization and life experiences. Those who grow up with parents who consider all alcohol use immoral, for example, will be less likely to drink as adults than those whose parents considered alcohol to be just another beverage.

At the same time, life choices affect life chances and vice versa. For example, those who choose to drive safely (a life choice) are more likely to avoid injury (a life chance). Conversely, those who live in poverty (a life chance) may choose to drive fast (a life choice) because doing so is a cheap source of fun and because they don't expect to live long anyway. As this theory suggests, life choices and life chances come together to create **habitual dispositions** toward health behaviors—routine, almost instinctual ways of thinking about whether certain behaviors are or are not worth adopting. These dispositions are crucial to the health lifestyles adopted by individuals and groups.

Finally, Cockerham notes, not only do dispositions affect health lifestyles, but health lifestyles affect dispositions. As people's ways of thinking about behaviors such as smoking change, so do their behaviors. And as their behaviors change, so do their dispositions.

Health Projects

One outgrowth of the health lifestyle currently common among middle- and upper-class Americans is the rise of the "health project." The idea of a health project draws on the idea of a **body project**. As originally developed, the latter term referred to the intense focus that many young women now bring to shaping their bodies and to the ways that those activities are now considered both important work and central to individual identity (Brumberg, 1997). Similarly, we can speak of a modern **health project**, common among many affluent Americans, that requires individuals to actively protect their health and defines this as important work, central to individual identity (Shilling, 2001). This health project reflects both the modern emphasis on appearance and the long-standing American emphasis on the virtues of hard work.

Embedded in the concept of the health project is the idea that good health comes not from God, nature, or genes, but rather from individual hard work. Similarly, the health project is based on the assumption that the body is both unfinished and highly malleable, so individuals can always choose to shape and control it (Dworkin and Wachs, 2009). Conversely, the health project suggests that those who don't take on this work are "slackers," less morally worthy than those who do so. For this reason, it's not at all unusual to hear lawyers, businesspeople, and others offhandedly mention their various athletic injuries, not only to elicit sympathy but also to subtly suggest their "moral" worth. These ideas are reinforced by a wide range of media (television shows, magazines, advertisements, and others) that constantly exhort us to work on our diets, "abs," and cholesterol levels (Dworkin and Wachs, 2009).

SOCIAL STRESS AND SOCIAL NETWORKS

Even among those who smoke, drink, or engage in other risky behaviors, some individuals are more likely to survive to old age than others. And even among those who do everything "right," some will die earlier than others, even if those others share their genes, physical environment, and living conditions. To understand this paradox, sociologists have looked at the concepts of social stress and social networks.

Social Stress

The term **stress** has three major meanings. First, stress refers to *situations* that make individuals feel anxious and out of balance. Second, stress refers to the *emotions* that result from exposure to such situations. Finally, stress refers to the *bodily changes* that occur in response to these situations and emotions. For example, a fight with a friend is a stress; can produce the emotion of stress; and can lead to the physical stress of tensed muscles, rapid heartbeats, and heavy breathing. Stress can be either acute (such as the death of a spouse) or chronic (such as long-term loneliness or financial difficulties resulting from a spouse's death). Importantly, stress is often cumulative. An individual's **cumulative stress burden**—the sum of acute and chronic stresses that one has experienced—is a powerful predictor of ill health (Thoits, 2010).

Stress is a natural, unavoidable, and sometimes beneficial part of life. Thousands of years ago, hunters experienced stress as they anxiously prepared to track wild animals. That emotional stress put physical stress on their bodies, but it also kept their minds focused on their tasks. If, for example, a wild animal suddenly attacked, a hunter might survive because the emotional stress resulted in the physical stress response known as the fight-or-flight syndrome. The same quick heartbeat we experience while fighting with a friend could have saved the life of someone fighting a lion, because these physical changes help our bodies produce additional energy and oxygen and hence respond more quickly and effectively to threats.

Although the fight-or-flight response works well for dealing with sudden threats such as rampaging lions, speeding cars, and last-minute quizzes, it is far less useful for dealing with chronic stresses such as poverty or an ill child. Each time the body responds to a threat, it uses muscles, energy, and other resources. Over the long run, such stresses can wear out the body; lead to heart disease, diabetes, and other illnesses; and encourage individuals to adopt unhealthy behaviors such as smoking tobacco or having sex without condoms (Avison and Thomas, 2010; Sapolsky, 2004).

The impact of the stress, however, depends heavily on the nature of the stress: Studying for a short quiz, for example, is less stressful than failing a final exam. Stress is particularly likely to affect health when it stems from a "fateful loss," is physically exhausting, or disrupts social support systems (Avison and Thomas, 2010). For example, an accountant who loses his job, has to work exhausting double shifts as a cashier to replace his lost income, and no longer has the time, money, or energy to hang out with friends may experience dangerous levels of stress. As this suggests, chronic stress is especially important, diminishing individuals' abilities to ward off infections, depression, and other health problems (Avison and Thomas, 2010; House, 2002; Siegrist, 2010).

But even when exposed to similar levels of stress, some individuals are more susceptible to illness than others. The likelihood that stress will affect health depends in part on how individuals *appraise* the stress and how they cope with the stress. In turn, both of these responses to stress depend on the *social resources* individuals bring to the situation (Avison and Thomas, 2010). For example, flunking an exam is far more stressful for a student who risks losing his scholarship than for other students. It will also be less stressful if the student copes by quickly seeking out a good tutor rather than by getting high or blaming her grade on an incompetent teacher. But the student's ability to respond effectively will also be determined in part by her social resources: Has she learned from a young age to turn to alcohol as a coping measure? Do her friends encourage her to continue trying or to drop out? Does she have the funds needed to hire a tutor and the contacts needed to find a good one? The answers to each of these questions will affect whether this acute stress leads to chronic stress and, in the end, to ill health.

Gender, Race, Class, and Social Stress

As this suggests, the likelihood of experiencing traumatic social stress depends in part on one's position in society (a topic described in more detail in Chapter 3). Men, for example, more often experience traumatic physical injuries on the job, but women more often experience chronic stress from holding jobs while raising children. Women, however, typically have a greater *cumulative* stress burden, which may explain their higher rates of disability (Thoits, 2010).

Minorities, too, have higher cumulative stress burdens and resulting ill health. That stress burden includes the emotional burden of living with racist discrimination and prejudice (Thoits, 2010). In addition, discrimination and prejudice increase the odds that such individuals will be poor and (even if not poor) will live in neighborhoods characterized by poverty, neglect, crime, and pollution—all factors that can contribute to stress, illness, and injury. Immigrants, too, may face similar problems, especially if they belong to a stigmatized minority or have entered the country illegally.

Similarly, individuals with lower incomes and education levels experience more stress overall than do more affluent, better-educated individuals (Thoits, 2010). Poverty exposes individuals to a wide range of stresses, including hunger, worries over bills, poor living conditions, and physically exhausting work. Moreover, poor people are far more likely to hold jobs that combine *high demand*—bosses constantly pressuring workers to produce more and faster—with *low control*—bosses who offer workers few choices, even over seemingly small issues such as when to take bathroom breaks. Such working conditions are particularly likely to result in dangerous levels of stress (Siegrist, 2010).

Social Networks

Whereas social stress can lead to illness regardless of individuals' behaviors, social networks can help to *prevent* illness (Smith and Christakis, 2008; Thoits, 2010). **Social networks** are the webs of social relationships that link people to each other, whether as friends, relatives, acquaintances, coworkers, or in some other way. Because most social networks are relatively homogeneous—primarily linking people who share ethnicity, social class status, personality traits, political views, and so on—they tend to amplify the advantages and disadvantages that different social groups already experience.

Social networks affect health in various ways. Among other things, social networks offer individuals access to social support, financial assistance, health information, and other forms of aid that can help them stay (or become) healthy. In general, people with smaller social networks; only weak ties to others in their networks; or networks that tie them to poor, marginalized individuals will be less able to avoid or recover from illness, injury, or substance abuse (Smith and Christakis, 2008).

Networks also affect health by exposing individuals to specific social norms. If, for example, several individuals in a social network smoke tobacco, others who join that network may conclude that smoking is acceptable and therefore start smoking (Kaplan et al., 2001). Conversely, joining new social networks can *reduce* individuals' health risks when, for example, a student from a poor family transfers to a middle-class school and gains access to a new social group that frowns on tobacco use or encourages athletic activity.

IMPLICATIONS

Recent years have seen an increasing tendency to blame individuals for their own health problems (a topic discussed further in Chapter 5). Yet as we have seen, patterns of disease stem from social conditions as much as, if not more than, they stem from individual behaviors or biological characteristics. As Marshall Becker, a sociologist and one of the researchers who has done the most to help elucidate why people engage in health-endangering activities, writes:

> I would argue, first, that health habits are acquired within social groups (i.e., family, peers, the subculture); they are often supported by powerful elements in the general society (e.g., advertising); and they have proven to be extremely difficult to change. Second, for most people, personal behavior is not the primary determinant of health status and it won't be very effective to intervene at the individual level without concomitant attempts to alter the broader economic, political, cultural, and structural components of society that act to encourage, produce, and support poor health (1993:4).

In sum, to improve the public's health we must look beyond individual behavior and personal troubles to structural issues and, in C. Wright Mills's terms, public issues. Such a change in focus will enable us to address the underlying causes of illness and to ensure that national health policy is driven by concern for the public rather than concern for special interests.

SUMMARY

1. *Epidemiology* refers to the distribution of illness in a population. Epidemiologists rely on concepts such as life expectancy, mortality and morbidity rates, incidence, and prevalence. *Incidence* refers to the number of new occurrences of an event (disease, births, deaths, etc.) within a specified population during a specified period. *Prevalence* refers to the total of both new cases existing in a population at a given time and older cases that are still surviving.
2. Infectious illnesses again have become a growing source of illness and death in the Western world, partly because of overuse of antibiotics, changing physical environments, and globalization. HIV/AIDS is an example of the resurgence of infectious diseases.
3. Sociologists suggest that to improve the population's health, we should look beyond individual behavioral choices to the manufacturers of illness: groups that promote illness-causing behaviors and social conditions.
4. Research suggests that ten factors account for at least 50 percent of all preventable deaths. These factors, in order of importance, are tobacco, medical errors, diet and exercise, alcohol, bacteria and viruses, toxic agents, motor vehicles, firearms, sexual behavior, and illegal drugs. The dangers posed by toxic agents reflect life in a *risk society*, one in which dangerous modern technologies and the risks they pose have become commonplace and accepted.
5. The health belief model predicts that individuals will be most likely to adopt healthy behaviors if they believe they are susceptible to a problem, believe the problem is serious, believe changing their behaviors will decrease the risk, and face no significant barriers to so doing. Health lifestyle theory offers a more comprehensive analysis of why healthy behaviors are adopted by emphasizing social structure as well as personal agency.
6. Many middle and upper class Americans now focus intently on protecting their health and consider this work—known as a *health project*—central to their identity.
7. Social stress has three meanings: (1) situations that make individuals feel anxious and out of balance, (2) the emotions that result from exposure to such situations, and (3) the bodily changes that occur in response to these situations and emotions. Social stress, especially when chronic, can cause physical and mental health problems.
8. The likelihood that stress will affect health depends on how individuals appraise the stress and on how they cope with the stress, both of which

depend on individuals' social resources. It also depends on individuals' position in society.

9. Social networks are the webs of social relationships that link people to each other. Health risks are greatest among those with small social networks, only weak ties to others in their networks, or networks that tie them to poor, marginalized individuals.

REVIEW QUESTIONS

1. What is the difference between morbidity and mortality and between incidence and prevalence?
2. What factors have caused the recent increases in infectious diseases?
3. How is globalization affecting rates of disease?
4. How have the "manufacturers of illness" increased deaths caused by tobacco? By alcohol? By toxic agents? By diet?
5. How have social forces and political decisions increased deaths caused by sexual behavior? By illegal drugs?
6. What system-level factors help to explain medical errors? How does medical culture keep doctors from identifying medical errors?
7. Think of someone you know who smokes or engages in another unhealthy behavior. Use the health belief model to explain what would have to change for him or her to change this behavior. Then use health lifestyle theory to explain why you do or don't have a generally healthy lifestyle.
8. What are the benefits of the modern health project? What problems does the health project cause for those who adopt it? For those who don't?
9. How does social stress affect health? How do gender, race, and class affect average levels of social stress?
10. How can social networks reduce individuals' health risks? How can they *increase* those risks?

CRITICAL THINKING QUESTIONS

1. What are the *political* consequences of focusing on how social factors cause illness rather than focusing on biological factors?
2. This text identifies tobacco and alcohol as two of the most important underlying causes of premature death in the United States. What social policies would help stop the manufacture of illnesses by alcohol and tobacco in the first place? (Be sure you do *not* confuse this with policies that would

stop individuals from using these substances or would treat the health consequences of using these substances.)

3. Think of something you do (or believe you should do) to protect your health such as wearing seat belts, wearing bike helmets, drinking alcohol only moderately, eating fruits and vegetables, or flossing your teeth. Use the health belief model to explain why you do or don't take these precautions. (If you can't think of an example from your own experience, use an example from a friend or relative's life.)
4. First, use *each* of the *four* elements of the health belief model to explain why so few Americans eat five portions of fruits and vegetables daily. Second, explain why policy makers who want to improve Americans' diets need to *additionally* pay attention to *one* element from health lifestyle theory (you can choose any element).
5. Think of someone you know whose health is poor. How might social stress have worsened his or her health? How might this person's health problems have increased his or her stress?
6. Think of a social network you belong to (such as a fraternity, religious organization, athletic team), and discuss the effect of that social network on your health and health behaviors.

CHAPTER

3

The Social Distribution of Illness in the United States

Marka/Alamy

LEARNING OBJECTIVES

After reading this chapter, students should be able to:

- Understand how social class affects health and illness.
- Compare the major health issues faced by different ethnic groups in the United States.
- Analyze the combined impact of poverty and ethnicity on health.
- Assess the impact of age on health and illness.
- Describe how sex and gender can affect health and illness.
- Evaluate how social capital can affect health and illness.

Meat and poultry processing is one of the most dangerous jobs in the United States, with injury and death rates several times higher than in other occupations. These are jobs that only the poor will take, and increasingly are jobs filled only by undocumented Hispanic immigrants, many of whom barely speak English. One worker interviewed by observers from the nonprofit Human Rights Watch (2005) said:

> *The [meat processing] line is so fast there is no time to sharpen the knife. The knife gets dull and you have to cut harder. That's when it really starts to hurt, and that's when you cut yourself. I cut my hand at the end of my shift, around 10:30 at night.... I went to the clinic the next day at 11:00 a.m. They gave me stitches and told me to come back at 2:30 before the start of my shift to check on the stitches. They told me to go back to work at 3:00. I never stopped working (Human Rights Watch, 2005:35).*

Another man, with fingers swollen and bent nearly into claws, said:

> *I hung the live birds on the line. Grab, reach, lift, jerk. Without stopping for hours every day. Only young, strong guys can do it. But after a time, you see what happens. Your arms stick out and your hands are frozen. Look at me now. I'm twenty-two years old, and I feel like an old man (Human Rights Watch, 2005:36).*

And a woman said:

> *I pull ribs with my fingers on the packing ribs line. My fingers and nails are in constant pain because the company won't give us hooks to pull the ribs, and they won't let us bring our own hooks. We need hooks to pull the meat more easily and to avoid injuries. But they say that meat gets lost using hooks, and using fingers pulls more meat, so no hooks (Human Rights Watch, 2005:45).*

Complaints from workers about conditions at processing plants are few because those who complain are usually fired or deported (after plant supervisors call immigration authorities). And medical care is usually unavailable except from company doctors whose jobs depend on minimizing rather than treating workers' health complaints.

Although the conditions faced by meat and poultry workers are extreme, they illustrate how social class and ethnicity can leave individuals vulnerable to illness, injury, and death. In this chapter, we explore how these factors as well as age, sex, and gender result in an unequal distribution of health and illness across the population.

SOCIAL CLASS

Overview

Social class refers to individuals' position within a society's economic and social hierarchy. Most often, it is measured by looking at individuals' education, income, or occupational status, with some researchers using only one of these indicators and some combining two or more. Other researchers have argued for additional measures, with wealth perhaps the most important. For example, imagine two students who work together at Starbucks, earning the same income. Now imagine that one receives a new wardrobe and a trip to Europe from her parents every summer, but the other receives only a bus ticket home. These students have the same income, education, and occupation but differ in social class because they differ in wealth.

The link between social class and ill health is strong and consistent (Adler and Rehkopf, 2008; Hadler, 2008; Lahelma, 2010). For example, the food, shelter, and clothing available to poor Americans 200 years ago differed greatly from that available to poor Americans now, which in turn differs greatly from that available to poor Brazilians these days. Even so, in each place and era and for almost all illnesses, poor persons experience more illness than wealthier persons do. Because of this very strong link between social class and health across time, place, and disease, some sociologists label social class a "fundamental cause" of disease (Link and Phelan, 2010; Phelan et al., 2004).

Fundamental-cause theory argues that even though the common diseases and their causes may change over time and place, in each situation those with greater access to resources will experience better health because those resources help protect their health. For example, Link and Phelan write:

> [A] person with many resources can afford to live in a high-status neighborhood where … enormous clout is exerted to ensure that crime, noise, violence, pollution, traffic, and vermin have been kept at a minimum and the best health-care facilities, parks, playgrounds, and food stores are conveniently located nearby. Once a person has used [social class-based] resources to locate in an advantaged neighborhood, a host of health-enhancing circumstances comes along as a package deal (2010:6).

In such circumstances, wealthier individuals can increase their odds of good health without even trying.

The impact of social class on health is obvious: Around the world, in every age group and racial or ethnic group, those with higher social class status have lower rates of **morbidity** and **mortality** (Adler and Rehkopf, 2008; Lahelma, 2010; Marmot, 2002, 2004). The relationship between social class and health holds true for all major and most minor causes of death and illness, regardless of how researchers measure social class (Wilkinson, 1996, 2005). For example, in the United States heart disease occurs three times as often and arthritis twice as often among low-income persons compared to more affluent persons. Moreover, these health differences appear across the entire income scale, with each group on the social class ladder having better health than the group just below it (Marmot, 2004; Wilkinson, 1996, 2005). And when researchers **control** for all known individual risk factors (such as obesity and smoking), the impact of social class on health declines only slightly (Wilkinson, 1996, 2005).

That impact begins at birth, with infant mortality significantly higher among the poor (CDC, 1995; Matthews and MacDorman, 2010). Similarly, poor children are considerably more likely than other children to fall ill, have a chronic health problem, or die (Federal Interagency Forum on Child and Family Statistics, 1999). The evidence linking social class to health, then, is extremely strong. However, some sociologists (most notably Richard Wilkinson) have argued that **income inequality**—the *gap* in income between a nation's poorest and wealthiest—rather than income itself, may best explain why some nations are healthier overall than others (Wilkinson, 1996, 2005). These theorists point, for example, to data showing that as income inequality declined in the United States from about 1960 to 1980, life expectancy rose among all social classes (Krieger et al., 2008; Ezzati et al., 2008). More recent research, however, has convinced most sociologists that income in and of itself has a greater impact on health than does income inequality (Mechanic, 2006). For example, since 1980, income inequality has soared in the United States, but life expectancy fell only for *poorer* Americans. This suggests that income affects health more than does income inequality (Ezzati et al., 2008; Krieger et al., 2008).

The Sources of Class Differences in Health

How can we explain the link between poverty and illness? One explanation is that illness causes poverty: As people become disabled or ill, their ability to earn a living or attract an employed spouse decline, and they fall to a lower social status than that of their parents. This explanation is known as **social drift theory**. Studies that have tracked Americans over time, however, have found that social drift explains only a small proportion of illness among the poor (Adler and Rehkopf, 2008; Marmot, 2002, 2004). Instead, and far more often, poverty causes illness.

But *how* does poverty cause illness? As Chapter 2 discussed, social stress is a major cause of illness. Thus, one important reason why poorer persons suffer worse health than do wealthier persons is because poorer persons experience *more stress* and have *less control* over that stress (Adler and Rehkopf, 2008; Link and Phelan, 1995; Phelan et al., 2004). For example, factory workers must keep pace with the production line but can't control the speed of the line and can't even choose when to take bathroom breaks. Numerous studies have found that workers who face high demands with little control over work conditions are particularly likely to experience stress, resulting in both physical and psychological illness (North et al., 1996; Marmot, 2004; Wilkinson, 2005).

As this suggests, stress amplifies the myriad health risks embedded in everyday aspects of lower-class life. First, the work available to poorly educated lower-class persons—when they can find it—can cause ill health or even death. A coal miner, for example, is considerably more likely than a mine owner to die from accidental injuries or lung disease caused by coal dust. In addition, lower-status workers typically experience both demanding work conditions and low control over those conditions.

Second, environmental conditions place poor people at risk of illness and death. Chemical, air, and noise pollution all occur more often in poor neighborhoods than in wealthier neighborhoods both because the cheap rents in neighborhoods blighted by pollution attract poor people and because poor people lack the money, votes, and social influence needed to keep polluting industries, waste dumps, and freeways out of their neighborhoods (Brulle and Pellow, 2006; Bullard, Warren, and Johnson, 2001). Such pollution can foster cancer, leukemia, high blood pressure, and other health problems as well as emotional stress.

Third, inadequate, overcrowded, and unsafe housing increases the risk of injuries, infections, and illnesses, including lead poisoning when children eat peeling paint; gas poisoning when families rely on ovens for heat; and asthma triggered by cockroach droppings, rodent urine, and mold (Reading, 1997; Brown et al., 2003).

Fourth, the food poor children eat—or don't eat—affects their lifetime risk of illness. Federal researchers currently estimate that more than 20 percent of US children sometimes go hungry (Seligman and Schillinger, 2010). These children get significantly more colds each year and are significantly more likely to experience poor health, lack sufficient iron, experience chronic headaches or stomachaches, or have a disability (Seligman and Schillinger, 2010). They are also more likely to miss school and to do poorly in school, thus increasing their chances—and their children's chances—of remaining poor.

The foods available to poor children and adults also increase their health risks (Seligman and Schillinger, 2010). This diet relies heavily on fatty or sweet foods that satisfy hunger and provide energy inexpensively but offer little nutrition. As Chapter 2 discusses, such a diet may lead to heart disease, diabetes, and other illnesses.

Lack of access to health care also fosters illness and disability among the poor, although its effect is relatively weak (Williams et al., 2010; Adler and Rehkopf, 2008). Despite recent health care reforms (discussed in Chapter 8), many Americans still lack health insurance. Others find it difficult to obtain health care because they can't afford transportation to the doctor, time off from work to visit a doctor, or drugs or services not covered by their insurance. Even when poor or near poor people have health insurance, they are more than twice as likely as other insured adults to forgo needed medical care (*Morbidity and Mortality Weekly Report*, 2010). In these circumstances, small health problems can quickly mushroom, as when an unfilled cavity leads to a deadly brain infection. The prestigious, federally run Institute of Medicine (2002) estimates that undertreatment and low-quality treatment leave uninsured Americans *25 percent* more likely than other Americans to die in any given year.

All these issues are pulled together by what is known as **cumulative inequality (CI) theory** (Ferraro and Shippee, 2009; Goosby, 2013). This theory argues that inequality primarily results from social systems, rather than individual choices, and that it causes health problems that accumulate over the lifetime. This theory helps to further explain why those who grow up poor are more likely than others to be ill, disabled, or dead before they turn 60. So, for example, if a poor child who lives in a rural area breaks a bone in her arm, her parents might try setting it on their own because they can't afford to pay a doctor, can't afford the gas to drive to a doctor, or fear they will lose their job if they take time off to go to a doctor, especially if the only feasible care is at a public clinic where they might have to wait for hours. If the wound becomes infected or the bone doesn't heal straight, the child might be left with chronic pain. That pain might make it harder for her to focus on her schoolwork and therefore less likely to finish high school, let alone college, especially if her parents need her to be earning a living, and her teachers and guidance counselors assume she isn't "college material." If the only work she can find is cleaning houses or taking care of a sick elderly person, the physical labor may worsen her pain and lead to depression. She might at that point conclude that her best options are taking antidepressants, using prescription narcotics to dull her pain, and smoking cigarettes to increase her energy, all of which can bring further health complications. And throughout all of this, it's likely that she has lived in low-quality housing with limited access to healthy foods. All of these problems add to her **cumulative stress burden**, which, as Chapter 2 described, can wear out the body, including its muscles and immune system. Thus inequality and its health effects accumulate over a lifetime.

However, CI theory also argues that these consequences are not set in stone. Although those who grow up poor have fewer ways to protect their health and more cumulative exposure to health risks and stresses, there are ways to make a difference. CI theory recognizes that individuals sometimes find the means to make healthier choices on their own, but suggests that to truly break the cycle of cumulative inequality, broader social change will be needed.

RACE AND ETHNICITY

Race is a social construction with almost no biological basis. Research on the human genome has found almost zero support for the concept of race: All humans share virtually all the same genes (Epstein, 2007). Moreover, when researchers sort individuals according to their genetic variations, the resulting categories don't match existing racial categories (Williams et al., 2010).

The social rules for identifying individuals' race also suggest that the term has little meaning. For example, although these days everyone considers Irish people to be white, a century ago that was not the case (Jacobson, 1998). Similarly, most Americans consider individuals to be nonwhite if they have any known African ancestors *even if most of their ancestors were European*. For this reason, from this point on this textbook uses the term *ethnicity*, which suggests cultural rather than biological differences, rather than the term *race*.

Although social class explains many observed health differences among ethnic groups, ethnicity nevertheless is an important and independent factor in predicting health status (Bradby and Nazroo, 2010). In this section, we look at health and illness among African Americans (13% of the US population), Hispanic Americans (17%), Asian Americans (5%), and Native Americans (1.0%). As Figure 3.1 shows, life expectancy is shortest among Native Americans (living on or near tribal lands) and longest among Asian Americans. The remainder of this section explores in more detail some reasons for these and other ethnic differences in health.

FIGURE 3.1 **Life Expectancy by Race/Ethnicity and Sex**

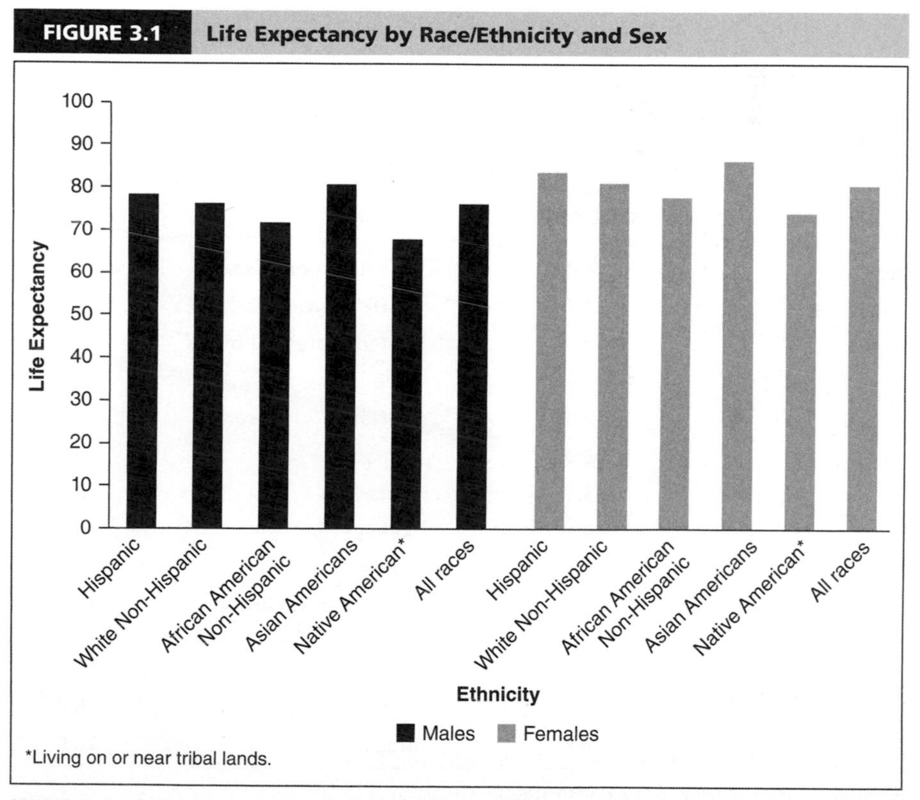

SOURCE: National Center for Health Statistics (2014a); Arias, Xu, and Jim (2014).

African Americans

The impact of ethnicity on health stands out vividly when we look at infant mortality (National Center for Health Statistics, 2014a). Not only is infant mortality twice as common among African Americans than among whites, but as Table 3.1 shows, it is also more common among African Americans than among citizens of Kosovo, Turkey, and Costa Rica.

As we have seen, poverty is a major cause of infant mortality. About two-thirds of African American children are either poor or near poor, and so poverty definitely helps explain high rates of African American infant mortality (National Center for Health Statistics, 2014a). However, even among the middle and upper classes, African American infants are more likely to die than are white infants.

The same holds true for other health problems: At all income levels, African Americans have higher mortality and morbidity rates than do whites (Williams et al., 2010). One explanation for this is racism. First, research consistently shows that the experience of racial discrimination is highly stressful and affects both physical and mental health (Bradby and Nazroo, 2010; Schnittker and McLeod, 2005; Williams et al., 2010). If, for example, pregnant women experience these stresses, they may be more likely to give birth prematurely. Second, racial discrimination by landlords, realtors, or mortgage bankers can leave even middle-class African Americans unable to obtain decent housing in neighborhoods free

TABLE 3.1 Infant Mortality Rates in Different Nations and US Ethnic Groups*

Location	Rate per 1,000 Births	Location	Rate per 1,000 Births
Japan	1.9	**US White non-Hispanic**	**5.3**
Singapore	2.0	**US Hispanic**	**5.4**
Sweden	2.3	**United States, all births**	**6.4**
Czech Republic	2.5	Chile	7.7
Greece	2.9	**US Native American****	**8.4**
Portugal	3.0	Romania	8.5
Austria	3.1	Costa Rica	8.7
Spain	3.1	Turkey	10.0
Italy	3.2	Kosovo	11.0
Germany	3.3	**US African American, non-Hispanic**	**12.2**
US Asian or Pacific Islander	**4.4**	Mexico	13.0
Cuba	4.6	Iran	16.0
Canada	4.8	Brazil	20.0

*By mother's ethnicity within the United States.

**Anywhere in the United States, not only near tribal lands.

SOURCE: Population Reference Bureau (2014); National Center for Health Statistics (2014a).

from pollution and violence (Williams and Jackson, 2005; Williams et al., 2010). Other middle-class African Americans prefer living in poorer, segregated neighborhoods rather than facing the daily hostility—or simply social discomfort—of white neighbors. Consequently, more-affluent African Americans sometimes live in conditions similar to those experienced by poorer African Americans, thus placing themselves and their families at risk.

The disparities in health status between African Americans and whites don't end in infancy. At each age, and for 13 of the 15 leading causes of death, African Americans have higher death rates than do whites. Thus, for example, compared with whites who smoke cigarettes, African American smokers are more likely to get lung cancer, to get lung cancer at younger ages, and to die from lung cancer (Williams et al., 2010). Although ethnic gaps in life expectancy have declined, white, non-Hispanic women still live an average of 4.2 years longer than do their African American counterparts, and white, non-Hispanic men live 6.4 years longer (Arias, 2010).

Table 3.2 shows the top causes of death for each major ethnic group in the United States. Heart disease, cancer, and stroke are among the top four causes for each group except Native Americans. Each of these diseases are linked to aging, and so it is not surprising that they should be major causes of death.

The table illustrates the impact of poverty on African Americans in two ways (Heron, 2013). First, African Americans are considerably more likely than non-Hispanic whites to die from homicide, which is often a marker of poverty and the despair it engenders. Second, African Americans are more likely than non-Hispanic whites to die of diabetes, which (as Chapter 2 described) is closely linked to poverty. In turn, diabetes largely accounts for high rates of kidney disease, a common consequence of long-term diabetes.

Yet kidney disease need not kill if individuals can receive transplanted kidneys. However, African Americans are significantly less likely than whites to receive transplants. Doctors less often refer African Americans to transplant programs, less often put African Americans on wait lists for donated kidneys, and more often reject African Americans as transplant patients because they lack transportation to hospitals and aftercare facilities (Epstein et al., 2000).

Hispanic Americans

Like African Americans, Hispanic Americans experience an array of diseases linked to poverty and despair, including fatal accidents, diabetes, and chronic liver disease (usually caused by alcohol abuse). Hispanic Americans are 2.5 times more likely than non-Hispanic whites to live in poverty and, except for Cubans, are half as likely to have completed college. Hispanics are also substantially more likely than whites to lack health insurance and to have no usual source of health care (National Center for Health Statistics, 2014a). In addition, cultural and language barriers as well as discrimination can make it difficult for Hispanics to take advantage of health care resources even when they can afford them.

That said, recent research using data from a wide variety of sources suggests that compared with white non-Hispanics, Hispanics enjoy comparable rates of infant mortality and *longer* life expectancies (see Figure 3.1 and Table 3.1). This

TABLE 3.2 Top Causes of Death by Ethnicity

Non-Hispanic Whites	Non-Hispanic African Americans	Hispanic Americans	Asian Americans	Native Americans
Heart disease	Heart disease	Cancer	Cancer	Heart disease
Cancer	Cancer	Heart disease	Heart disease	Cancer
Chronic lower respiratory diseases	Stroke	Accidents	Stroke	Accidents
Stroke	Diabetes	Stroke	Accidents	Diabetes
Accidents	Accidents	Diabetes	Diabetes	Chronic liver disease
Alzheimer's disease	Kidney disease	Chronic liver disease	Influenza and pneumonia	Chronic lower respiratory diseases
Diabetes	Chronic lower respiratory diseases	Chronic lower respiratory diseases	Chronic lower respiratory diseases	Stroke
Influenza and pneumonia	Homicide	Alzheimer's	Kidney disease	Suicide
Kidney disease	Septicemia	Kidney disease	Influenza and pneumonia	Kidney disease
Suicide	Alzheimer's	Influenza and pneumonia	Suicide	Influenza and pneumonia

SOURCE: Heron (2013).

surprising finding is known as the **Hispanic paradox**. The three most common explanations for this paradox are migration effects, cultural differences, and problems with the data (Arias, 2010). First, some researchers argue that Hispanic Americans have unusually good health because only the healthy migrate and because migrants who fall ill often return home (Palloni and Arias, 2004). Second, some researchers argue that strong social networks—both within immigrant communities and within Hispanic families—protect Hispanic Americans from disease. This hypothesis is supported by data suggesting that the Hispanic health advantage declines with each generation after immigration (Williams and Sternthal, 2010). Conversely, other researchers argue that Hispanic Americans only appear to be healthier than white non-Hispanics because the former are often inaccurately identified as the latter on death certificates (Smith and Bradshaw, 2006). At present, data are insufficient to fully support any of these three explanations (Arias, 2010).

At any rate, health status varies enormously among Hispanic Americans. One group at especially high risk is migrant farm workers (Azevedo and Bogue, 2001; Greenhouse, 2001). Farm work is physically hazardous, with long days of repetitive stooping and bending, heavy lifting, and exposure to toxic pesticides (Gwyther and Jenkins, 1998; Sandhaus, 1998). Access to clean water and sanitary toilets is often limited, and workers are routinely exposed to weather extremes. Living conditions for migrant workers are often poor, with many individuals crowded together in rooms that are poorly heated or cooled, with insufficient water and toilets and low wages that make it difficult to obtain nutritious foods. Finally, lack of transportation, cultural differences, and communication problems make it difficult for laborers and their families to obtain good health care. As a result, life expectancy is substantially reduced among migrant workers and their families, and chronic health problems, miscarriages, infant mortality, and infectious diseases (including tuberculosis, typhoid, and hepatitis) are several times more common than among the rest of the population (Gwyther and Jenkins, 1998; Sandhaus, 1998).

Immigrants from the poorer countries of Central America, too, are especially at risk, whether farm workers or not. Because these migrants must cross more than 2,000 miles plus at least two national borders to reach the United States, immigration from Central America is more dangerous and expensive than it is from Mexico. As a result, undocumented Central Americans are more likely than undocumented Mexicans to stay in low-paying, dangerous occupations such as palm tree trimming or roof building because they especially fear the scrutiny and potential for deportation that might accompany the search for new jobs.

Native Americans

As is true with any ethnic group, Native Americans are highly diverse. Native Americans in the United States belong to more than 500 different tribes, each with a distinct language and culture. About 20 percent live in rural areas, 20 percent on reservations or other trust lands, and 60 percent in suburbs or cities (Office of Minority Health, 2014).

Life expectancy among Native Americans has increased steadily since the 1950s. However, it is still considerably lower than among other Americans, especially when we look only at those who live on or near tribal lands (Arias, Xu, and Jim, 2014). (Statistics that include individuals who live away from tribal lands can be misleading because many of these people have few genetic or cultural ties to Native American peoples.) Similarly, Native Americans are considerably more likely than other Americans to live with disabilities and to report unmet needs for health care (Barnes et al., 2010).

Infant mortality is a particularly crucial marker of the health problems faced by Native Americans. Compared to non-Hispanic whites, rates of infant mortality are 60 percent higher among Native Americans overall (see Table 3.1) and 240 percent higher among those living on South Dakota's large and exceptionally poor Pine Ridge Reservation (Indianz.com, 2012).

These differences in infant mortality grow even starker when we separate **neonatal infant mortality** (deaths occurring during the first 27 days after birth) from **postneonatal infant mortality** (deaths occurring between 28 days and 11 months after birth). *Neonatal* infant mortality rates are essentially the same among Native Americans and whites, but *postneonatal* infant mortality is almost three times higher among Native Americans. Most postneonatal deaths stem from poverty, malnutrition, maternal tobacco use, poor living conditions, and lack of health care for ill infants (Tomashek et al., 2006).

For Native Americans who survive past infancy, heavy alcohol use stands out as an especially serious health risk (see Table 3.2). Although alcohol-related deaths among Native Americans have decreased in recent years, deaths from liver disease, suicide, and accidents (often caused by alcohol use) remain strikingly common. Native Americans (like other poor populations) are also more likely than white Americans to die from diabetes (Barnes et al., 2010; Heron, 2013).

Asian Americans

Overall, Asian Americans enjoy far better health than do other American minority groups (see Figure 3.1 and Table 3.2). The largest Asian American groups (Chinese, Japanese, and Filipino) have life expectancies and infant mortality rates equal or superior to those of white Americans.

These statistics, however, tell only part of the story. Since 1975, a substantial portion of Asian immigration has come from the war-torn countries of Southeast Asia. These immigrants typically have far lower income and education levels than those of established Asian Americans. In addition to having the health problems that always accompany poverty, they may experience unavoidable dietary changes, culture shock, tropical diseases for which diagnosis and treatment can prove elusive, and the long-lasting traumas of warfare and refugee life.

In addition, Southeast Asians living in the United States typically have less access to health care than do other Asian Americans (Association of Asian Pacific Community Health Organizations, 2008). Rates of health insurance coverage are low, and even those who have insurance sometimes find that linguistic or cultural barriers make it difficult to communicate with health care workers or obtain quality health care. As a result, Southeast Asians are less likely than are other Americans to use Western health care (although some continue to use traditional Asian healers and therapies).

Writer Anne Fadiman poignantly describes the communication barriers between new immigrants and their doctors and the problems these barriers create for both groups in her prize-winning book *The Spirit Catches You and You Fall Down: A Hmong Child, Her American Doctors, and the Collision of Two Cultures* (1997). Fadiman describes the completely divergent worldviews of American doctors and Hmong patients in Merced, California, where many Hmong refugees from Laos have settled:

> Most Hmong believe that the body contains a finite amount of blood that it is unable to replenish, so repeated blood sampling [for lab tests] … may be fatal. When people are unconscious, their souls are at large, so anesthesia may lead to illness or death. If the body is cut or disfigured, or if it loses any of its parts, it will remain in a condition of perpetual imbalance, and the damaged person not only will become frequently ill but may be physically incomplete during the next reincarnation; so surgery is taboo. If people lose their vital organs after death, their souls can't be reborn into new bodies and may take revenge on living relatives, so autopsies and embalming are also taboo….
>
> Not realizing that when a man named Xiong or Lee or Moua walked into the Family Practice Center with a stomachache he was actually complaining that the entire universe was out of balance, the young doctors of Merced frequently failed to satisfy their Hmong patients. How could they succeed? … They could hardly be expected to "respect" their patients' system of health beliefs (if indeed they ever had the time and the interpreters to find out what it was), since the medical schools they had attended had never informed them that diseases are caused by fugitive souls and cured by (sacrificing) chickens. All of them had spent hundreds of hours dissecting cadavers … but none of them had had a single hour of instruction in cross-cultural medicine. To most of them, the Hmong taboos against blood tests, spinal taps, surgery, anesthesia, and autopsies—the basic tools of modern medicine—seemed like self-defeating ignorance. They had no way of knowing that a Hmong might regard these taboos as the sacred guardians of his identity, indeed, quite literally, of his very soul (Fadiman, 1997, 33, 61).

Growing recognition of problems like these has spurred medical schools to incorporate training in working with culturally diverse populations in their programs, as we will consider in more detail in Chapter 11.

Case Study: Environmental Racism

One health issue that cuts across America's minority communities is **environmental racism**. Environmental racism refers to the disproportionate burden of environmental pollution experienced by ethnic minorities, from Hispanic farm workers exposed to dangerous pesticides to Navajo communities poisoned by deadly uranium mines and inner-city African Americans plagued by asthma-inducing air pollution (Bullard et al., 2001; Brulle and Pellow, 2006; Taylor, 2014). The most important of these environmental hazards, because it is so widespread and devastating, is lead—found in polluted air, contaminated soil, and the paints and pipes of older residences. Among children under age five and known to have high levels of lead in their blood, 17 percent are white non-Hispanic, 16 percent are Hispanic, and 60 percent are African American (Meyer et al., 2003). Compared with whites, minorities are exposed more often to dust and soot, carbon monoxide, ozone, sulfur, and sulfur dioxide, as well as to pesticides, emissions from hazardous waste dumps, and other hazardous substances. Exposure to environmental pollution is more highly correlated with race than with any other factor, including poverty (Brulle and Pellow, 2006).

Environmental racism is a consequence of "everyday" racism. Racial discrimination keeps members of minority groups in segregated communities and enables industrialists, with the tacit approval of government bureaucrats and politicians, to place environmental hazards in those communities without worrying that residents will have the political power or financial resources to resist (Taylor, 2014). Poverty and lack of other job opportunities can even encourage minority communities to welcome polluting industries for the jobs they will bring. This doesn't mean, however, that those who make decisions about where to locate environmental hazards *intend* to discriminate against minorities—certainly those who make these decisions would argue that they decide solely on economic and technical considerations—only that their actions have the effect of discriminating.

Currently, dozens of grassroots organizations of African Americans, Hispanic Americans, Asian Americans, and Native Americans are working to fight for environmental justice (Brulle and Pellow, 2006), as are numerous national civil rights and environmental organizations. These organizations have had considerable success, both locally and nationally, in keeping polluting companies out of local communities and getting the federal Environmental Protection Agency to consider environmental justice in its decisions (Brulle and Pellow, 2006).

AGE

Overview

Not surprisingly, age is the single most important predictor of mortality and morbidity. As noted in Chapter 2, until the twentieth century deaths during the first year of life were common in the United States. Although far less common now, infant mortality remains an important issue because so many years of productive life are lost when an infant dies and because infant mortality so often is caused by preventable social and environmental conditions.

Once individuals pass the danger zone during and immediately after birth, mortality rates drop precipitously. Those rates begin to rise significantly beginning at about age 40 and escalate with age. For those who survive past age 65,

chronic illnesses rather than **acute illnesses** comprise the major health problems, often bringing years of disability in their wake.

The American population is aging steadily, with the population above age 85 growing the fastest. Although most middle-aged and older persons are relatively healthy, rates of illness, disability, and mortality nevertheless are rising as the population ages. Similarly, both the total costs for health care and the percentage of health care dollars spent on the elderly—already greatly disproportionate to the size of that population—are bound to increase. At the same time, as young persons become a smaller proportion of the population, the pool of persons who can provide or pay for the care needed by the elderly is shrinking. Consequently, in the future, it will become more difficult to provide services to all the elderly persons who will need health care or assistance with daily tasks such as shopping or cooking.

These problems are amplified by the **feminization of aging**—the steady rise in the proportion of the population who are female in each older age group so that women comprise a larger proportion of the elderly than of the young and middle aged. Because elderly women more often than elderly men are poor and lack a spouse who can or will care for them and because (as we will see in the next section) women, in general, experience more illness than men do, the feminization of aging will increase the costs of providing health and social services to the elderly.

SEX AND GENDER

Overview

Both sex and gender strongly affect health status. **Sex** refers to the biological categories of male and female, to which we are assigned based on our chromosomal structure, genitalia, hormones, secondary sexual characteristics such as facial hair, and so on; those with two X chromosomes and a vagina are sexually female, and those with one X chromosome, one Y chromosome, and a penis are sexually male. (Later in this section, we will consider those who don't fit neatly into these categories.) In contrast, **gender** refers to the social categories of masculine and feminine and to the social expectations regarding masculinity and femininity that we are expected to follow based on our assigned sex. Because these categories are social, they vary across time and across culture.

Basic epidemiological data show that both sex and gender affect health (Bird and Rieker, 2008; Read and Gorman, 2010; Rieker, Bird, and Lang, 2010). For example, before the twentieth century, complications of pregnancy and childbirth often cut short women's lives, so on average, women died younger than did men. These days, however, American women (regardless of race) live longer than men do, as Figure 3.1 showed—even though the same diseases (including heart disease, cancer, and cerebrovascular disease) eventually kill most people. The *differences* between men's and women's life expectancies suggest that sex may directly affect health, but the *changes* in these differences across time suggest that gender affects health: Women now live longer than men not because their biology has changed, but because their social position has changed.

But mortality differences tell only part of the story. If we look only at life expectancy, we might conclude that women are biologically hardier than men. When we look at morbidity rates, however, the picture blurs. At each age, men have higher rates of mortality and of fatal diseases, even though women have higher rates of morbidity and of nonfatal disease (Bird and Rieker, 2008; Read and Gorman, 2010). Arthritis, for example, is the most common chronic, nonfatal condition among both men and women older than age 45 years, but it strikes women about 50 percent more often than it does men. In addition, at each age, women experience a 20 percent to 30 percent greater incidence of *acute* conditions (not including health problems related to their reproductive systems). In sum, women live longer than men but experience more illness and disability, whereas men experience relatively little illness but die more quickly when illness strikes.

How can we explain these paradoxical findings? Some researchers have hypothesized that women's higher rates of illness are more apparent than real—that women don't actually experience more illness than men but simply *label* themselves ill and seek health care more often. Most research, however, suggests that the health differences between men and women are real (Bird and Rieker, 2008; Read and Gorman, 2010). These differences stem from both the biological differences of sex and the socially reinforced differences of gender.

Sex does seem to offer females some biological health benefits (Bird and Rieker, 2008; Read and Gorman, 2010). Around the globe, more females than males survive at every stage of life from fetus to old age as long as they receive adequate nutrition. Although the exact mechanisms through which this works are unknown, some theorize that estrogen and other "female" hormones (which in fact also occur in males but in lower proportions) somehow protect the heart and other organs from fatal disease.

Gender, too, protects women from fatal disease and injury (Bird and Rieker, 2008; Read and Gorman, 2010; Rieker, Bird, and Lang, 2010). Most importantly, because of differences in male and female gender roles, women less often abuse alcohol, drive dangerously, participate in dangerous sports, and so on. They are also less likely than men to work in dangerous industries such as agriculture or commercial fishing. Less importantly, gender roles more often bring women than men into routine contact with medical care. Unlike men, who are socialized to downplay physical problems as signs of weakness, women are more comfortable seeking health care when they experience problems. In addition, because women often must obtain health care for children or elderly parents and must seek obstetric or gynecological care for themselves, women are more likely than men to meet with health care providers. As a result, women are more likely to have health problems identified and treated early enough to make a difference.

Sex and gender may also help explain why, despite women's lower rates of mortality, they have higher rates of morbidity than do men. Research on this topic, however, is far less conclusive (Barker, 2005). Most commonly, theories suggest that women are more susceptible to nonfatal illnesses because of their hormones (a sex effect) or because of their relatively high stress levels coupled with low control over their lives (a gender effect). The latter theory gains support from an article published in the prestigious *American Sociological Review* and based on national data collected from **random samples** over a 30-year period (Schnittker, 2007). The article found that women's self-reported health (a measure of morbidity) had improved considerably from 1974 to 2004 and that almost all of the improvement was explained by women's increased educational attainment, which in turn increased their employment and income. Increased education apparently gives women more power over their lives and therefore improves their health.

These changes are part of a broader move toward gender convergence (Annandale, 2010). **Gender convergence** refers to the growing similarities in expectations for how men and women should behave in their everyday lives. This gender convergence may well result in greater convergence in men and women's patterns of health, illness, and mortality. For example, women's new freedom to join the Marines may increase their health risks, whereas men's new freedom to put family ahead of careers may reduce their stress levels and so improve their health.

A Sociology of Intersex

So far, we have been talking about sex as if it were a *binary category*—one with two and only two conditions: male or female. However, up to 2 percent of babies are born with genitalia that appear neither clearly male nor clearly female (Blackless et al., 2000). Such babies are referred to as **intersex**: having characteristics of both sexes. Intersexuality refers to biological sexual characteristics, and it is not the same as homosexuality (which refers to same-sex sexual desires and practices) or to transsexuality (which refers to the perception that one belongs to the opposite of one's biological sex).

Intersexuality can be caused by hormonal factors, chromosomal factors, or both. During their first eight weeks of development, the only sex differences among fetuses are their chromosomes (XX among females, XY among males). After that point, the production of male hormones leads some fetuses to develop male genitalia; the same fetal tissue becomes female genitalia in the absence of these hormones. A slightly different hormonal balance produces fetuses that have both male and female external genitalia (penis, testicles, clitoris, vagina, labia) or internal genitalia (gonads, uterus, fallopian tubes). This can happen for many reasons. For example, some fetuses inherit unusual hormonal patterns or sex chromosome patterns (such as XO or XXY rather than the typical male XY or female XX), and others are affected by environmental pollutants that their mothers absorbed while pregnant.

The social response to intersex individuals varies greatly across cultures. Some cultures revile them and kill such babies at birth. Other cultures assume that three or more sexes occur naturally in the population and consider intersex a normal human variation. These cultures typically integrate intersex individuals into normal social life. Still others assign special, valued roles to intersexed individuals. Modern Western culture, however, generally supports hiding intersexuality, stigmatizing it, or eliminating it in some way.

Beginning in the 1950s, surgery and hormonal manipulation became the standard medical practice for handling intersex children in the United States (Kessler, 1998). Doctors urged parents of intersex children to have the children surgically reassigned to be either male or female as early in infancy as possible on the assumption that this would help children develop into the "appropriate" gender. Decisions about which sex to assign reflected doctors' cultural assumptions about gender: Children were assigned to be boys if doctors considered their penises sufficiently large and were assigned to be girls if their internal organs would allow them to give birth. Boys with penises considered too small had their penises surgically removed and artificial vaginas constructed even if their hormonal and chromosomal makeup were indisputably male. Girls with clitorises considered unattractively large had their clitorises surgically removed or reduced, even though this meant removing healthy organs and impairing their ability to ever experience sexual pleasure. To assist the children in adopting their assigned sex, parents were instructed to socialize them strictly to their new gender, hide their history from them, and place them on a steady (if secret) diet of sexual hormones to change the children's bodies to better match their assigned sex.

Currently, surgery is performed on about 1 to 2 of every 1000 babies, with lifelong hormonal injections following (Blackless, 2000). This treatment became the norm because doctors assumed it was the most humane option, although no research was available on its psychological, social, or physical consequences. Since the 1990s, however, this standard medical treatment has come under considerable attack, both from scholars and from activists who themselves experienced sex reassignment as children. Opponents of sex reassignment point out that this treatment is based not on scientific evidence, but on gender beliefs: that small penises are "unmanly;" that large clitorises are frightening; that children need strict socialization into "appropriate" gender behaviors; and that a vagina need only permit penile penetration, not provide natural lubrication, elasticity, or the possibility of female sexual pleasure (Kessler, 1998; Preves, 2003). Moreover, opponents argue, sex reassignment reinforces children's sense of difference, reduces their ability to ever enjoy sexual pleasure, and depends on webs of deception among children, parents, and doctors that create their own psychological nightmares (Kessler, 1998; Preves, 2003).

At this point, there is insufficient evidence to say whether sex assignment more often helps or harms these children. Surgical intervention remains the norm, but doctors increasingly hold off on surgery for six months to give parents time to consider other options (Navarro, 2004). Another change is that some intersex activists are now working closely with doctors and medical societies to create better models of care. This cooperation resulted in a joint document which supported the medicalization of intersex by creating the new diagnosis of Disorders of Sex Development and somewhat *demedicalized* intersex by stating that surgery should not be performed solely for cosmetic reasons—except for girls whose clitorises are exceptionally large (Accord Alliance, 2014).

Case Study: Intimate Partner Violence and Health

One health issue in which gender plays an especially critical role is violence by intimate partners. Although neither health care workers nor the general public typically thinks of partner violence as a health problem, it is a major cause of injury, disability, and death among American women, as among women worldwide.

According to data collected by federal researchers from a large-scale, national, random sample, about 25 percent of women (compared with 8% of men) have been raped or physically assaulted by a spouse, ex-spouse, lover, or date at some point during their lives (Tjaden and Thoennes, 2000). These numbers, of course, don't include the approximately 1200 women who are killed each year by intimate partners and so are unable to answer survey questions (National Center for Injury Prevention and Control, 2009).

Women whose romantic partners are male are about three times more likely to suffer an attack than those whose romantic partners are female. In addition, Native American women report experiencing violence considerably more often, and Asian American women considerably less often, than do other women. This last finding, however, may reflect differences in the likelihood of *reporting* violence more than differences in *rates* of violence (Tjaden and Thoennes, 2000).

The consequences of violence are also more severe for women than for men. Women are about three times as likely as men to be killed during an attack and twice as likely to be seriously injured, with about one-third of women requiring emergency health care afterward. Extrapolating from these data, the researchers estimate that more than a half million women per year seek care at hospital emergency departments for injuries resulting from assault by intimate partners (Tjaden and Thoennes, 2000).

That assaults by men far surpass assaults by women should not surprise us. Before 1962, US courts consistently ruled that women could not sue their husbands for violence against them—in essence declaring violence against wives a man's legal and even moral right. Even after that date, most police refused to arrest men for such violence, and most courts refused to prosecute, a situation that did not begin to change for more than a decade.

Intimate violence against women continues to exist because it reflects basic cultural and political forces in our society and, indeed, around the world (Dobash and Dobash, 1998). Through religion, schools, families, the media, and so on, women often are taught to consider themselves responsible for making sure that their personal relationships run smoothly. When problems occur in relationships, women are taught to blame themselves even if their husbands or romantic partners respond to those problems with violence. Men, meanwhile, often receive the message—from sources ranging from pornographic magazines to religious teachings that give husbands the responsibility to "discipline" their wives—that violence is an acceptable response to stress and that women are acceptable targets for that violence. Although most men resist these messages, enough men absorb these messages to make woman battering a major social problem. Moreover, women typically have less access to money than do their male partners and so often find themselves financially unable to leave if a relationship turns violent.

Violence against women occurs most often among men who believe that their power within the family is threatened, such as men who have less education than their wives (Tjaden and Thoennes, 2000). In addition, violence occurs most often among men who have a high need for power and who support traditional gender roles. Taken together, these data tell us that intimate violence against women is not only an individual response to social stress but also a form of **social control**: a way of reinforcing social expectations and power relationships. Specifically, intimate violence against women operates as social control by reinforcing men's power over women and women's inferior position within society. Consequently, as long as gender inequality remains the norm, woman battering will persist.

Recognition of intimate partner violence as a health risk has led various health-related organizations to enter the fight against it. The Centers for Disease Control and Prevention (CDC), is the federal agency responsible for tracking and preventing the spread of diseases in the United States. During the past decade, it has begun funding research on the causes, consequences, and prevention of intimate partner violence. In addition, the US Public Health Service has developed violence prevention programs, trained health professionals and others in violence prevention, and encouraged health care workers to learn how to identify battered women in hospital emergency departments. Similarly, the American College of Obstetricians and Gynecologists now requires medical schools to teach how to identify and respond to battered women and publishes materials designed to aid health professionals in doing so.

SOCIAL CAPITAL

Social capital refers to the resources available to an individual through his or her social network (Song et al., 2010). It is typically measured by some combination of the number of people with whom one has close personal relationships and the resources one can access through those relationships. Those resources can take many forms. For example, your social networks might (or might not) offer you access to expert advice, useful skills, or a good apartment to rent. Although one's social capital is typically linked to one's social class, each affects health independently. In addition, social capital also varies depending on one's ethnicity, gender, and age, among other factors. Thus social capital pulls together the advantages (and disadvantages) built into all of the social statuses discussed in this chapter.

Social capital affects health in several ways (Song et al., 2010). Among other things, individuals with higher social capital typically have better access through their networks to high-quality information. For example, individuals who have doctors as friends are more likely to learn of the dangers of popular new drugs. In addition, social capital can offer individuals the emotional and practical support needed to preserve their health, such as encouragement to stop smoking or a loan to cover the costs of an operation. At a broader level, social capital can protect health by providing power and political influence. For example, because

CONTEMPORARY ISSUES

Ethnicity, Class, Gender, and Hotel Maids

If you have ever stayed in a nice hotel, you probably enjoyed the fresh sheets, clean sinks, and other amenities provided by hotel maids. Yet maids' work is largely invisible—and far more dangerous than most realize. Moreover, those dangers have multiplied in recent years. Luxury hotels now often use three sheets per bed rather than two, requiring maids to lift the mattress eight times simply to make a bed. Meanwhile, mattresses have grown thicker and heavier, making them considerably more difficult to lift. The number of pillows and blankets per bed also has increased along with the physical labor needed to deal with those linens. Yet the number of rooms maids are expected to clean each day has increased. As a result, maids typically lift more weight daily than federal guidelines allow and consequently face risks of back injuries equal to those of construction workers (Greenhouse, 2006).

Maids also risk injury when they scrub sinks and tubs or wash floors. The repetitive motion required for these tasks can injure joints and muscles, as can the need to bend, crouch, or reach up to clean various surfaces. Toxic chemicals used in cleaning can also place maids at risk.

Almost all maids are female, immigrants, nonwhite or Hispanic, and poor or near poor. Each of these statuses reduces the odds that they can find other jobs or negotiate for better working conditions. Moreover, each of these statuses exposes maids to health risks outside of the work environment that can multiply the risks they face at work. The union UNITE HERE (www.unitehere.org), which represents hotel workers as well as workers in several other industries, has sponsored an ongoing boycott and campaign aimed at changing dangerous work conditions. Their hope is that by working together, maids will gain greater power to change their lives than they ever could have as individuals.

affluent neighborhoods typically include not only doctors but also lawyers, politicians, and business leaders, social networks in those neighborhoods have the resources needed to fight against highways, polluting factories, or anything else that might harm the health of individuals in those neighborhoods.

IMPLICATIONS

Far from being purely biological conditions reflecting purely biological factors, health and illness are intimately interwoven with social position. In the United States, as elsewhere, those who are poor or are targets of racial discrimination die younger than others do. Sex and gender have more complex health consequences: Women enjoy longer life spans than men do, but they are subject to more illness and disability. Importantly, each of these factors (along with age) interacts with the others, leaving some individuals at much greater risk of illness and injury than others; *Contemporary Issues: Ethnicity, Class, Gender, and Hotel Maids* explores the health risks faced by women who are disadvantaged by social class, ethnicity, and immigration status, as well as by gender.

Given that social forces as well as biological factors affect health, understanding social trends can help us predict future health trends. For example, as women's social roles have changed, their rates of tobacco use and lung cancer have approached those of men, and their ability to protect themselves from the health consequences of male violence has increased. Similarly, if economic and ethnic inequality either increase or decrease, we are likely to see changes in the health status of currently disadvantaged economic and ethnic groups.

SUMMARY

1. The causes and types of illness are not randomly distributed among the US population, but rather vary dramatically according to social class, ethnicity, age, sex, and gender.
2. Social class very strongly affects rates of mortality and morbidity. Poor persons are substantially more likely than others to experience illness and disability and to die young. These social class differences primarily reflect lower class persons' exposure to environmental hazards, unsafe working conditions, inadequate housing, poor nutrition, and psychological stress.
3. Ethnicity also affects health status. Infant mortality rates are especially high among African Americans and Native Americans. African Americans, Hispanic Americans, and Native Americans are all more likely than white Americans to die of conditions linked to poverty and despair, including liver disease and diabetes. In contrast, health is generally excellent among Asian Americans except for recent poor immigrants from Southeast Asia.
4. Environmental racism refers to the disproportionate burden of environmental pollution experienced by racial and ethnic minorities.
5. For those who survive infancy, mortality rates rise significantly beginning at about age 40. For those who survive past age 65, chronic illnesses rather than acute illnesses comprise the major health problems.
6. The American population is aging steadily. As a result, future years will see higher rates of illness, disability, and mortality as well as increased health care costs.
7. The United States is experiencing the feminization of aging. That is, each age cohort has a higher percentage of women than the next younger cohort. As a result, in the future more Americans will likely need health care, and fewer will be able to afford it.
8. Although men have higher rates of *fatal* diseases and die younger, women experience higher levels of *nonfatal, chronic* conditions. These differences stem from both *sex* differences (such as hormone levels) and *gender* differences (such as levels of risk taking).
9. The term *intersex* refers to individuals born with characteristics of both sexes. Until recently, intersex was treated solely as a medical and surgical problem, based on social ideas about gender.

10. Social capital refers to the resources available to an individual through his or her social network. Although social capital is closely linked to social class, both independently affect health.

REVIEW QUESTIONS

1. How and why does social class affect people's health?
2. What are the special health problems of migrant farm workers?
3. How does ethnicity affect health separately from social class? How does social class affect health separately from ethnicity?
4. How and why do the particular health problems of African Americans, Hispanic Americans, Native Americans, and Asian Americans differ from those of whites?
5. What is environmental racism?
6. What are the health care consequences of an aging population and of the feminization of aging?
7. Why do men have higher mortality rates than women but lower morbidity rates?
8. What are the sources and consequences of intimate partner violence? Why do some health care workers consider it to be a serious health problem?
9. Everyone has social capital, regardless of their social class. Give an example of the social capital that a working-class person might have, and how that social capital might help him or her to stay healthy.

CRITICAL THINKING QUESTIONS

1. Explain why poor persons become ill more often and die younger than wealthier persons.
2. Assume that over the next 20 years both men and women increasingly adopt behavior patterns now associated with the other gender. What changes would you expect to see in the health of men and women? Explain your answer.
3. Assume that 20 years from now, African Americans are as likely as whites to graduate from college. Why and in what ways would you expect the health of the African American population to improve? Why and in what ways would you expect it to remain the same?

CHAPTER

4

Illness and Death in the Less Developed Nations

Trip/Alamy

LEARNING OBJECTIVES

After reading this chapter, students should be able to:

- Explain the differences between more, less, and least developed nations.
- Assess the ways that globalization affects health around the world.
- Understand the changing patterns of disease in less developed nations.
- Identify the main types and causes of disease in the less developed nations.

Mahabouba Muhammad grew up in a small village in Ethiopia. Many Ethiopian girls receive little education and have few rights, but Mahabouba's situation was particularly poor: Her parents had divorced and left her with an aunt who treated her like a servant. As a result, Mahabouba eventually ran away to the nearest town to find work as a maid in exchange for room and board:

> *"Then a neighbor told me he could find better work for me," Mahabouba recalled. "He sold me for eighty birr [ten dollars]. He got the money, I didn't. I thought I was going to work for the man who bought me, in his house. But then he raped me and beat me.... I was about thirteen."*
>
> *The man, Jiad, was about sixty years old and had purchased Mahabouba to be his second wife. In rural Ethiopia, girls are still sometimes sold to do manual labor or to be second or third wives....*
>
> *[Jiad and his first wife] wouldn't let Mahabouba out of the house for fear she might run away. Indeed, she tried several times, but each time she was caught and thrashed with sticks and fists until she was black, blue, and bloody. Soon, Mahabouba was pregnant, and as she approached her due date, Jiad relaxed his guard over her. When she was seven months pregnant, she finally succeeded in running away....*
>
> *Unable to afford a midwife when she went into labor, Mahabouba tried to have the baby by herself. Unfortunately, her pelvis hadn't yet grown large enough to accommodate the baby's head, a common occurrence with young teenagers. She ended up in obstructed labor, with the baby stuck inside her birth passage. After seven days, Mahabouba fell unconscious and at that point someone summoned a birth attendant. By then the baby's head had been wedged there for so long that the tissues between the baby's head and Mahabouba's pelvis had lost circulation and rotted away. When Mahabouba recovered consciousness, she found that the baby was dead and that she had no control over her bladder or bowels. She also couldn't walk or even stand, a consequence of nerve damage that is a frequent by-product of [obstructed pregnancies] (Kristof and WuDunn, 2010:93–94).*

Mahabouba's story—rape, beatings, pregnancy too young, unattended childbirth—is all too common in much of the world. As this suggests, the sources and patterns of illness and death in poorer countries differ dramatically from those found in more affluent countries—and often reflect social conditions as well as biological forces. In this chapter, we first compare some of these differences. We then focus on explaining the main sources of death and disease (including illness and death in childbirth), focusing on the role played by social, economic, and political conditions and forces.

SETTING THE STAGE: KEY CONCEPTS

A few key concepts are needed to understand disease patterns around the world. This section lays out those concepts.

Understanding Development Patterns

In making international comparisons, politicians, social scientists, medical researchers, and others typically divide the world into two broad groups: the ***more* developed nations** and the ***less* developed nations**. Essentially, this division reflects the economic status of the various nations. The *more developed* nations are primarily defined by their relatively high gross national income (GNI) per capita compared with the *less developed* nations. In addition, the more developed nations are characterized by diverse economies made up of many different industries, whereas the less developed nations have far simpler economies, in some cases still relying heavily on extractive industries such as mining or logging or a few agricultural products such as rubber or bananas. These economic differences—primarily resulting from centuries of exploitation by political and economic powers in the more developed nations—have left the less developed nations with high infant and maternal mortality, low life expectancies, and damaging levels of infectious and parasitic diseases.

That said, the less developed nations also differ substantially from each other. Sociologists and other researchers use the term **least developed nations** to refer to those less developed nations that suffer from the *least* diverse economy and *lowest* GNIs and life expectancies. For example, life expectancy in Ethiopia is only 63 years, and gross national product (GNP) per capita is only $1,350 (Population Reference Bureau, 2014). Table 4.1 compares life expectancies and infant mortality rates in the least, less, and more developed nations. As is common in the field, except when directly comparing the less and least developed nations, this textbook uses the former term to refer to both groups.

Although dividing the globe into least, less, and more developed nations is a useful analytic tool, it is important to recognize that development level is a scale, not a dichotomy. Mexico and Thailand, for example, fall near the border between the more and less developed nations: Each has both complex industries and traditional agricultural crops, and each enjoys infant mortality rates and life expectancies approaching those found in the United States. And although

TABLE 4.1 Life Expectancy and Infant Mortality by Development Level

Country	Life Expectancy at Birth	Infant Mortality per 1,000 Births
Most Developed		
Japan	83	1.9
Italy	82	3.2
France	82	3.6
Germany	80	3.3
Denmark	80	3.0
United States	79	5.4
Less Developed		
Mexico	74	13
China*	75	15
Philippines	69	23
Thailand	75	11
Bolivia	67	39
India	66	44
Least Developed		
Haiti	63	59
Ethiopia	63	50
Somalia	55	80
Sierra Leone	45	92
Afghanistan	61	74

*Does not include Hong Kong, which only became part of China in 1997 and operates under a separate political structure.

SOURCE: Population Reference Bureau (2014).

infectious and parasitic diseases remain more common in Mexico and Thailand than in the United States, chronic diseases are now the most common cause of death in all three nations (WHO, 2010).

This terminology also should not keep us from recognizing that social conditions and hence health patterns vary from community to community and from social group to social group within each nation. Thus, conditions in central Detroit in some ways resemble those in Bangladesh, whereas conditions in wealthy sections of Bangkok resemble those in Beverly Hills. Within the less developed nations, the income gap—and consequently the "health gap"—between rich and poor has increased in the past two decades. These growing gaps in income and health largely stem from "structural adjustment" policies that were heavily promoted by international organizations based in the most developed nations. These policies pressed developing nations to cut back programs such as food subsidies and low-cost health care in exchange for economic aid from international nonprofit organizations (Kolko, 1999; Peabody, 1996).

Finally, although the terms *least*, *less*, and *more developed* imply linear progression from one status to the other, this is not necessarily the case. For example, economic and health conditions worsened in Eastern Europe after the collapse of the Soviet Union and in southern Africa after the start of the HIV/AIDS epidemic.

Understanding Globalization

Although it is important to understand development stages and disease patterns within individual nations, it is equally important to understand that *diseases respect no national borders*. Because of **globalization**, diseases and disease-causing conditions spread rapidly from less to more developed nations and vice versa (Quammen, 2013). For example, air pollution from China is now causing heart disease and asthma in the western United States, and the recycling of used US electronics equipment in China is releasing toxic acids and metals in China's drinking water (Markoff, 2002; Polakovic, 2002).

Because the United States and Mexico share the same water, air, and, to a growing extent, economies where the two nations meet, US citizens need to be especially concerned about health conditions in Mexico. For example, only one-third of the sewage generated by the more than 1 million residents of Juarez, a large city just south of El Paso, Texas, is appropriately treated (Schmidt, 2000). As a result, human wastes drain from Juarez into the Rio Grande, which provides water for drinking and farms to El Paso as well as Juarez. This untreated sewage has made gastrointestinal disease a leading cause of infant mortality in both cities. Diseases such as cholera and hepatitis could easily take root in these areas and spread into the interiors of both countries. As this example suggests, those who live in the more developed nations have a vested interest in understanding health and illness in the less developed nations.

Understanding Global Health

As this discussion of globalization suggests, dealing with health issues in one nation at a time has inherent limitations. This problem has led to new interest in what is referred to as *global health* (Farmer, Kim, and Kleinman, 2013). **Global health** refers to the ways that health and illness transcend borders—along with people, goods, health providers, floods, crops, and so on. The idea of global health emphasizes that disease can be spread or prevented not only by national governments but also by myriad other players, from the World Bank to local nonprofits, to small peddlers who move drugs, needles, food, and other supplies across borders. Finally, the concept of global health emphasizes the similarities as well as differences in health problems around the world and the importance of developing equitable solutions to those problems. As this suggests, the term is primarily used by those who take a critical stance toward health and society.

EXPLAINING DEATH AND DISEASE IN LESS DEVELOPED NATIONS

In this section, we look at the main types and causes of diseases in the less developed nations, including malnutrition, infectious diseases, maternal mortality, and war.

Chronic Disease

In a major change from past generations, chronic disease (especially heart disease and strokes) is rapidly emerging as a common cause of death in the less developed nations. Table 4.2 shows the leading causes of death around the world (WHO, 2014b). However, residents of less developed nations who have chronic diseases are far less likely to have access to appropriate treatment than are residents of the more developed nations.

TABLE 4.2 Leading Causes of Death around the World

Least Developed Nations	Less Developed Nations*		More Developed Nations
	Lower Income	Higher Income	
Lower respiratory infections	Coronary heart disease	Stroke and other cerebrovascular diseases	Coronary heart disease
HIV/AIDS	Stroke and other cerebrovascular disease	Coronary heart disease	Stroke and other cerebrovascular diseases
Diarrheal diseases	Lower respiratory infection	Chronic obstructive pulmonary disease	Trachea, bronchus, lung cancers
Stroke and other cerebrovascular diseases	Chronic obstructive pulmonary disease	Trachea, bronchus, lung cancers	Alzheimer's disease and other dementias
Coronary heart disease	Diarrheal diseases	Diabetes	Chronic obstructive pulmonary disease
Malaria	Maternal mortality	Lower respiratory infections	Lower respiratory infections
Maternal mortality	HIV/AIDS	Road traffic accidents	Colon and rectum cancers
Tuberculosis	Diabetes	Hypertensive heart disease	Diabetes mellitus
Infant mortality	Tuberculosis	Liver cancer	Hypertensive heart disease
Malnutrition	Cirrhosis of the liver	Stomach cancer	Breast cancer

*For these data, the World Health Organization divides the less developed nations into two groups, based on gross national incomes.
SOURCE: World Health Organization (2014b).

Ironically, the rise in chronic disease reflects in part the *problems* caused by rising incomes. As new middle classes have emerged in countries such as China and India, tobacco use, alcohol use, automotive travel, and obesity have all increased, causing deaths from lung cancer, alcohol-related disease and injuries, fatal accidents, diabetes, and heart disease. Moreover, these nations still have millions of poor citizens, so they are burdened by the economic, social, and health costs of both "diseases of wealth" such as diabetes and "diseases of poverty" such as tuberculosis (Yach et al., 2004).

Poverty, Malnutrition, and Disease

The primary cause of low life expectancies in the less developed nations is poverty. In Chapter 3, we saw that wealthy Americans experience less illness and live longer than do poorer Americans. In the same way, wealthier nations have lower rates of illness and mortality than do poorer nations. The average life expectancy is 61 years in the least developed nations, 69 years in the less developed nations, and 79 years in the more developed nations—an 18-year difference all told (Population Reference Bureau, 2014).

In large part, poverty causes disease and death by causing chronic malnutrition. Malnutrition causes disease and death by damaging the body's immune system, leaving individuals more susceptible to all forms of illness and contributing to both infant and maternal mortality. In addition, malnutrition leads to numerous health problems, including brain damage caused by iodine deficiency, blindness caused by vitamin A deficiency, and mental retardation caused by anemia. As Table 4.2 shows, malnutrition remains among the top ten causes of death in the least developed nations.

The Roots of Chronic Malnutrition Given the link between malnutrition, illness, and death, the importance of investigating the roots of chronic malnutrition is clear. At first thought, we might easily assume that malnutrition in less developed nations that have not yet experienced the **epidemiological transition** results naturally from overpopulation combined with insufficient natural and technological resources. Yet on a global level, farmers now grow twice as much food as is needed to feed the world's population (Holt-Giménez and Peabody, 2008; Lappé, Collins, and Rosset, 1998; UNICEF, 2014).

Nor can malnutrition be blamed on population density (Lappé et al., 1998). The Netherlands, for example, is one of the most densely populated countries in the world, yet chronic malnutrition no longer occurs there. Similarly, malnutrition has largely disappeared from Costa Rica but remains common in nearby Honduras, even though the latter has twice as much cropland per person.

If overpopulation, lack of food, population density, and lack of cropland don't explain chronic malnutrition, what does? The answer lies in the social distribution of food and other resources: *Malnutrition occurs most often in countries where resources are most inequitably distributed.* In other words, malnutrition occurs not in countries where resources are scarce, but in countries where a few people control many resources while many people have access to very few resources

(Dreze and Sen, 1989; Lappé et al., 1998). Similarly, within each country, malnutrition occurs most often among those groups—typically females and the poor—with the least access to resources (Messer, 1997). In essence, then, malnutrition is a disease of powerlessness.

If powerlessness causes malnutrition, then eliminating power inequities should eliminate malnutrition. Evidence from China (officially known as the People's Republic of China) and Costa Rica supports this thesis. In the past, both nations adopted socialistic strategies for redistributing resources somewhat more equitably. By giving farmland to formerly landless peasants, extending agricultural assistance to owners of small farms, working to raise the status of women, and so on, they made chronic malnutrition almost unknown within their borders. On the other hand, China has not proved immune to *acute* malnutrition caused by famines. According to Nobel Prize-winning economist Amartya Sen, famines occur only when (1) natural events reduce harvests *and* (2) nondemocratic governments (such as China's) can ignore citizen's basic needs because politicians know they can't be voted out of office (Sen, 1999). *Contemporary Issues: Linking Sanitation and Malnutrition* further illustrates how power inequalities can continue to breed malnutrition, even when countries begin to develop and food becomes more widely available.

The Role of International Aid In less developed nations that are democratically run, international aid—both food aid and development projects—has helped improve citizens' standard of living and health status. But in *nondemocratic* nations, aid often has the opposite effect (Calderisi, 2006; Easterly, 2006; World Bank, 1998). In such nations, small, powerful elites often take control of food aid, sell it on the black market, and pocket the profits. Poor people can't afford to buy food aid sold in the marketplace, so it doesn't help them at all.

CONTEMPORARY ISSUES

Linking Sanitation and Malnutrition

Over the last 20 years, India has experienced an economic boom. As a result, many more children across the country receive what should be enough calories and nutrients to foster healthy growth. Yet an estimated 65 million children under age five are malnourished, including one-third of those from wealthy families (Harris, 2014).

How could this be? Until very recently, nutrition researchers assumed, quite reasonably, that malnourishment resulted solely from lack of healthy food. In the last few years, however, researchers have increasingly concluded that lack of proper sanitation is a major cause of malnourishment in densely populated countries like India (Harris, 2014). Because of continuing inequities in how public services are distributed around the nation, about half of all Indians lack toilets and must defecate outdoors. As a result, Indian children are constantly fighting infections caused by parasites and germs carried by rains down streets and alleys and into water supplies used for bathing and drinking. As a result, their bodies lack the energy needed for them to develop physically and mentally. Unfortunately, the resulting physical frailty and mental retardation are permanent.

Like international food aid, internationally sponsored development projects have had mixed impacts on malnutrition and on health in general (Calderisi, 2006; Easterly, 2006; World Bank, 1998). According to the politically conservative World Bank, carefully designed projects, sensitive to local conditions and culture and located in countries with democratic governments, open trade, social safety nets, and conservative economic policies can reduce malnutrition and its root causes. In Pakistan, for example, school enrollment of girls soared in 1995 when local communities received development money to open new schools only if they increased girls' enrollment rate (World Bank, 1998).

On the other hand, when aid projects are not built around local needs and culture, the results can be harmful. For example, large dam projects around the world have brought electricity to urban elites and to factories run by multinational corporations, while flooding and destroying agricultural fields and bringing plagues of waterborne diseases to rural dwellers (Basch, 1999:280–281; Farmer, 1999). Similarly, agricultural development projects have often encouraged men to grow cash crops, leading them to take over farmlands that women had previously used to grow food. But in many countries, men consider feeding the family to be a woman's responsibility, and so men use their profits to purchase tobacco or other high-status goods for themselves rather than to purchase food for their families. As a result, malnutrition increases among women and children (Lappé et al., 1998).

Infectious and Parasitic Diseases

One indirect result of malnutrition and, more broadly, of poverty is a high rate of infectious and parasitic disease. As Table 4.2 shows, such diseases account for more deaths in the less developed nations (and far more in the least developed nations) than in the more developed nations.

As in Europe and the United States before the twentieth century, the high rates of infectious and parasitic diseases in the less developed nations reflect the dismal circumstances in which many people live. As we've already seen, malnutrition leaves individuals far more susceptible to a wide range of diseases. In addition, overcrowding promotes the spread of airborne diseases such as tuberculosis, and contamination of the water supply with sewage spreads waterborne diseases such as cholera and intestinal infections. Similarly, poor housing and lack of clean water for bathing result in frequent contact with disease-spreading rats, fleas, and lice.

The infectious and parasitic diseases that cause the most deaths in the less developed nations are HIV/AIDS, tuberculosis, diarrheal diseases, and malaria.

HIV/AIDS **HIV/AIDS** now kills more persons than any other infectious or parasitic disease in the less developed nations. Heterosexual intercourse remains the major mode of HIV transmission (as it has been from the start), although illicit intravenous drug use and blood transfusions are also sources of infection (in the absence of funds to purchase sterile needles or medical equipment). Transmission from childbearing women to their babies, however, has declined sharply now that large-scale, internationally funded programs provide antiretroviral drugs to pregnant women (UNAIDS/WHO, 2010).

Still, in the hardest-hit countries (most located in sub-Saharan Africa) up to one-quarter of adults are infected with HIV/AIDS (Population Reference Bureau, 2014). Life expectancies have increased a bit due to antiretroviral drugs, but remain under 50 years (Population Reference Bureau, 2014). Sub-Saharan Africa still accounts for about two-thirds of all new infections among adults and virtually all new infections among children (UNAIDS/WHO, 2010). HIV is also spreading rapidly in Eastern Europe and Central Asia, primarily among individuals who inject heroin and their sexual partners.

As stunning as these numbers might appear, they understate the impact of HIV/AIDS. Unlike most illnesses, HIV/AIDS commonly strikes at midlife, normally the most economically productive years. In the hardest-hit countries, agricultural production is declining steeply, causing food shortages. Moreover, unlike most diseases, HIV/AIDS has struck not only the poor but also the middle and upper classes (because of their greater access to sexual partners; reduced commitment to traditional, more conservative sexual norms; and residence in cities, where the disease is more common) (Fortson, 2008; WHO, 2009a). Deaths among teachers, doctors, businesspeople, and the like have crippled schools and the economy in numerous countries. The resulting increase in unemployment and poverty is sending ripples of illness and death throughout these countries. In addition, the deaths of many young mothers have produced a corresponding rise in deaths among children who lose their only (or best) provider.

Poverty primarily explains why HIV/AIDS has hit Africa especially hard. In addition, the epidemic has been stoked by labor migration, women's low status, and sexual behavior patterns.

Labor migration is common across Africa because the need to earn a living draws African men from small villages to cities and other areas where factories, mines, and plantations offer jobs. These men often must live apart from their wives and families for weeks, months, or even years at a time. Such conditions foster the use of prostitutes and consequently foster the spread of sexually transmitted diseases (STDs), including HIV/AIDS. In turn, some migrants may eventually carry these diseases back to their villages (Hunt, 1996; UNAIDS/WHO, 2010).

Meanwhile, health conditions also deteriorate among women and children left in rural villages (Hunt, 1996). The loss of men's labor makes it more difficult for women to grow sufficient crops to feed themselves and their children, leaving them increasingly malnourished and susceptible to disease. Faced with these conditions, women's only option is to seek employment in cities, where many find that they must trade sex for cash or other favors to survive, even if doing so increases their risk of HIV/AIDS (Hunt, 1989; Simmons, Farmer, and Schoepf, 1996).

As this suggests, girls' and women's low social and economic status also fosters the spread of HIV/AIDS in Africa. In countries where girls and women have low status, they may face physical violence if they ask a husband or other sexual partner to use a condom, find that male teachers demand sex as a requirement for attending school, or be pressured into marriages or sexual relationships with older men who are more likely to be infected. In addition, because their low status often keeps them from accessing medical care, women are more likely than men to have untreated STDs, which can produce open sores and thus increase the chances of infection for any woman who is exposed to HIV.

Sexual behavior patterns also play a role in the epidemic. Current research suggests that risks of infection are greater in Africa not because the average *number* of sexual partners is high there, but because long-term, *concurrent* sexual partners are more common (UNAIDS/WHO, 2010). In Western countries, individuals typically have **serial sexual partners**—one after another—such as a first marriage followed by a brief sexual relationship or two and then a second marriage. In contrast, in parts of Africa (especially sub-Saharan Africa), individuals often have long-term **concurrent sexual partners**: multiple sexual relationships *during overlapping time periods*.

Concurrent partnerships increase the chances of spreading HIV/AIDS for two reasons. First, around the world, people typically use condoms early in relationships but stop doing so if the relationship continues. Consequently, persons in long-term concurrent relationships are more likely than those in short-term monogamous relationships to reach the point where they stop using condoms (Mah and Halperin, 2010). Second, HIV/AIDS is most easily transmitted only when an individual is healthy enough to have an active sex life and has a "high viral load" (i.e., has many HIV cells in his or her body). If an individual hits that peak transmission point while he or she has concurrent sexual partners, all of those partners—and their partners—will be at risk.

Tuberculosis Each year, tuberculosis infects about 9 to 10 million people and kills about 1 million (WHO, 2014c). The disease is most common in Asia, followed by Africa. Tuberculosis is particularly devastating because, like HIV/AIDS, it typically hits people during their prime work years, so it sharply curtails family incomes.

Because of a consolidated, worldwide effort to make powerful treatment available even in poor regions, rates of tuberculosis have been falling for the past two decades in most of the world (WHO, 2014c). However, because HIV/AIDS makes individuals more susceptible to other infections, tuberculosis continues to increase in those African nations where HIV/AIDS is most common.

Diarrheal Diseases In the more developed nations, diarrhea typically causes only passing discomfort. In the less developed nations, diarrheal diseases are the second leading cause of death among children younger than age five (WHO, 2009b).

Diarrhea is a symptom, not a disease, and can result from infection with any of several bacteria, viruses, or parasites. Diarrhea kills by causing dehydration and electrolytic imbalance. It also leads to malnutrition when affected children not only eat less but also absorb fewer nutrients from the foods they do eat. In turn, malnutrition leaves children susceptible to other fatal illnesses. Conversely, other illnesses can leave children susceptible to both diarrheal diseases and malnutrition.

Diarrheal diseases (including dysentery, cholera, and infection with *Escherichia coli*) occur when individuals ingest contaminated water or foods. The likelihood of severe diarrhea is greatest when families lack refrigerators, sanitary toilets, sufficient fuel to cook foods thoroughly, or safe water for cooking and cleaning. WHO estimates that about 2 billion people lack access to "improved" water supplies, and many more lack access to truly safe water (WHO/UNICEF, 2014). The *number* of persons without safe water is greatest in Asia, but the *percentage* of those without safe water is highest in sub-Saharan Africa.

Survival rates for children with diarrheal diseases in less developed nations have improved rapidly in recent years. Before the 1960s, those suffering from diarrheal diseases could be treated only by using expensive intravenous fluids, thus making treatment unfeasible for many in the less developed nations. Since then, however, scientists have developed saline solutions and peanut butter pastes that keep children alive at least as well as do more expensive treatments.

Malaria Each year, between 250 and 500 million people become infected with malaria, and about 1 million—mostly African children—die from the resulting anemia, general debility, or brain infections (Shah, 2010). In addition, millions more find themselves unable to work because of continuing malarial chills and fevers, or die because malaria leaves them susceptible to other fatal illnesses.

Malaria poses the greatest threat to pregnant women, infants, and young children. Among pregnant women, malaria increases the risks of miscarriage, anemia, and premature labor, each of which increases the risk of potentially fatal hemorrhaging. Infants born to malaria-infected women typically have lower than average birth weight and hence a higher chance of death or disability.

Malaria is caused by protozoan parasites belonging to the genus *Plasmodium*. Malaria is transmitted only by *Anopheles* mosquitoes and consequently exists only where those mosquitoes live. The disease cycle begins when a mosquito bites an infected individual and ingests the parasite from the individual's blood. The parasite reproduces in the mosquito's stomach and then migrates to the mosquito's salivary glands. The next time the mosquito bites someone, it transmits the parasite to that person.

Because of this transmission cycle, eliminating *Anopheles* mosquitoes will eliminate malaria. Since the 1940s, anti-malaria campaigns have depended heavily on pesticides to kill mosquitoes. Although such campaigns initially work well, over time, pesticide-resistant mosquitoes evolve, and the pesticides lose their potency (Shah, 2010). As a result, nations must constantly search for new and more toxic pesticides, each of which can endanger birds, fish, and insects that benefit humans. Because of these problems, some recent campaigns have instead focused on encouraging the use of insect repellents, mosquito netting, and screens to prevent infection. These campaigns also have focused on encouraging the use of drugs, such as chloroquine and mefloquine, which can both prevent and treat malaria. Unfortunately, because these drugs can cause debilitating side effects and cost more than many residents of developing nations can afford, infected individuals often stop taking the drugs before they are cured. This continual undertreatment of malaria, like the undertreatment of tuberculosis, has encouraged the evolution of drug-resistant malaria.

Infant Mortality

Like infectious and parasitic diseases, and as Table 4.1 on page 75 shows, infant mortality is many times higher in the less developed nations than in the more developed nations (Population Reference Bureau, 2014). The most common causes of infant mortality in poorer nations are malnutrition and infections (particularly respiratory infections and diarrheal diseases). Because we examined these factors earlier in this chapter, we focus here on two other important sources of infant mortality: women's status and infant formula manufacturers.

The Role of Women's Status The low status of women plays a critical role in infant mortality in less developed nations. In these countries, infant mortality occurs most often among babies born to underfed, overworked mothers, many of whom suffer from untreated illnesses (WHO, Reproductive Health and Research Department, 2004).

These conditions reflect women's low status. Throughout the less developed nations, girls and women often spend long hours in heavy labor (Messer, 1997). Yet they typically receive less food and less health care (including immunizations) than do boys and men (Kristof and WuDunn, 2010; Messer, 1997). As a result, girls often enter their childbearing years already ill and malnourished—a situation that worsens as pregnancies further stress their bodies and drain their energy. Map 4.1 shows the prevalence of intimate partner violence against women around the world. Such violence is an excellent marker of women's low social status, and in itself can cause health problems that make safe childbearing difficult.

Similarly, infant mortality is highest among infants born to very young or very old mothers and to infants born less than 18 months after a sibling. This situation occurs most commonly in cultures that expect women to marry at young ages and that judge women's worth by how many children they have (especially male children). In part, these cultural values reflect the economic realities of agricultural life: In agricultural societies, children produce more economic resources than they consume, so a family with many children is more likely to survive than a family with few children. In addition, in the absence of pension systems, individuals can guarantee their security in old age only by having sons given that daughters generally are expected to take care of their husbands' parents rather than their own. For this reason, it is common for families in some parts of Asia to let girl babies die by giving them less food or medical care, to use medical technologies to identify and abort female fetuses, or to kill girl babies outright (Kristof and WuDunn, 2010; Zhu, Lu, and Hesketh, 2009). This situating is discussed in *Ethical Debate: The Ethics of Prenatal Sex Selection.*

In sum, research suggests that if women's social status were higher, they would enter their childbearing years with healthier bodies, wait longer before

MAP 4.1 Lifetime Prevalence of Intimate Partner Violence against Women*

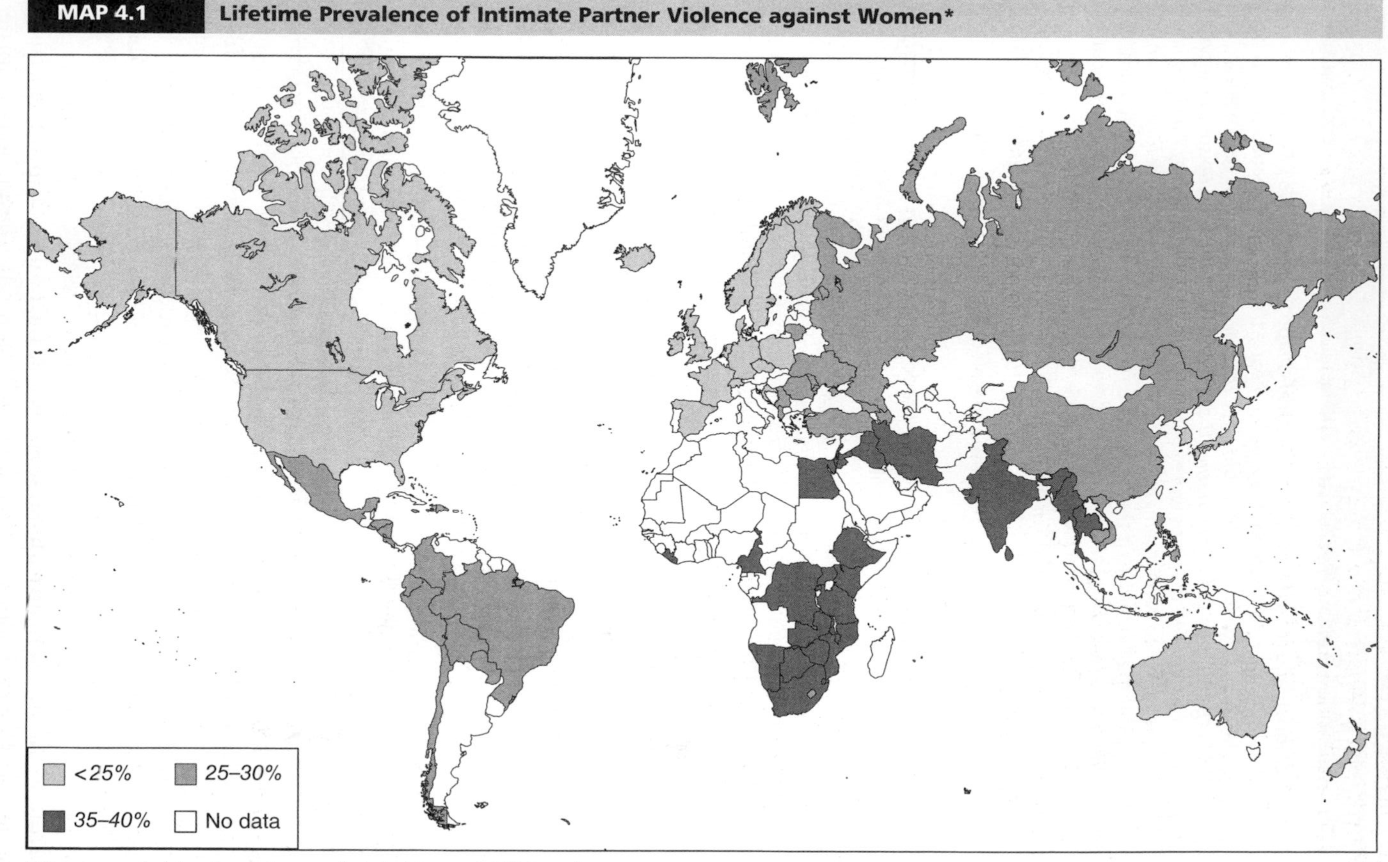

*15 years and older. Percentages calculated for each WHO region.

SOURCE: World Health Organization, Department of Reproductive Health and Research (2013).

ETHICAL DEBATE

The Ethics of Prenatal Sex Selection

Zhang Zhiquan and his wife, Mei, live in a rural village in the People's Republic of China. Growing up in rural China, they learned early that couples needed sons to prosper and to care for them in their old age. They also learned that sons were essential for passing on the family name, that wives who produced no sons deserved mockery and abuse, and that girls were so useless that in the past many rural families did not even bother to name them. When Mei became pregnant, therefore, they had to decide what they would do if the baby were female. In the past, if they felt unable or unwilling to raise a daughter, their only options would have been to kill the baby or give her up for adoption—choices that some families still make. Now, however, they have one additional option: having a health care worker identify the fetus's sex through ultrasound or amniocentesis and perform an abortion if the fetus is female.

Half a world away, the same issues of sex preselection and selective abortion arise, although in a different form:

Sharon and James Black live in Denver, Colorado, with their two young daughters. Because they both believe that children need a parent home at the end of the school day, Sharon works only part time as a secretary, and James works two jobs so they can make ends meet. Sharon has just learned she is pregnant again. Although they had only planned on having two children, James always wanted a son with whom he can share his interests in sports and automobiles. Having another child, however, will further strain their finances and make it difficult for Sharon to return to full-time work for several more years. Consequently, continuing the pregnancy does not seem worthwhile unless they know that the fetus is male.

Is prenatal sex selection ethically justified in these cases? Although the circumstances differ enormously, for both families, the birth of a daughter would bring substantial economic hardship. For both families, too, a daughter would enter life unwanted and already having failed to meet her parents' expectations. In addition, for the Chinese family and possibly (although to a lesser extent) the American family, the birth of another daughter might lower the wife's status and strain the marriage. Given these circumstances, wouldn't it be best for all concerned if the families use the available medical technology to determine their fetuses' sex and to abort them if they are female?

For hundreds of thousands of couples in Asia and a growing number in the West, the answer, resoundingly, is yes. In rural China, for example, 126 boys are born

having babies, wait longer between babies, and have fewer babies in total, with each of these factors lowering the infant mortality rate. For all of these reasons, many researchers and public health workers have suggested that the most effective way to reduce infant mortality is to improve the status of women, thereby increasing their power to make decisions for themselves. This at least partly explains why infant mortality is so much lower in Costa Rica and China than in some other countries at similar levels of development.

The Role of Infant Formula Manufacturers A final cause of infant mortality in the less developed nations is the use of infant formula. Researchers estimate that 13 percent of all deaths before age five could be prevented if infants were breastfed during their first six months of life (UNICEF, 2005).

for every 100 females overall (Lawn, Cousens, and Zupan, 2005; Zhu et al., 2009). That ratio increases to 146 boys to 100 girls for second births to a couple.

Those who support prenatal sex selection argue that selective abortion causes little harm, whereas the birth of unwanted girls in poorer nations can financially strain families; leave mothers open to ridicule or even physical abuse; and result in child neglect, abuse, or abandonment. Those who oppose prenatal sex selection argue that it does more harm than good because it reinforces the low status of females. Although in rare circumstances families use medical technologies to ensure that their babies are female (such as families with a history of hemophilia, a disease that affects only males), in the less developed nations, prenatal sex selection almost always means selecting males. However, in the more developed nations, the preference for sons has declined substantially or even reversed (Andersson, Hank, and Ronsen, 2006; Edgar et al., 2006).

When families select male fetuses over female fetuses, they proclaim male babies preferable. Moreover, when health care workers help families to select male babies, the workers in essence validate this preference. Finally, when health care workers assist in prenatal sex selection—whether helping families to select males or females—they reinforce the idea that males and females are inherently different. After all, if male and female personalities, interests, and aptitudes were more similar than different, why would families need to choose one over the other?

In sum, to assess the ethics of prenatal sex selection, we need to weigh the potential benefits and costs for families and for society as a whole.

Sociological Questions

1. What social views and values about medicine, society, and the body are reflected in prenatal sex selection?
2. Which social groups are in conflict over this issue? Whose interests are served by allowing prenatal sex selection? By forbidding it?
3. Which of these groups has more power to enforce its view? What kinds of power do they have?
4. What are the intended consequences of permitting prenatal sex selection? What are the unintended social, economic, political, and health consequences of this policy?

In the less developed nations, several factors contribute to the especially high rates of death and disease among infants who are not breastfed. First, in addition to the inherent nutritional limitations of breast milk substitutes, bottle-feeding itself can expose infants to tremendous risks. Infant formula is typically sold as a powder that must be mixed with water and then transferred to a bottle before it can be used. In most of the less developed nations, this water contains dangerous infectious organisms. Those organisms can be killed if the water and bottle are boiled, but many families don't understand how or why they should do so. Moreover, throughout the less developed nations, many women and children already spend hours each day getting water and firewood and lack the time and energy to get the extra supplies needed to sterilize water and bottles.

Second, infant formula is not free. To cut the costs, families often stretch infant formulas by diluting them with water. Babies fed diluted formula in essence starve to death while filling their stomachs.

Finally, by altering the hormonal levels in a woman's body, breast-feeding serves as a moderately effective contraceptive. Breast-feeding thus helps women to space out pregnancies and gives each baby a better chance for survival.

Given all the benefits of breast-feeding, why don't more women in less developed nations breastfeed? Part of the answer lies in traditional cultural beliefs, such as the conviction that children require certain traditional foods for health or that it is unsafe for men to have sex with breast-feeding women (Dettwyler, 1995). Part of the answer lies in practical economic and social issues, such as the difficulty of meshing breast-feeding with paid work. And part of the answer lies with multinational food corporations (based in the more developed nations) that have mounted massive advertising campaigns to convince women that bottle-feeding is healthier. One particularly pernicious strategy is to dress saleswomen as nurses and send them to villages and maternity hospitals to convince women that bottle-feeding is healthier and more "modern."

Maternal Mortality

Although maternal mortality is now rare in the more developed nations, it remains the primary cause of death among women of reproductive age in the less developed nations. For example, in Afghanistan 1 out of every 32 women die from childbirth complications, compared to 1 out of every 4,600 women in the United Kingdom (Population Reference Bureau, 2014).

How can we account for the tremendous toll maternal mortality takes in the less developed nations? Patricia Smyke, writing for the United Nations, explains:

> If you ask, "Why do these women die?" the technical response is: "The main causes of maternal death are hemorrhage, sepsis (infection), toxemia, obstructed labor and the complications of abortion." But looking beneath those immediate causes, one must ask why they occurred or why they were fatal. The answer to that is: lack of prenatal care; lack of trained personnel, equipment, blood or transport at the moment the obstetrical emergency arose, or earlier, when it might have been foreseen and avoided; lack of family planning to help women avoid unwanted pregnancies, too many or too closely spaced births, or giving birth when they were too young or too old; preexisting conditions like malaria, anemia, fatigue and malnutrition that predispose to obstetrical complications; problems arising from female circumcision. From that list of intermediary causes one must go deeper still to identify the cultural and socioeconomic factors that put young girls, almost from birth, on this road to maternal death: … low status of women and discrimination against them; poverty; lack of education; local customs; and government policies that give low priority to the needs of women (Smyke, 1991: 61–62).

Box 4.1 Female Genital Cutting

According to UNICEF (2013), more than 125 million girls and women across Africa, as well as in Malaysia, Indonesia, Yemen, and elsewhere, have experienced the ordeal of female genital cutting. Female genital cutting is a brutal and sometimes fatal procedure, far different from male circumcision. In most cases, the entire clitoris and labia minora are removed. In addition, in about 20 percent of cases, the clitoris, labia minora, and parts of the labia majora are removed, and the sides of the vulva are stitched together, leaving only a small opening for urine and menstrual fluid to escape. Most commonly, a midwife or other lay healer does the cutting using a razor blade, knife, or piece of broken glass.

Those who support female genital cutting believe it makes women more docile and reduces their sex drive, making them better wives and less likely to disgrace their families by engaging in premarital or extramarital sexual relationships. In addition, supporters believe that women who have been cut are cleaner, healthier, more fertile, and prettier. In countries where female genital cutting is the norm, these beliefs reduce marriage prospects for those who are uncut and so put pressure on parents to have their daughters cut, even if the parents disapprove of the practice.

Female genital cutting substantially impairs the health of young girls and women. Given the unsanitary conditions in which it is usually performed, the cutting can cause life-threatening shock, hemorrhage, infections, or tetanus. Those who survive often experience pain during intercourse and chronic urinary, vaginal, or pelvic infections, sometimes resulting in infertility. If they do become pregnant, scar tissue and the narrowed vaginal opening can make it difficult for a baby to emerge, causing women to die from hemorrhage and babies to die from brain damage. To avoid these problems, families increasingly turn to sympathetic nurses and doctors who are willing to perform the procedure (UNICEF, 2013).

To date, most nations where female genital cutting occurs officially oppose the practice, and several have outlawed it. Beliefs and practices are changing, and rates of female genital cutting have declined somewhat, with substantial declines in a few countries (UNICEF, 2013).

Maternal mortality occurs most often when malnutrition or malaria causes anemia and as a result causes women to hemorrhage during birth. Maternal mortality is especially common among women who give birth in unsanitary conditions or who give birth before age 20, after age 35, or more than three times. Finally, maternal mortality is more common among women who have endured genital cutting; Box 4.1 provides further details on this dangerous practice.

Another cause of maternal mortality in the less developed nations—accounting for 13 percent of deaths—is unsafe abortion, most commonly because of infections caused by unsterile instruments, hemorrhage when instruments pierce the uterus, or poisoning when women try to perform abortions on themselves by swallowing toxic chemicals (Sedgh et al., 2007). Abortion is a technically simple procedure, far safer than childbirth when performed by trained professionals working in sterile conditions with proper tools. However, most of the less developed nations have restricted or outlawed abortion because of cultural traditions, religious beliefs, a desire by political elites to increase population, or political pressure from US anti-abortion forces. Yet a comprehensive global study published in the prestigious medical journal *The Lancet* found that outlawing abortion has no effect on the number of women who *get* abortions but greatly increases the number who *die* from abortions (Sedgh et al., 2007).

Maternal mortality has declined significantly during the past 30 years, primarily because of decreases in the number of births per woman and increases in women's education, income, and access to skilled birth attendants (whether doctors or not) (Hogan et al., 2010). Mortality is now concentrated in countries torn by war (including Afghanistan and Ethiopia) and the southern African countries most severely affected by the HIV/AIDS epidemic.

Respiratory Diseases

Finally, respiratory diseases, such as emphysema, are also major killers in the less developed nations, as in the more developed nations. As with all disease in the less developed nations, poverty and malnutrition increase individual susceptibility to illness. In addition, long periods spent cooking over open fires in closed rooms expose millions of women to cancer-causing toxins equivalent to smoking several packs of cigarettes daily. Meanwhile, those who live in cities such as Caracas and Calcutta risk their health daily because of pollution from automobiles and industries. Unfortunately, in some less developed nations, government officials lack the political or economic power to control polluting industries, and in other nations, officials profit from and hence promote these industries. Equally important, officials in less developed nations sometimes believe that pollution and the attendant morbidity and mortality are short-term costs they must pay to industrialize and to improve their nation's health in the long run.

To these factors must be added the growing role of tobacco, which, in the less developed nations as in the more developed nations, is a major cause of respiratory disease, heart disease, and cancer (WHO, 2014a). Tobacco use has grown steadily in the less developed nations since 1964, when the US Surgeon General declared tobacco a cause of lung cancer, US sales of cigarettes plummeted, and tobacco manufacturers (most based in the United States) turned to the less developed nations for new markets.

Manufacturers now devote enormous sums to advertising tobacco in those nations. In countries where direct advertising of tobacco on television or radio is restricted, manufacturers instead sponsor cultural and athletic events, especially those oriented toward youths.

War

The most unnatural cause of death and disease in the less developed nations (and elsewhere) is war (Geiger and Cook-Deegan, 1993; Toole and Waldman, 1993). Political and economic instability, combined with environmental degradation, have made the less developed nations particularly vulnerable to war. Wars can not only wipe out a generation of soldiers but can also take astoundingly high tolls among civilians. For example, in the first 18 months after coalition forces led by the United States invaded Iraq in 2003, about 100,000 Iraqi civilians—most of them women and children—were killed by military forces (Roberts et al., 2004).

Civilians are killed not only by bombs and guns but also by forced labor; malnutrition after soldiers burn crops, kill farm animals, and force farmers off their lands; and diseases that spread when refugees are forced into overcrowded, unsanitary camps and when soldiers destroy water, sewage, and health care facilities. During six years of warfare in Darfur, for example, 80 percent of those who died—many of them women and children—were killed not by guns or bombs, but by diseases that spread when warfare led to economic, social, and ecological destruction (Olivier and Debarati, 2010).

Survivors, too, pay a huge price, often including both long-lasting disability and the psychological trauma of losing one's family, community, and work. The traumas are particularly high for victims of mass rape, a common tool of warfare that has been used extensively in recent years in the Sudan and the Congo; those who survive can find themselves not only infertile or permanently disabled by their injuries but also stigmatized and sometimes abandoned by families and neighbors. Finally, an estimated 300,000 children in more than 30 less developed nations are serving as soldiers (UNICEF, 2010). Mortality rates are very high, as are the health risks experienced by those who survive. These children are exposed to all the horrors and dangers of warfare and to increased risks of malnutrition, disease, injuries from land mines, sexual abuse, and substance abuse, while losing opportunities for education and normal family life that might protect their mental and physical health as adults.

Mike Goldwater/Alamy

Whenever children serve as soldiers, they risk terrible injuries, psychological trauma, and of course death.

Disasters

The devastation wrought by earthquakes, tsunamis, floods, and other natural disasters in the less developed nations is impossible to miss: 131,000 people confirmed dead in 2004 when a tsunami hit Indonesia; more than 60,000 confirmed dead after a 2008 earthquake in Sichuan, China; and so on. Like war, natural disasters typically result in public health disasters as crops and jobs are lost; sewer, water, and health care facilities are destroyed; and health care workers are scattered or killed.

Although humans can't prevent natural disasters, they can greatly reduce—or increase—their toll (Revkin, 2005). Schools, homes, and other built structures can be retrofitted or built to withstand most earthquakes at costs far less than the cost of replacing or repairing damaged or destroyed structures. Dams, nuclear power plants, and other dangerous structures can be located away from vulnerable flood plains or tectonic faults. And disaster preparedness programs can be developed that will warn people of impending disasters, offer means of escape, and secure public health infrastructures afterward. For example, despite the extraordinary violence of the tsunami and earthquake that hit Japan in 2011, deaths were far lower than they otherwise would have been because the country had prepared so well for natural disasters. However, such preparation requires not only technical knowledge but also both the money and the political will to act on that knowledge (Revkin, 2005). The less developed nations are particularly vulnerable to disasters both because they lack the necessary funding and because in many cases they are ruled by small elites who have no real commitment to protecting the citizenry.

The earthquake that struck Haiti in January 2010 illuminates these points. The earthquake's impact was particularly devastating not only because of its power but also because of the population's poverty and the government's corruption. When the earthquake struck, most of the population was already living in poverty, and half in extreme poverty (*New York Times,* 2010). Government corruption had siphoned money into politicians' pockets and away from building hospitals, roads, clean water systems, and earthquake-proof housing. As a result, many died when buildings were crushed, aid workers couldn't reach the injured, few hospital beds were available, and vulnerable water systems made it easy to spread cholera.

Structural Violence

To a large extent, the causes of ill health discussed in this chapter can be summed up using the concept of structural violence. **Structural violence** refers to social arrangements that are deeply embedded in the politics, culture, or economy of a society *and* that harm individuals or keep them from reaching their full potential (Farmer et al., 2006; Farmer et al., 2013) (see *Key Concepts: Structural Violence*). This concept is particularly useful for explaining why poor populations are especially likely to fall ill and for protecting the health of those vulnerable populations.

KEY CONCEPTS

Structural violence

Structural violence refers to social structures and institutions that are embedded in a society's politics, culture, or economy and that either harm individuals or keep them from reaching their full potential.

Type of Violence	Committed by	Effects	Visibility of Effects	Example
"Ordinary" violence	Individuals	Injury, death	Obvious, concrete	Bar fight Father who beats his child
Structural violence	Social structures, social institutions, or individuals working on behalf of social structures or institutions	Injury, illness, death, economic harm, social harm	Often hard to recognize	Armies that massacre minority groups Laws that set the minimum wage so low that poor parents can't provide healthy food to their children

The case of HIV/AIDS in Rwanda provides a useful example. A small sliver of Rwandans is wealthy, whereas most are very poor. Meanwhile, across the social classes, women continue to have far less power than do men. In addition, Rwanda experienced a massive, genocidal war during the 1990s, which caused untold deaths and injuries and forced much of the population to flee to refugee camps. All these factors reflect and reinforce structural violence, and have contributed to an epidemic of HIV/AIDS in Rwanda.

As this suggests, effective interventions need to take structural violence into consideration (Farmer et al., 2006). For example, training doctors in the best ways to treat HIV/AIDS will have little effect if doctors work primarily in cities and only the wealthy can afford to visit them. Instead, it may be more effective to train lay workers to provide basic, low-cost treatment. Similarly, any efforts devoted to reducing poverty in Rwanda would attack structural inequality at its roots and increase the odds that Rwandans could purchase both condoms and HIV/AIDS treatment.

IMPLICATIONS

One of the major threads throughout this chapter is the important role poverty plays in causing illness and death in the less developed nations. Consequently, reducing poverty in these nations should raise them to the health levels found in the more developed nations. So, too, can a variety of relatively inexpensive public health measures; deaths among children in poor countries have fallen precipitously in the last 20 years due primarily to interventions such as distributing insecticide-treated mosquito nets and increasing vaccination rates (UNICEF, 2014). Conversely, the situation in Russia demonstrates how a more developed nation can slide toward health levels lower than those found in some less developed nations (Feshbach, 1999; Feshbach and Friendly, 1992).

After the collapse of the Soviet Union in 1991, political and economic upheaval in Russia led poverty to spread across the country and living conditions to deteriorate. Increasingly across this vast territory, people lived in inadequately heated, overcrowded, and ramshackle housing. Moreover, public awareness that the government could no longer guarantee citizens a minimum standard of living demoralized people, encouraging many to find solace in drugs and especially alcohol.

To these problems were added those caused by environmental degradation. In past decades, the Soviet Union expanded its economic base as rapidly as possible, with little regard for the human or environmental toll. The Soviet government rarely established and almost never enforced regulations designed to protect the environment from industrial pollution. As a result, industries wreaked far greater environmental havoc in the Soviet Union than elsewhere in Europe, polluting farmlands and waterways beyond repair and leaving radioactivity, lead, and other dangerous toxins behind. Similarly, the emphasis on increasing agricultural yields as quickly as possible led to overplowing and resultant soil erosion as well as to overuse of herbicides, chemical fertilizers, and pesticides that have poisoned the water, the land, and food crops.

Since about 2000, median incomes in Russia have risen, and living conditions have improved. The nation's health also seems to have begun to improve, but years of environmental damage, social turmoil, and poor living conditions continue to take a toll in human lives. Infant mortality remains twice as high in agricultural areas where pesticides were used heavily compared with other regions, and **incidence** rates for numerous infectious diseases have increased, especially for drug-resistant tuberculosis. In addition, the collapse of the social structure and economy during the 1990s contributed to a proliferation of STDs, with rates of both syphilis and AIDS skyrocketing. Partly as a result, deaths caused by alcohol, suicide, violence, and accidents remain far more common in Russia than in Western Europe or the United States (WHO Regional Office for Europe, 2006). For all of these reasons, even though life expectancy has increased steadily for the last decade, it remains 11 years lower for men in Russia than for men in the United States (Population Reference Bureau, 2014).

In sum, no natural progression leads countries toward an increasingly healthy citizenry. Rather, as the political and economic fortunes of a country shift and as the natural environment improves or declines, so too will the health of its population. Only through continued commitment to eliminating poverty, reducing inequality, and protecting the environment can a nation guarantee that it will keep whatever health gains it has achieved.

SUMMARY

1. The *more developed nations* are nations that have relatively high GNI per capita and diverse economies composed of many different industries. The *less developed nations* are those nations with relatively low GNI per capita and relatively simple economies. The *least developed nations* are the worst-off subset of the less developed nations.
2. Compared with the more developed nations, the less developed nations have higher infant and maternal mortality; lower life expectancies; and a greater burden of infectious and parasitic diseases, especially HIV/AIDS, tuberculosis, diarrheal diseases, and malaria.
3. In a major change from past generations, chronic disease (especially heart disease) is rapidly emerging as a common cause of death in the less and even least developed nations.
4. The main reason for low life expectancy in the less developed nations is chronic malnutrition. Chronic malnutrition occurs most often in undemocratic countries where a few people control most resources. Within countries, malnutrition occurs most often among those groups with the least access to resources—typically poor women and their children. International aid can increase malnutrition when it increases power inequities.
5. HIV/AIDS has hit parts of Africa especially hard, primarily because of poverty. In addition, the epidemic has been stoked by labor migration (which takes men away from their families and increases their use of prostitutes), women's low status, and concurrent sexual partners (i.e., having more than one long-term sexual partner at a time).
6. Infant mortality is a far more common cause of death in the less developed nations. The most common killers of infants in the less developed nations are malnutrition and infections. In addition, the low status of women and the mass marketing of infant formula by multinational corporations have contributed to infant mortality.
7. Maternal mortality is the primary cause of death among women of reproductive age in the less developed nations. Because of their low status, girls are married off young, bear children before their bodies have matured enough to do so safely, receive too little food to nourish their fetuses or their own bodies, and lack access to birth control or safe abortions.
8. Political and economic instability, combined with environmental degradation, leave the less developed nations particularly vulnerable to war. Wars typically kill far more civilians than soldiers, primarily by causing famines and spreading illnesses.
9. The less developed nations are particularly vulnerable to disasters both because they lack the necessary economic funds to build earthquake-safe infrastructures and because in many cases they are ruled by small elites who have no real commitment to doing so.

10. *Structural violence* refers to social structures and institutions that are embedded in a society's politics, culture, or economy and that either harm individuals or keep them from reaching their full potential.

REVIEW QUESTIONS

1. How does poverty contribute to illness in less developed nations?
2. How do international politics and multinational corporations contribute to illness in less developed nations? How do undemocratic governments contribute?
3. How does the low status of women contribute to maternal mortality in less developed nations? To infant mortality?
4. How are the effects of natural disasters amplified by the political and economic conditions in less developed nations?

CRITICAL THINKING QUESTIONS

1. For the past five years, you have worked as a public health worker in a poor, urban, minority neighborhood in the United States. You have just accepted an exchange agreement to work for three years in Cape Town, South Africa. What parallels will you expect to see between these two settings in terms of the nature and sources of health problems and the best ways for dealing with health problems?
2. Identify the three changes you think would contribute most to improving the health of people in the less developed nations. Justify your choices.
3. Identify three *selfish* reasons why Americans (individuals, corporations, government, voluntary organizations) should care about illness and death in *less* developed nations.
4. Use the concept of *structural violence* to help explain why infant mortality is so much higher in poor nations than in wealthy ones.

PART
II

The Meaning and Experience of Illness

Our commonsense understandings of the world tell us that illness is a purely biological condition, definable by objectively measured biological traits. As we will see in Part II, however, definitions of illness vary considerably over time and across social groups. In Chapter 5, we explore the social meanings of illness and consider how ideas about the nature and causes of illness have changed historically, from biblical explanations that attributed illness to punishment for sin to modern explanations that attribute illness to risky lifestyles. We also examine how defining something as an illness can act as a form of social control.

Whereas Chapter 5 discusses the meaning of illness in the abstract, Chapter 6 looks at the consequences of chronic illness, chronic pain, and disability for individuals. Beginning with a discussion of how Western society historically has treated those who have chronic illnesses and disabilities, we then consider the modern experience of illness and of chronic pain, including the processes involved in responding to initial symptoms, searching for mainstream or alternative therapies, and coming to terms with a changed body and self-image.

In Chapter 7, we examine parallel questions regarding mental illness. We begin by exploring what it means to call something a mental illness. Then we look at how and why mental illness is distributed among social groups; how Western society historically has treated persons with mental illnesses; and how individuals experience mental illness, from initial symptoms to treatment, to social status after treatment.

CHAPTER

5

The Social Meanings of Illness

Chuck Savage/Cosbis/Cusp/Glow Images

LEARNING OBJECTIVES

After reading this chapter, students should be able to:

- Understand how cultural explanations for illness have changed over the centuries.
- Compare the medical and sociological models of illness.
- Assess the impact of medicalization.
- Use and critique the concept of a sick role.

According to Lunesta.com, a website aimed at the general public and run by the company that manufactures the popular insomnia drug Lunesta:

> *Approximately 20 million adults in the U.S. suffer from insomnia—a medical condition in which difficulty falling asleep and/or staying asleep has a negative impact on the next day ... Symptoms [include] difficulty falling asleep, waking up frequently during the night, difficulty returning to sleep, waking up too early in the morning, unrefreshing sleep, daytime sleepiness, [or] difficulty concentrating.... It [insomnia] is a serious medical condition that can affect your mind and body [and cause] daytime fatigue, irritability, decreased feelings of well-being, decreased ability to concentrate, decreased ability to problem solve, [and] difficulty in making decisions.*

To encourage readers to seek (drug) treatment for sleep problems, the website not only stresses the need to seek medical help but also offers suggestions on what to say when seeking help from doctors; information on getting Lunesta at discounted prices; and web links to national nonprofit organizations (partly funded by pharmaceutical manufacturers) that focus on identifying and treating sleep problems.

Does this seem like a reasonable way to deal with sleep problems—many of which are caused by nonmedical issues such as working odd hours, job stress, an over-firm mattress, or watching late-night television? According to many doctors, the answer is yes. Diagnoses with insomnia have skyrocketed in recent years, following intensive marketing campaigns for these drugs—even though research suggests that they provide only a few extra minutes of sleep per night but can cause traffic accidents, crippling falls, and amnesia-like episodes during which individuals may eat, walk, engage in sex, or do other activities they would not consciously have chosen (Moloney, Konrad, and Zimmer, 2011).

This raises the question: What do we mean when we say that something is a disease? In this chapter, we examine the meaning of illness and disease. We look first at how people have explained illness across history and then at the medical and sociological models of what illness is and means. Then we consider how medicine can act as an institution of social control, highlighting the process through which behaviors or conditions become defined as illnesses, the consequences of these definitions for individuals and society, and the potential consequences of our growing reliance on genetic explanations for disease.

EXPLAINING ILLNESS ACROSS HISTORY

Throughout history, people have feared illness. To relieve their anxiety and make the world seem less frightening, they typically have sought explanations for why illness occurs and why it strikes some rather than others. Most often, these explanations defined illness as a deserved punishment for sinful or foolish behaviors and blamed individuals for their own illnesses (Brandt and Rozin, 1997; Weitz, 1991). Such explanations provide psychological reassurance by reinforcing people's belief in a "just world" in which punishment falls only on the guilty (Meyerowitz, Williams, and Gessner, 1987). For example, both the Jewish and Christian Bibles describe leprosy as punishment for sin. As a result, throughout the Middle Ages, Christian society required anyone diagnosed with leprosy to participate in a special mass for the dead, in which a priest would shovel dirt on the individual's feet to symbolize his or her civil and religious death. From then on, the individual was legally prohibited from entering public gathering places, washing in springs or streams, drinking from another's cup, wearing anything other than the special "leper's dress," touching anything before buying it, and so on (Richards, 1977:123–124).

By the early nineteenth century, pre-scientific ideas about illness had begun to erode as the idea grew, especially among the elite, that scientific principles controlled the natural order. According to the new scientific thinking, illness occurred when biological forces combined with personal susceptibility. Doctors (still lacking a concept of germs) argued that illness occurred when persons whose constitutions were naturally weak or had been weakened by unhealthy behaviors came in contact with dangerous **miasma**, or air "corrupted" by foul odors and fumes. As a result, whereas earlier theories had blamed illness on *immoral* behavior, these new theories blamed illness on *unhealthy* behavior.

As the history of cholera shows, however, these new ideas still allowed the healthy to blame the ill for their illnesses. Cholera first appeared in the Western world in about 1830, killing its victims suddenly and horrifyingly through overwhelming dehydration brought on by uncontrollable diarrhea and vomiting. Cholera is caused by waterborne bacteria generally transmitted when human wastes contaminate food or drinking water. It most often strikes poor persons because they are the most likely to lack clean water and to be weakened by insufficient food, clothing, or shelter.

To explain why cholera had struck, and why it struck the poor especially hard, early nineteenth-century doctors asserted that cholera could attack only individuals who had weakened their bodies through improper living (Risse,

1988; Rosenberg, 1987). According to this theory, the poor caused their own illnesses, first by lacking the initiative required to escape poverty and then by choosing to eat an unhealthy diet, live in dirty conditions, or drink too much alcohol. Thus, for example, the New York City Medical Council concluded in 1832 that "the disease in the city is confined to the imprudent, the intemperate, and to those who injure themselves by taking improper medicines" (Risse, 1988:45). Conversely, doctors (and their wealthy patrons) assumed that wealthy persons would become ill only through gluttony, greed, or by "innocently" inhaling some particularly noxious air. This theory of illness allowed the upper classes to adopt the new scientific explanations for illness while retaining older moralistic assumptions about ill people and avoiding any responsibility to aid the poor or the ill. In sum, instead of believing that immorality directly caused illness, people now believed that immorality left one *susceptible* to illness.

Despite the tremendous growth in medical knowledge about illness during the past 200 years, popular explanations for illness have remained remarkably stable. Theories connecting illness to sin continue to appear, as do theories that conceptualize illness as resulting from poorly chosen (although not necessarily sinful) behaviors and attitudes (Brandt and Rozin, 1997; Tesh, 1988). Parents still act as if colds are caused by playing in the rain rather than exposure to viruses, and public health authorities more often focus on urging people to exercise than on addressing the conditions (such as dangerous neighborhoods, lack of gyms, or the need to hold down two jobs) that keep people from exercising. Similarly, author Rhonda Byrne, in her hugely popular book *The Secret*, argues that individuals "attract" health, wealth, sickness, or poverty to themselves simply by thinking about these conditions (Byrne, 2006:130–132). In sum, theories of illness that focus on individual responsibility continue to reinforce existing social arrangements and to help us justify our tendency to reject, mistreat, or simply ignore those who have illnesses.

MODELS OF ILLNESS

But what do we mean when we say something is an illness? The answer is far from obvious. Most Americans are fairly confident that someone who has a cold or cancer is ill. But what about women whose bones have become brittle with age; men who have bald spots or enlarged prostates; or young boys who have trouble learning, drink excessively, or enjoy fighting? Depending on whom you ask, these conditions and actions may be defined as normal human variations, illnesses, bad character, or bad behavior. As this suggests, defining illness is not a simple task. In this section, we explore how doctors and sociologists approach these issues.

The Medical and Sociological Models of Illness

The **medical model of illness** refers to what doctors typically mean when they say something is an illness. This medical model is not accepted in its entirety by

KEY CONCEPTS

Medical and Sociological Models of Illness

Medical Model	Sociological Model
Illness is an objective label: All educated people agree on what is normal and what is illness. *Example:* Female sexual dysfunction (FSD) is a biological disease characterized by lack of sexual responsiveness.	Illness is a subjective category: Educated people sometimes disagree on what should be labeled illness. *Example:* FSD is a label given to women who are distressed by their lack of sexual responsiveness with their current sexual partner.
Illness is nonmoral: Conditions and behaviors are labeled illness scientifically without moral considerations or consequences. *Example:* Labeling FSD an illness and labeling individuals as having FSD are neutral biological statements that don't reflect moral judgments of the condition or individual.	Illness is a moral category: Conditions and behaviors are labeled illness when they are considered bad or abnormal. *Example:* We label sexual nonresponsiveness an illness because we find it repugnant, and we typically look down on those who have FSD.
Illness is an apolitical label: Politics have no impact on who or what is labeled illness. *Example:* FSD was first identified by doctors through scientific research.	Illness is a political label: Some groups have more power than others to decide what is an illness and who is ill. *Example:* The concept of FSD was promoted by pharmaceutical companies to sell drugs.
Each illness results from a unique biological cause. *Example:* FSD results from a biochemical imbalance best treated with a drug.	Illness results from a combination of social and biological causes. *Example:* Women's sexual problems often reflect psychological and interpersonal as well as biological problems.

all doctors, but it is the dominant conception of illness in the medical world. In contrast, the **sociological model of illness** offers a very different way of thinking about what illness means in practice (rather than ideally). This model is most often adopted by sociologists who take a critical approach, and is also sometimes adopted by health care workers who share sociologists' concerns about how social forces affect health and health care. *Key Concepts: Medical and Sociological Models of Illness* compares these two models using as an example female sexual dysfunction (FSD), a recently developed and still contentious diagnosis.

The medical model of illness begins with the assumption that illness is an objective label given to anything that deviates from normal biological functioning (Mishler, 1981). Most doctors, if asked, would explain that polio is caused by a virus that disrupts the normal functioning of the neurological system; menopause is a "hormone deficiency disease" that, among other things, impairs the body's normal ability to regenerate bone; and men develop urinary problems when their prostates grow excessively large and unnaturally compress the urinary tract. By extension, the medical model assumes that each illness has specific features that any doctor can recognize (Mishler, 1981).

In contrast, the sociological model of illness begins with the statement that illness (as the term is actually used) is a *subjective* label, which reflects personal and social ideas about what is normal as well as scientific reasoning (Weitz, 1991). Sociologists point out that ideas about normality differ widely across both individuals and social groups. A height of 4 feet 6 inches would be normal for a Pygmy man, but not for an American man. Drinking three glasses of wine a day is normal for Italian women but could lead to a diagnosis of alcoholism in American medical circles. In defining normality, therefore, we need to look not only at individual bodies but also at the broader social context.

Moreover, even within a given group, "normality" is a range and not an absolute. The median height of American men, for example, is 5 feet 9 inches, but most people would consider someone several inches taller or shorter than that as still normal. Yet medical authorities routinely decide what is normal and what is illness based not on objective markers of health and illness, but on arbitrary, statistical cutoff points—deciding, for example, that anyone in the fourth percentile for height or the fiftieth percentile for cholesterol level is ill (Cohen and Cosgrove, 2009; Hadler, 2008).

Similarly, sociologists note, the process of assigning diagnoses to individuals is far from objective. For example, African American patients with chest pain often receive diagnoses of indigestion, whereas white patients more often receive diagnoses of heart disease. Meanwhile, French doctors often attribute patients' headaches to liver problems, whereas US doctors more often attribute them to neurological causes (Hoffman and Tarzian, 2001; Nelson, Smedley, and Stith, 2002; Payer, 1996).

Because the medical model assumes illness is an objective, scientifically determined category, it also assumes that moral judgments play no role in labeling conditions or behaviors as illnesses. Sociologists, on the other hand, argue that illness is inherently a moral category because deciding what is illness always means deciding what is good or bad. When, for example, doctors label menopause a "hormonal deficiency disease," they label it an undesirable deviation from normal. In contrast, many women consider menopause both normal and desirable, and enjoy the freedom from fear of pregnancy that menopause brings (Avis and McKinlay, 1991). In the same manner, when we define cancer, polio, or diabetes as illnesses, we judge the bodily changes these conditions produce to be both abnormal and undesirable rather than simply normal variations in functioning, abilities, and life expectancies. (Conversely, when we define a condition as healthy, we judge it to be normal and desirable.)

Similarly, whenever we label someone ill, we suggest there is something undesirable about that person. By definition, an ill person is one whose actions, abilities, or appearance don't meet social **norms**: expectations within a given culture regarding proper behavior or appearance. Such a person is typically considered less whole and less socially worthy than those deemed healthy. Illness, then, like virginity or laziness, is a **moral status**: a social condition that we believe indicates the goodness or badness, worthiness or unworthiness, of a person.

From a sociological standpoint, illness is not only a moral status (such as crime or sin) but also a form of **deviance** (Parsons, 1951). To sociologists, labeling something deviant does not necessarily mean that it's immoral. Rather, deviance refers to behaviors or conditions that socially powerful persons within a given culture *perceive*, whether accurately or inaccurately, to be immoral or to violate social norms. We can tell whether behavior violates norms (and therefore whether it's deviant) by seeing if it results in **negative social sanctions**. This term refers to any punishment, from ridicule to execution. (Conversely, **positive social sanctions** are rewards, ranging from token gifts to knighthood.) These social sanctions can be enforced by parents, police, teachers, and peers, as well as doctors. Later in this chapter, we'll look at some of the negative social sanctions imposed against ill persons.

For the same reasons that the medical model doesn't recognize the *moral* aspects of illness labeling, it doesn't recognize the *political* aspects of that process. Although doctors sometimes participate actively in these political processes—arguing, for example, that insurance companies should cover treatment for newly labeled conditions such as fibromyalgia—few doctors recognize how politics underlie the illness-labeling process in general. In contrast, sociologists point out that any time a condition or behavior is labeled an illness, some groups will benefit more than others, and some groups will have more power than others to enforce the definitions that benefit them. As a result, open political struggles often emerge around illness definitions (a topic we'll return to later in this chapter). For example, US vermiculite miners who were constantly exposed to asbestos dust in their work and who now have strikingly high rates of cancer have fought with insurance companies and doctors in clinics, hospitals, and the courts to have "asbestosis" labeled an illness. Meanwhile, the mining companies and the doctors they employed have argued that no such disease exists and that the high rates of cancer in mining communities are merely coincidences (Schneider and McCumber, 2004).

Finally, the medical model of illness assumes that each illness has not only unique symptoms but also a unique biological cause (Mishler, 1981). Modern medicine assumes, for example, that tuberculosis, polio, and other infectious diseases are each caused by a unique microorganism. Similarly, doctors continue to seek limited and unique causes of heart disease and cancer, such as high-cholesterol diets and exposure to toxins. Yet even though illness-causing microorganisms exist everywhere and environmental health dangers are common, relatively few people become ill as a result. By the same token, although cholesterol levels and heart disease are strongly correlated among middle-aged men, many men eat high-cholesterol diets without developing heart disease, and others eat low-cholesterol diets but die of heart disease anyway. Belief in unique biological causes discourages medical researchers from asking why individuals respond in such different ways to the same health risks and encourages researchers to search for **magic bullets**—a term used by Paul Ehrlich, discoverer of the first effective treatment for syphilis, to refer to drugs that almost miraculously prevent or cure illness by attacking one specific etiological factor.

In sum, to sociologists who work from a critical perspective, illness is a **social construction**, something that exists in the world not as an objective condition, but *because we have defined it as existing.* This doesn't mean that the virus causing measles does not exist or that it doesn't cause a fever and rash. It does mean, though, that when we talk about measles as an illness, we have organized our ideas about that virus, fever, and rash in only one of the many possible ways. In another place or time, people might conceptualize those same conditions as manifestations of witchcraft, as a healthy response to the presence of microbes, or as some other illness altogether. To sociologists, then, *illness*, like *crime* or *sin*, refers to biological, psychological, or social conditions subjectively defined as undesirable by those within a given culture who have the power to enforce such definitions.

MEDICINE AS SOCIAL CONTROL

In everyday life, we use the word *medicine* to refer to the drugs that doctors prescribe. But we can also use the word *medicine* to refer to the world and culture of doctors. For example, we might say that modern medicine is an exceedingly complex enterprise or that modern medicine primarily focuses on treating disease rather than on looking for environmental causes of illness. Even more broadly, sociologists refer to medicine as an *institution*. Sociologists use the term **institution** to refer to enduring social structures that meet basic human needs, such as the family, religion, and education. When we talk of medicine as an institution, we refer to the world and culture of doctors as well as to the economic, social, and political underpinnings of that world. We might, for example, talk about how the power of medicine *as an institution*—doctors, hospitals, the medical way of thinking about the world, and so on—has grown over the past century. Importantly, sociologists (and others) increasingly talk about *biomedicine* as an institution. **Biomedicine** refers to the ways in the which medicine, science and technology often now work together as one social institution—an institution that can increase or reduce the power of medicine as an institution (Clarke et al., 2003, 2010). (For simplicity's sake, we will primarily use the term *medicine* in this book.)

One of the central concepts in the sociology of health and illness is the idea that medicine is, among other things, an institution of social control. **Social control** refers to the formal and informal methods used by a social group to ensure that individuals conform to social norms and to protect the existing balance of power among groups. When we say that medicine is an institution of social control, we are saying that medicine is a basic structure of our society that sometimes serves to "keep people in line." For example, doctors have the power to decide whether someone is a malingerer who should be shunned or is truly ill and deserves sympathy. In such a situation, doctors act as **social control agents**: individuals or groups (such as parents and religious leaders) that enforce social norms. *Contemporary Issues: Citizenship and Biomedicine* explores how the broader institution of biomedicine now plays a powerful role in deciding who can become a US citizen.

CONTEMPORARY ISSUES

Citizenship and Biomedicine

Becoming a US citizen is a long and difficult process. Each year, however, the United States allows some refugees and immigrants to bypass the usual procedures if they are the spouse, parent, or young child of a US citizen. Increasingly, however, the federal government has mandated that individuals must first prove their family relationship through expensive genetic testing (Dove, 2013; Lakhani and Timmermans, 2014).

The government, of course, has both a need and an obligation to prevent immigration fraud. At the same time, relying on DNA testing leads to a different set of problems (Dove, 2013; Lakhani and Timmermans, 2014). Most basically, it enshrines biomedical information as uniquely accurate and meaningful: more important than legal documents, a parent's sworn statement, or a child's obvious desire to be reunited with the adults he or she loves. Similarly, it implicitly declares that we are all defined by our genes. In addition, it defines family in narrow terms, omitting stepchildren and adopted children. Yet in many other cultures, it is common for adults to informally adopt nieces, nephews, cousins, or others when parents die or are unable to care for them. This is especially true in war-torn areas like those where most refugees were born. Relying on genetic testing also has the effect of denying legitimacy to polygamous families, in which all wives may consider all children born into the family to be their own. Finally, mandated genetic testing can rip apart—rather than unite—families when it reveals that a child is not genetically related to a man long assumed to be his or her father.

In the next sections, we will see how doctors, as agents of social control, both decide which conditions or behaviors should be labeled illness and press sick people to get well and return to normal social roles.

Creating Illness: Medicalization

The process through which a condition or behavior becomes defined as a medical problem requiring a medical solution is known as **medicalization** (Conrad, 2007). For example, during the nineteenth century most Americans considered chronic drunkenness to be a sin, but by the mid-twentieth century many instead considered it a form of mental illness. Similarly, over the last few decades various natural conditions and processes, such as uncircumcised penises, male balding, aging, loss of sexual desire, and pregnancy have all increasingly come to be seen as medical problems (Armstrong, 2000; Conrad, 2007; Hartley, 2006; Rosenfeld and Faircloth, 2005). The term *medicalization* also refers to the process through which the definition of an illness is *broadened*. For example, when doctors expanded the definition of osteoporosis to include anyone with low bone density, rather than only individuals who had experienced unusual bone fractures, the number of persons diagnosed with osteoporosis almost doubled (Grob and Horwitz, 2009).

For medicalization to occur, one or more organized social groups must have both a vested interest in medicalization and sufficient power to convince others (including doctors, the public, and insurance companies) to accept their new definition of the situation. Not surprisingly, doctors often play a major role in medicalization because medicalization can increase their power, the scope of their practices, and their incomes. For example, during the first half of the twentieth century, improvements in the standard of living coupled with various public health measures substantially reduced the number of seriously ill children. As a result, the market for pediatricians declined, and their focus shifted from serious illnesses to minor childhood illnesses and well-baby care. Pediatrics thus became less well paid, interesting, and prestigious. To increase their market while obtaining more satisfying and prestigious work, some pediatricians have expanded their practices to include children whose behavior concerns their parents or teachers and who are now defined as having attention-deficit hyperactivity disorder (ADHD) (Conrad, 2007). Doctors have played similar roles in medicalizing crooked noses, obesity, drinking during pregnancy, impotence, and numerous other conditions (Grob and Horwitz, 2009; Loe, 2004).

In other instances, however, doctors have actively opposed medicalization (Swoboda, 2008). This by definition is the case with any **contested illness**: distressing and painful symptoms that affected individuals believe constitute an illness even though many doctors disagree. For example, fibromyalgia is characterized by many common symptoms, including pain, dizziness, insomnia, and headache. Moreover, no blood test or X-ray can identify an individual as having fibromyalgia. As a result, many doctors question whether it really is a disease. The same is true for chronic fatigue syndrome, multiple chemical sensitivity, and Gulf War syndrome, among others. In these cases, consumers often press for medicalization to get validation for their experiences, stimulate research on treatments and cures, and get health and disability insurance coverage for their problems (Barker, 2005, 2008; Conrad, 2007). The rise of the Internet has made it much easier for such consumers to find each other, reaffirm each other's sense that they suffer from a real illness, and lobby for medicalization.

Another major force in battles over medicalization is **managed care organizations (MCOs)**. MCOs (discussed in detail in Chapter 8) are health insurance providers that restrain costs (and ideally improve quality of care) by monitoring closely the health services given to patients. MCOs either support or oppose medicalization, depending on which tactic best protects their interests (Conrad, 2007). For example, in the past, MCOs typically rejected requests for gastric bypass surgeries to help obese patients lose weight, implicitly arguing that obesity was a personal rather than a medical issue. More recently, MCOs have started approving these surgeries in hopes of reducing their long-term costs for obesity-related disease.

The final major force behind medicalization is the pharmaceutical industry (Conrad, 2007). The industry has a vested economic interest in medicalization whenever it can sell a drug as a treatment (Cohen and Cosgrove, 2009; Conrad, 2007; Rothman and Rothman, 2003). For example, in 1985, the pharmaceutical company Genentech patented a genetically engineered human growth hormone designed to increase height in children with pituitary gland defects. Such defects, however, are rare. To expand the market for its drug, Genentech sponsored in-school screening programs that identified the shortest three percent of students and then informed the students' parents that the students might benefit from hormone treatment. That treatment, however, carried significant side effects and only increased height in children with pituitary defects.

Case Study: Working Together to Medicalize Attention-Deficit Hyperactivity Disorder (ADHD) Successful medicalization often depends on the interwoven interests and activities of multiple interest groups. The history of ADHD illustrates this process.

Scientists have yet to discover any biological markers (such as viruses or genes) for ADHD (Furman, 2009). As a result, doctors instead diagnose individuals with ADHD if they are overactive, impulsive, or easily distractible but show no evidence of brain damage (Conrad 2007).

The diagnosis only became popular in the 1960s, following a massive advertising campaign for Ritalin (methylphenidate). That campaigned aimed to convince both doctors and the public that ADHD was a real disease and that Ritalin was a safe treatment. Ritalin is indeed safer than other amphetamines and can, in the short term, improve individuals' concentration, impulse control, and discipline. But it may cause addiction, loss of appetite, sleep deprivation, headache, stomachache, and cancer (Davis, 2007; Vastag, 2001), and does *not* improve individuals' chances of graduating high school, holding a job, avoiding drug abuse, or avoiding trouble with the law (Diller, 1998).

Despite these problems, pediatricians proved a ready audience for this marketing campaign, which promised a way to boost their flagging income and prestige. Teachers, too, began recommending that certain students get tested (and treated) for ADHD, in part because the drugs could make students more manageable, and the diagnosis could shift blame for student problems away from the teachers themselves (Diller, 1998). Meanwhile, parents hoped that diagnosis with ADHD would move blame for their children's school or behavior problems away from both their children and their parenting skills. Finally, parents hoped that diagnosis would give their children the benefits federally guaranteed to any children with disabilities: special educational services plus protection against suspension or expulsion for disciplinary problems related to ADHD (Conrad 2007; Diller, 1998).

Currently, 11 percent of US schoolchildren—and almost 20 percent of high school boys—have been diagnosed with ADHD, and sales of drugs to treat it have more than doubled in the last few years (Schwarz and Cohen, 2013). In addition, diagnoses have spiked among toddlers, even though official definitions of ADHD limit it to children age four or over (Schwarz, 2014).

Unintended Consequences of Medicalization In some circumstances, medicalization can be a boon, leading to social awareness of a problem, sympathy toward those diagnosed with an illness, and the development of helpful treatments. Persons with epilepsy, for example, lead far happier and more productive lives now that their seizures are treated with drugs rather than treated as signs of demonic possession. But defining a condition as an illness does not necessarily improve the social status of those who have that condition. Those who use alcohol excessively, for example, continue to experience social rejection even when alcoholism is labeled a disease. Moreover, medicalization also can lead to new problems, known by sociologists as **unintended negative consequences** (Conrad, 2007).

First, once a situation becomes medicalized, doctors become the only experts considered appropriate for diagnosing the problem and for defining appropriate responses to it. As a result, the power of doctors increases while the power of other social authorities (including judges, the police, religious leaders, legislators, and teachers) diminishes. For example, now that troublesome behavior by children is increasingly diagnosed as ADHD, parents, teachers, and the children themselves have lost credibility when they disagree with this diagnosis. Similarly, doctors are now given considerable authority to answer questions such as who should receive abortions or organ transplants, how society should respond to drug use, and whether severely disabled infants should receive experimental surgeries, while the authority of the church and family members to answer these questions has diminished.

As this suggests, medicalization significantly expands the range of life experiences under medical control. For example, the natural process of aging is increasingly regarded as a medical condition. Doctors now scrutinize all aspects of the aging body and recommend psychological tests to measure mental decline, hormones to improve virility, cosmetic surgery for wrinkles, and more (Conrad, 2007).

Second, once a condition is medicalized, medical treatment may seem the only logical response to it. For example, if woman battering is considered a medical condition, then doctors need to treat women and the men who batter them. However, if woman battering is considered a social problem stemming from male power and female subordination, then it makes more sense to arrest the men, assist the women financially and emotionally, and work for broader structural changes to improve all women's status and options.

Third, when doctors define situations in medical terms, they reduce the chances that these situations will be understood in *political* terms. For example, China, Pakistan, and other countries have removed political dissidents from the public eye by committing them to mental hospitals. By so doing, these governments discredited and silenced individuals who might otherwise have offered powerful dissenting voices. In other words, medicalization allowed these governments to **depoliticize** the situation: to define it as a medical rather than a political problem.

Fourth, and as the examples of China and Pakistan illustrate, medicalization can justify involuntary treatment. Yet treatment sometimes harms more than it helps. For example, since the 1980s, US doctors have legally forced small numbers of women to submit to cesarean deliveries, in which babies are surgically removed from their mothers' uteruses rather than delivered naturally through the vagina (Daniels, 1993; Roth, 2003). In these cases, doctors argued successfully that childbirth is a dangerous medical condition rather than a natural process, that doctors are better qualified than pregnant women to judge fetuses' needs, and that fetuses' right to health is more important than women's right to control their own bodies. Yet the rate of cesarean section in the United

States is twice that recommended by the World Health Organization, suggesting that doctors are far too ready to perform this potentially life-threatening surgery. *Ethical Debate: Medical Social Control and Fetal Rights* explores how the growing acceptance of the idea of "fetal rights" is affecting the lives of pregnant women.

Medicalization and the "Potentially Ill" In addition to creating new illnesses, medicalization has also led to labeling increasing numbers of individuals as "potentially ill" (Boyer and Lutfey, 2010; Conrad, 2005; Scott, Wood, and Gray, 2005). The **potentially ill** are individuals identified as having an above average risk of illness, whether because of age, stress level, tobacco use, family history, medical test results, or other factors.

The risks faced by the potentially ill vary substantially. Some learn that they carry a gene guaranteed to cause a fatal disease. Many more, however, learn that they have a condition such as high cholesterol that may increase their risk of illness. The numbers of such individuals continues to increase as corporations develop more tests for risk factors and as doctors (often reimbursed per test) adopt such tests. Similarly, the ranks of the potentially ill have expanded as pharmaceutical companies have encouraged both doctors and consumers to expand their ideas about health risks and to adopt treatments for those risks. For example, pharmaceutical companies have worked not only to broaden the definition of osteoporosis but also to create a new category, osteopenia, for those at *risk* of osteoporosis. Because *osteoporosis* refers to the risk of bone fractures caused by low bone density, osteopenia is essentially the *risk* of a *risk* of a health problem.

As this suggests, the health benefits of learning that one is potentially ill depend on the magnitude of the identified risk and the effectiveness of available treatments (Scott et al., 2005). Those benefits, however, must also be balanced against the psychological distress caused when people without any symptoms learn that illness might strike at any moment (Marteau and Richards, 1996). In addition, some of these individuals experience the stigma of illness without any of the benefits that those who have illnesses may receive, such as legal protection from discrimination.

The Rise of Demedicalization The problems inherent in medicalization have fostered a (much smaller) countermovement of **demedicalization** (Conrad, 2007). A quick look at medical textbooks from the late 1800s reveals many "diseases" that no longer exist. For example, nineteenth-century medical textbooks often included several pages on the health risks of masturbation. One popular textbook from the late nineteenth century asserted that masturbation caused "extreme emaciation, sallow or blotched skin, sunken eyes, … general weakness, dullness, weak back, stupidity, laziness, … wandering and illy defined pains," as well as infertility, impotence, consumption, epilepsy, heart disease, blindness, paralysis, and insanity (Kellogg, 1880:365). Today, however, medical textbooks describe masturbation as a healthy part of human sexuality.

Like medicalization, demedicalization often begins with lobbying by consumer groups. For example, medical ideology now defines childbirth as an

ETHICAL DEBATE

Medical Social Control and Fetal Rights

In October 2006, 20-year-old Tiffany Hitson gave birth to a healthy baby girl. The next day, after traces of marijuana and methamphetamine were found in her baby during routine testing, Hitson was arrested for "chemically endangering" her child. Hitson spent most of the next year in state prison.

Since her arrest, doctors have helped prosecutors to charge eleven other poor or working-class women in the same rural, Alabama county with child endangerment after drugs were detected in a newborn's system or after authorities learned that a pregnant woman was using drugs (Nossiter, 2008). So far, none of the women have had the money or the confidence to risk a trial and so all instead have pleaded guilty.

Over the last decade, hundreds of pregnant women—most poor and nonwhite—have faced criminal sanctions or been forced to endure medical treatments (Eckholm, 2013). These actions reflect a growing tendency among doctors, lawyers, and the general public to view mother and fetus as separate beings, with separable and sometimes conflicting rights, and to see the fetus rather than the mother as obstetricians' primary patient (Roth, 2003). Given that doctors have an ethical and a legal obligation to protect children from parents who abuse them, should doctors have a similar obligation to protect fetuses even if it means superseding mothers' wishes?

Those who argue in favor of medical intervention find it illogical to protect children from bodily harm *after* birth but to deny them protection that might ensure their health *before* birth. Children born prematurely, addicted to drugs, or with birth defects because their mothers did not follow medical advice may lead short, painful lives or may survive with mental or physical disabilities. In addition, caring for these children costs hospitals and taxpayers money. Those costs alone, one could argue, give the medical and legal systems the right to intervene when women endanger their fetuses.

Others, however, have raised several objections to placing **fetal rights** above mothers' wishes. First, critics question whether medical intervention really is

inherently dangerous process, requiring intensive technological, medical assistance. Since the 1940s, however, some American women have attempted to redefine childbirth as a generally safe, simple, and natural process and have promoted alternatives ranging from natural childbirth classes to hospital birthing centers, to home births assisted only by midwives. Similarly, and as described in Chapter 7, gay and lesbian activists have at least partially succeeded in redefining homosexuality from a pathological condition to a normal human variation. More broadly, innumerable books, magazines, television shows, and popular organizations now exist that focus on teaching people to care for their own health rather than (or in addition to) relying on medical care.

Genetic Research and Social Control

The potential for medicine to act as a form of social control continues to grow as scientists learn more about human genes—and as the public increasingly believes necessary. For example, almost all well-structured research studies on mothers' drug use during pregnancy have found that it causes little if any long-term harm to children (Roth, 2003; Singer et al., 2002). Second, drug withdrawal also can endanger fetuses, as can threats of legal sanctions that discourage pregnant women from seeking health care altogether. Third, opponents argue that doctors can't necessarily make better decisions than mothers do because they can't understand fully the circumstances in which mothers make those decisions. For example, many women continue to use drugs during pregnancy only because they can't obtain access to treatment programs, which usually are expensive, have long waiting periods, and won't accept pregnant women.

Finally, some argue, the concept of fetal rights put an undue burden on women (Roth, 2003; Toscano, 2005). Although we require parents to guard their children's health and welfare, we don't require them to donate kidneys, bone marrow, or even blood for their children's sake. Why, then, should we require women—and only women—to protect their fetuses? After all, when fathers drink alcohol, use illicit drugs, smoke cigarettes, or work in legal or illegal chemical laboratories, both sperm and fetuses can be harmed, but no man has ever been charged with fetal abuse. This has led some to conclude that the idea of fetal rights is aimed more at controlling women rather than at protecting children.

Sociological Questions

1. What social views and values about medicine, society, and the body are reflected in the concept of "fetal rights"? Whose views are these?
2. Which social groups are in conflict over this issue? Whose interests are served by promoting fetal rights? Whose interests are harmed?
3. Which of these groups has more power to enforce its view? What kinds of power do they have?
4. What are the intended consequences of this policy? What are the unintended social, economic, political, and health consequences of this policy?

that genes hold the key to health and illness. The shift toward increasingly defining genes as the cause of human disease, behavior, and differences is known as **geneticization** (Shostak, Conrad, and Horwitz, 2008). Geneticization is a form of medicalization.

Genes affect health in two ways: by *causing* "true" genetic diseases and by increasing individuals' *predisposition* to develop disease. True genetic diseases, such as hemophilia, only occur if an individual has a specific gene (Williams and Sternthal, 2010). Although doctors can't cure these diseases, genetic testing does provide affected individuals with the opportunity to learn whether they, their children, or (for pregnant women) their fetuses carry a disease-causing gene.

Individuals who learn through testing that they are not at risk certainly benefit by gaining peace of mind and the ability to plan their futures. Others, though, learn that they are guaranteed to eventually develop a genetic disease, a prospect that some find overwhelming (Marteau and Richards, 1996). It is hard, for example, to imagine how it can help individuals to learn at age 21 that by their forties they will develop Huntington's disease, a devastating neurological disorder that invariably causes progressive insanity, total disability, and death.

Even these individuals, though, gain some options and benefits. They can commit to living life to the fullest, make wills, or otherwise plan for their futures. They also can choose to avoid becoming pregnant; to abort any fetuses that carry the defect; or to continue a pregnancy to term, knowing that the fetus carries the defect and hoping that this foreknowledge will better prepare them for the birth of an ill or disabled child. Finally, they can try to have a healthy child who is biologically theirs by surgically removing the woman's eggs from her body, mixing them with the man's sperm in the laboratory, having a doctor test the resulting fetuses for genetic defects, and implanting only nondefective fetuses in the woman's uterus. This strategy is rare because the physical, financial, and psychological costs are extremely high, and the odds of success are low.

In most cases, however, instead of directly *causing* disease, genes merely increase the *probability* of disease. In these cases, disease occurs when genes *combine* with environmental factors, in a process known as an **epigenetic effect** (Landecker and Panofsky, 2013). For example, stressful conditions can "turn on" illness-causing genes and weaken the effectiveness of illness-preventing genes, thus increasing risks of depression, heart disease, and other illnesses (Shanahan, Bauldry, and Freeman, 2010). This explains why identical twins may not get the same genetic disease, even though their genes are identical (Roberts et al., 2012).

Genetic testing for this second sort of "genetic" disease can benefit individuals if it motivates those who test positive to take potentially health-preserving actions. For example, women who learn that they carry the BRCA-1 gene, and thus have an increased risk of breast cancer, might choose to adopt a low-fat diet or to have their breasts removed prophylactically.

On the other hand, researchers so far have had little success in treating genetic factors (Wade, 2010). Yet individuals identified through genetic testing as having an illness or being at high risk for illness may lose their jobs, health insurance, or life insurance, even if they are healthy and despite laws banning such discrimination (Council for Responsible Genetics, 2001). Moreover, genetic tests can't tell how soon or how severely an individual with a particular gene will contract a given disease. Increasingly, too, tests are identifying genetic anomalies whose effects, if any, are unknown. As a result, individuals often must make life plans with little knowledge of what their futures (or their fetuses' futures) will be like.

Social Control and the Sick Role

So far, we have looked at how medicine functions as an institution of social control by defining individuals as sick or defective. Medicine can also work as an institution of social control by pressuring individuals to *abandon* sickness, a process first recognized by Talcott Parsons (1951).

Parsons was one of the first and most influential sociologists to recognize that illness is deviance. From his perspective, when people are ill, they can't perform the social tasks normally expected of them. Workers stay home, homemakers tell their children to make their own meals, students ask to be excused from exams. Because of this, either consciously or unconsciously, people can use illness to evade their social responsibilities. To Parsons, therefore, illness threatened social stability.

Parsons also recognized, however, that allowing some illness can *increase* social stability. Imagine a world in which no one could ever "call in sick." Over time, production levels would fall as individuals, denied needed recuperation time, succumbed to physical ailments. Morale, too, would fall while resentment would rise among those forced to perform their social duties day after day without relief. Illness, then, acts as a kind of pressure valve for society—something we recognize when we speak of taking time off work for "mental health days."

From Parsons's perspective, then, the important question was how did society control illness so that it would increase rather than decrease social stability? His emphasis on social stability reflected his belief in the broad social perspective known as **functionalism**. Underlying functionalism is an image of society as a smoothly working, integrated whole, much like the biological concept of the human body as a homeostatic environment. In this model, social order is maintained because individuals learn to accept society's norms and because society's needs and individuals' needs match closely, making rebellion unnecessary. Within this model, deviance—including illness—is usually considered **dysfunctional** because it threatens to undermine social stability.

Defining the Sick Role Parsons's interest in how society allows illness while minimizing its impact led him to develop the concept of the **sick role**. The sick role refers to social expectations regarding how society should view sick people and how sick people should behave. According to Parsons, the sick role as it currently exists in Western society has four parts. First, the sick person is considered to have a legitimate reason for not fulfilling his or her normal social role. For this reason, we allow people to take time off from work when sick rather than firing them for malingering. Second, sickness is considered beyond individual control, something for which the individual is not held responsible. This is why, according to Parsons, we bring chicken soup to people who have colds rather than jailing them for stupidly exposing themselves to germs. Third, the sick person must recognize that sickness is undesirable and work to get well. So, for example, we sympathize with people who strive to recover from illness and question the motives of those who seem to revel in the attention illness brings them. Finally, the sick person should seek and follow medical advice. Typically, we expect sick people to follow their doctors' recommendations regarding drugs and surgery, and we question the wisdom of those who don't.

Parsons's analysis of the sick role moved the study of illness forward by highlighting the social dimensions of illness, including identifying illness as deviance and doctors as agents of social control (Shilling, 2001). It remains important partly because it was the first truly sociological theory of illness. Parsons's research also has proved important because it stimulated later research on interactions

KEY CONCEPTS

Evaluating the Sick Role Model

Elements of the Sick Role	Model Fits Well	Model Fits Poorly
Legitimate reason for not fulfilling obligations	Appendicitis, cancer	Undiagnosed chronic fatigue
Individual not held responsible	Measles, hemophilia	Herpes, lung cancer
Individual should strive to get well	Tuberculosis, broken leg	Diabetes, epilepsy
Individual should seek medical help	Strep throat, syphilis	"24-hour flu," cold

between ill people and others. In turn, however, that research has illuminated the weaknesses of the sick role model.

Critiquing the Sick Role Model Many recent sociological writings on illness—including this textbook—have adopted a **conflict perspective** rather than a functionalist perspective. Whereas functionalists envision society as a harmonious whole held together largely by socialization, mutual consent, and mutual interests, those who hold a conflict perspective argue that society is held together largely by power and coercion, as dominant groups impose their will on others. Consequently, whereas functionalists view deviance as a dysfunctional element to be controlled, conflict theorists view deviance as a necessary force for social change and as the conscious or unconscious expression of individuals who refuse to conform to an oppressive society. Conflict theorists, therefore, have stressed the need to study not only deviants but also social control agents.

The conflict perspective has helped sociologists to identify the strengths and weaknesses in each of the four elements of the sick role model (see *Key Concepts: Evaluating the Sick Role Model*). That model declares that sick persons are not held responsible for their illnesses. Yet, as we saw earlier in this chapter, society often *does* hold individuals responsible for their illnesses (Freidson, 1970). In addition, ill persons are not always considered to have a legitimate reason for abstaining from their normal social tasks. Certainly, no one expects persons with end-stage cancer to continue working, but what about people with arthritis or those labeled malingerers because they can't obtain a diagnosis after months of pain, increasing disability, and visits to doctors (Glenton, 2003; Ziporyn, 1992)?

Other aspects of the sick role model are equally problematic. The assumption that individuals will attempt to get well fails to recognize that much illness is **chronic** and by definition not likely to improve. Similarly, the assumption that sick people will seek and follow medical advice ignores the many people who lack access to medical care or who can't afford to take time off from work or purchase medications when ill. In addition, it ignores the many persons, especially those with chronic rather than **acute** conditions, who have found mainstream health care of limited benefit and who therefore rely mostly on their own experience and knowledge and that of other nonmedical people. Similarly (and understandably), it could not anticipate the ways the Internet has enabled lay people—both sick and well—to seek health information on their own and, occasionally, to challenge or ignore medical advice as a result (Shilling, 2001). Finally, the concept of a (singular) sick role ignores how gender, ethnicity, age, and social class affect the response to illness and to ill people. For example, women are both *more* likely than men are to seek medical care when they feel ill and *less* likely to have their symptoms taken seriously by doctors (Council on Ethical and Judicial Affairs, 1991; Steingart, 1991).

In sum, the sick role model is based on a series of assumptions about both the nature of society and the nature of illness. In addition, the sick role model confuses the experience of *patienthood* with the experience of *illness* (Conrad, 1987). The sick role model focuses on the interaction between the ill person and the mainstream health care system. Yet interactions with the medical world form only a small part of the experience of living with illness or disability, as the next chapter shows. For these among other reasons, research using Parsons's conception of the sick role has declined over time.

IMPLICATIONS

The language of illness and disease permeates our everyday lives. We routinely talk about living in a "sick" society or about the "disease" of violence infecting our world, offhandedly labeling anyone who behaves in a way we don't understand or don't condone as "sick."

This metaphoric use of language reveals the true nature of illness: behaviors, conditions, or situations that powerful groups find disturbing and believe stem from internal biological or psychological roots. In other times or places, the same behaviors, conditions, or situations might have been ignored, condemned as sins, or labeled crimes. In other words, illness is both a social construction and a moral status.

In many instances, using the language of medicine and placing control in the hands of doctors offers a more humanistic option than the alternatives. Yet, as this chapter has demonstrated, medical social control also carries a price. The same surgical skills and technology for cesarean sections that have saved the lives of so many women and children now endanger the lives of those who have cesarean sections unnecessarily. At the same time, forcing cesarean sections on women potentially threatens women's legal and social status. Similarly, the development of tools for genetic testing has saved many individuals from the anguish of rearing children doomed to die young and painfully but has cost others their jobs or health insurance.

In the same way, then, that automobiles have increased our personal mobility in exchange for higher rates of accidental death and disability, adopting the language of illness and increasing medical social control bring both benefits and costs. These benefits and costs will need to be weighed carefully as medicine's technological abilities grow.

SUMMARY

1. Throughout history, explanations for illness have commonly blamed ill persons for their illnesses. Such explanations encourage policy makers to ignore how social and environmental factors can foster illness.
2. Illness is a social construction—not something that simply exists in the world as an objective condition, but something that exists *because we have defined it as existing*. To sociologists, the term *illness* refers to biological, psychological, or social conditions that are subjectively defined as undesirable by those who have the power to enforce their definitions.
3. Illness is a moral status and a form of deviance. We label individuals ill when they don't meet our social norms for behavior, ability, or appearance.
4. The medical model of illness assumes that illness is an objective label, applied scientifically, without moral judgment or political bias. That model also assumes that each illness is caused by unique biological forces.
5. The sociological model of illness regards illness as a social construction, a moral category, and a political label, and emphasizes that what is labeled illness changes over time and space.
6. Medicine is an institution of social control. The institution of medicine acts as social control whenever it defines behaviors and conditions as deviant and pressures individuals to seek health care and strive to get well.
7. The process through which a condition or behavior becomes defined as an illness requiring a medical solution is known as medicalization; the reverse process is known as demedicalization. Four groups that often play prominent roles in fights over medicalization are doctors, consumers, the pharmaceutical industry, and managed care organizations.
8. Medicalization can reduce stigma, increase social awareness, and encourage medical research. It can also cause unintended negative consequences, such as increasing the power of doctors at the expense of other social groups, depoliticizing dissent, and justifying medical—and only medical—treatment.
9. Contested illnesses are combinations of distressing and painful symptoms that affected individuals believe constitute an illness even though many doctors disagree. Examples include fibromyalgia and multiple chemical sensitivity.
10. Genetic research and testing have increased the potential for medicine to act as a form of social control, especially because of geneticization: the shift toward assuming that genes cause human disease, behavior, and differences. Genetic testing brings both benefits and problems to individuals and society.

11. The "potentially ill" are individuals identified as having an above average risk of illness, whether because of age, stress level, tobacco use, family history, medical test results, or other factors.
12. The sick role model refers to social expectations regarding how society should view sick people and how sick people should behave. The sick role has four parts. First, sickness is considered beyond individual control. Second, sick persons are considered to have legitimate reasons for not fulfilling their normal social roles. Third, sick persons are expected to recognize that sickness is undesirable and are therefore expected to work to get well. Finally, the sick role assumes that sick persons should seek and follow medical advice.
13. Critics of the sick role model challenge each of the four parts of that model. They note that the model best fits acute rather than chronic illness, and suggest that the model confuses the experience of being a *patient* with the much broader experience of *illness*.

REVIEW QUESTIONS

1. What does it mean to say that illness is a social construction and a moral status?
2. How have explanations for illness changed over time, and how have explanations for illness blamed ill people for their illnesses?
3. What is the medical model of illness, and what are some of the problems with that model?
4. What is medicalization, why does it occur, and what are some of its consequences?
5. Who are the potentially ill? What are the consequences of being labeled potentially ill?
6. How can genetic research and testing lead to social control? What is geneticization?
7. What is the sick role model, and what are some of the problems with that model?

CRITICAL THINKING QUESTIONS

1. Do the four characteristics of the "sick role" apply to persons who have high cholesterol but no known evidence of heart disease? Do they apply to persons who learn that they have a gene that carries with it a high chance of developing breast cancer? Explain your answers.
2. Psychiatrists apply the diagnosis of premenstrual dysphoric distress syndrome to women who each month experience depression and anger before

menstruating. How might women benefit from psychiatry's decision to label this condition a disease? How might women be harmed by it?

3. Researchers have identified a gene that, if present, indicates that a person has a significant risk of developing Alzheimer's disease at a young age. Alzheimer's disease causes people to gradually lose their memory and mental abilities. Imagine that you are a family practice doctor. Explain to a concerned patient two arguments for and two arguments against getting tested for the gene.

CHAPTER

6

The Experience of Disability, Chronic Pain, and Chronic Illness

Photo by Tommy Hindley/Professional Sport/Popperfoto/Getty Images

LEARNING OBJECTIVES

After reading this chapter, students should be able to:

- Critique the medical and sociological models of disability.
- Describe the nature and distribution of disability and chronic pain.
- Understand how individuals who experience disability, pain, or illness respond to their diagnoses.
- Analyze how individuals who experience disability, pain, or illness manage their health care.
- Describe how individuals who experience disability, pain, or illness manage or fight against stigma.

Award-winning author A. Manette Ansay began college as a piano performance major. Within a few months, however, her arms and hands grew so painfully inflamed that she had to drop out. Over the next few years, her legs as well as her arms began to weaken, but doctors could find neither explanation nor treatment for her problems.

By age 22, Ansay usually relied on a wheelchair, although some days she could walk—shakily—using crutches. On one such day, she went with her mother for lunch at a local McDonald's:

> *While my mother goes around the corner to place her order, I continue my Frankenstein lurch toward the nearest open table, telling myself not to be paranoid, nobody is staring, and so what if they are, so you're on crutches, so who cares?...*
>
> *"My goodness, what happened to you?" A woman with three young children sits in a nearby booth. "Were you in an accident or something?"*
>
> *I shake my head, keep going. These are the questions I've grown to hate, even without suspecting, yet, that they'll follow me for the rest of my life like a complicated name, an alias I must live by. What's wrong with you, what happened to you, what's the matter? Sometimes they're prefaced with, Do you mind if I ask you a personal question? Often they're followed by a long account of another person's health complaint: an accident, a bout of cancer, a recent diagnosis.*
>
> *I take another step, another. The children stare, following their mother's example.*
>
> *"What happened?" she repeats. "Did you break your legs?"*
>
> *I sit down facing the opposite direction, expressionless, pretending I haven't heard. There are two kinds of pain: the kind that can be protected—the lump in the breast, the loved one's death, the broken heart—and the other kind, the*

visible kind, the kind that, in my case, is the first thing people see. It's right there, out in the open, where anyone might choose to poke at it, probe it, satisfy their grim curiosity.

"What's wrong with that lady?" one of the children asks.

"Nothing a smile wouldn't cure," the woman says, in a voice I am meant to hear (Ansay, 2001:34–35).

As Ansay's story illustrates, living with chronic illness, pain, or disability affects one's *social* as well as medical status. In this chapter, we begin by exploring the meaning, history, and distribution of disability. Then we discuss chronic pain, which falls on the border between disability and illness. Finally, we look broadly at the experience of chronic illness, pain, and disability, including the search for an accurate diagnosis, for treatments that preserve one's quality of life, and for a coherent and valued sense of self.

UNDERSTANDING DISABILITY

Defining Disability

As explained in Chapter 5, the meaning of the term *illness* is far from obvious. The same is true for the term **disability**. Typically, when people think about disability, they think of it as something wrong—a deficit—within an individual mind or body that should be cured if possible. This way of thinking about disability is referred to as the **medical model of disability** because it is common among doctors (although certainly there are doctors who do not take this view).

At first glance, the medical model of disability seems perfectly reasonable. After all, isn't a disability something that an individual *has*, a defect in his or her body? According to many people with disabilities, the answer is no. Instead, they argue, their disabilities primarily stem not from their physical differences, but from the way others respond to those differences and from the choices others have made in constructing the social and physical environment. For example, a man whose energy waxes and wanes unpredictably during the day might be able to work 40 hours per week on a flexible schedule but not within a rigid 9-to-5 schedule. Similarly, a woman who uses a wheelchair might find it impossible to work in an office where furniture fits only persons who walk and are of average height, but she might have no problems in an office with more adaptable furniture. Disability activists argue that this is not a matter of providing special benefits for people with disabilities, but rather of compensating for the unacknowledged benefits that existing arrangements offer those who walk, such as chairs to sit in, stools for reaching high shelves, and carpeted floors that make walking easier but wheeling more difficult.

This approach reflects a **sociological model of disability** in its emphasis on social forces and public issues rather than on individual physical variations and troubles. In the rest of this chapter, the term *disability* refers to restrictions or lack of ability to perform activities resulting largely or solely from either (1) social responses to bodies that fail to meet social expectations *or* (2) assumptions about the body reflected in the social or physical environment.

These two models of disability—the medical model and the more sociological model used by disability activists—have strikingly different implications. As Paul Higgins (1992:31) notes, "To individualize disability [as the medical model does] is to preserve our present practices and policies that produce disability. If disability is an internal flaw to be borne by those 'afflicted,' then we don't question much the world we make for ourselves. Our actions that produce disability go unchallenged because they are not even noticed."

Individualizing disability therefore exemplifies the broader process of **blaming the victim**, through which individuals (in this case, people with disabilities) are blamed for causing the problems they experience (Ryan, 1976); an example is the common belief that women wouldn't be battered if they didn't provoke their husbands. In contrast, the sociological model of disability challenges us to look at disability from a very different perspective. If we conclude that the problem resides primarily in social attitudes and the built environment, then we can solve the problem most efficiently by changing attitudes and environments rather than by "rehabilitating" people with disabilities.

People with Disabilities as a Minority Group

When we start thinking of disability as primarily a result of social attitudes and built environments rather than of individual deficiencies, strong parallels emerge between people with disabilities and members of **minority groups** (Hahn, 1985). A minority group is defined as any group that, because of its cultural or physical characteristics, is considered inferior and subjected to differential and unequal treatment and that therefore develops a sense of itself as the object of collective discrimination (Wirth, 1985).

Few would argue with the assertion that we differentiate disabled persons from others based on physical characteristics. But can we also argue, as the definition of a minority group requires, that people with disabilities are considered inferior and are subject to differential and unequal treatment?

Unfortunately, yes. Even a cursory look reveals widespread prejudice and discrimination against people with disabilities. **Prejudice** refers to unwarranted suspicion, dislike of, or disdain toward individuals because they belong to a particular group, whether defined by ethnicity, religion, or some other characteristic. Prejudice toward people with disabilities is obvious: Throughout history, most societies have defined those who are disabled as physically or even morally inferior and have considered disabilities a sign that either the individual or his or her parents behaved sinfully or foolishly (Albrecht, 1992).

Prejudice typically expresses itself through **stereotypes**, or overly simplistic ideas about members of a given group. Nondisabled people typically stereotype disabled people as bitter, menacing, and unattractive or as asexual, dependent, mentally incompetent, and pitiable (Basnett, 2001; Ryan et al., 2005). Ironically, because medical training especially values quick, technological cures, doctors may be especially likely to develop negative attitudes toward people who live with long-standing disabilities (Basnett, 2001). (Medical culture is discussed more fully in Chapter 11.)

Stereotypes about people with disabilities are reflected and reinforced in the popular media (Chivers and Markotic, 2005). In book and film characters, from Captain Hook in *Peter Pan* to Freddie Krueger in *Nightmare on Elm Street* and the Penguin in *Batman*, the media have equated physical deformity with moral deformity. The media also often portray disabilities as pitiful and thus something to be avoided at all costs (as when Jake in the film *Avatar* chooses to leave his entire life and universe behind for the chance to walk again in an alien body). Although contemporary media sometimes do present more positive images, such as stories about people with disabilities who have "heroically" compensated for their physical disabilities, who have chosen to live "saintly" lives, or whose innocence can help the rest of us learn to live better lives (*Riding the Bus with My Sister*, for instance), these stories, too, typically ignore the social nature of disabilities and instead offer simplistic stories about individual character. Exceptions to these rules—films such as *The Fault in Our Stars* and television shows such as *Glee* and *Game of Thrones*—remain rare, although they have become far more common in recent years.

All too often, prejudice against persons with disabilities results in **discrimination**: unequal treatment grounded in prejudice. As recently as the first decades of the twentieth century, American laws forbade those with epilepsy, leprosy, Down syndrome, and other conditions from marrying and mandated their institutionalization or sterilization (Shapiro, 1993:197; Trent, 2005). Discrimination continues into the present day. In a national survey conducted in 2010, almost half of people with disabilities reported encountering job discrimination, most often in the form of lower pay for the same work or being considered ineligible for a job because of their disability (Harris Interactive, 2010).

To fit the definition of a minority group, however, members of a group not only must experience prejudice and discrimination but also must believe that they belong to a group that shares a common experience. In fact, 79 percent of people with disabilities report feeling a sense of community with other such individuals (Harris Interactive, 2010).

In the United States, laws now offer at least some protection against discrimination for people with disabilities. Currently, the federal Education for All Handicapped Children Act requires school districts to educate all children regardless of disability in the least restrictive environment feasible. In addition, the **Americans with Disabilities Act (ADA)** outlaws discrimination and requires accessibility in employment, public services, and public accommodations (including restaurants, hotels, and stores). To date, the ADA has had limited impact, primarily because courts have narrowly defined who is disabled and thus qualifies for its protection (Gostin, Feldblum, and Webber, 1999). In 2009, however, Congress substantially broadened the definition of disabilities under the ADA, and so it should become easier for individuals to qualify for ADA protection.

The Social Distribution of Disability

According to US government researchers, approximately 12 percent of noninstitutionalized persons living in the United States have a disability. These researchers define disability as a chronic health condition that makes it difficult for individuals to perform activities considered appropriate for persons of a given age—play and study for children, work for adults, or basic activities such as shopping and dressing for elderly adults (National Center for Health Statistics, 2010b).

The proportion of the US population living with disabilities has grown significantly over time. A few decades ago, most people with paraplegia, babies born prematurely, persons with serious head injuries, and soldiers with major wounds died quickly. Now most live, although often with serious disabilities. For example, because of advances in body armor and medical care, far fewer US soldiers have died of wounds in Iraq and Afghanistan than in previous wars, but far more have survived with brain damage and amputated limbs (Glasser, 2005). In addition, average survival times for various common chronic conditions, such as hypertension and cardiovascular disease, have increased, leaving more people living with disabilities. Finally, as the proportion of the population older than age 65 has increased—and in the absence of meaningful attempts to remove the social and physical barriers that can prevent individuals from living independent lives—so has the proportion living with disabilities.

Table 6.1 shows the distribution of disability across the population, measured by limitations in "basic life activities" such as shopping, dressing oneself, or

TABLE 6.1 Percentage of Americans with Basic Activity Limitations*

	Ages 18–64	Ages 65 and Older
Total	24%	59%
Sex		
Male	21	50
Female	27	61
Income		
Very poor	35	70
Poor	29	67
Middle Income	23	57
Upper income	18	44
Ethnicity		
White, non-Hispanic	24	56
African-American, non-Hispanic	26	61
Hispanic	20	58
Asian	13	49

*Ability to perform activities needed to maintain an independent life, such as shopping, bathing, or working (for those younger than age 65).

SOURCE: National Center for Health Statistics (2014a).

working (for those under age 65). Age stands out as a major predictor of disability, affecting the majority of adults older than age 65 but only about one-quarter of younger persons. Sex also predicts disability, with women significantly more likely than men to report activity limitations (for reasons discussed in Chapter 3). Income is also directly related to disability: Disability is most common among the poor and becomes less and less common as income rises. Finally, ethnicity also affects rates of disabilities, largely because of its relationship to poverty. According to official statistics, Asian Americans have the lowest risk, and African Americans the highest risk. (It is likely that Native Americans are also at high risk, but data are not available.)

UNDERSTANDING CHRONIC PAIN

Chronic pain, which affects 40 percent of Americans (Institute of Medicine, 2014), falls on the border between disability and chronic illness. As writer Melanie Thernstrom (2010:5), who herself lives with chronic pain, explains:

> Ordinarily, pain is protective—a finely wired system warning the body of tissue damage or disease and enforcing rest for the bone to knit or the fever to run its course. This is known as acute pain; when the tissue heals, the pain disappears. When pain persists long after it has served its function, however, it transforms into the pathology of chronic pain. Chronic pain is the fraction of pain that nature can't heal, that does not resolve over time, but worsens. It can begin in many ways—as trivial as a minor injury or as grave as cancer or gangrene. Eventually, the tissue heals, the diseased limb is amputated, or the cancer goes into remission, and yet the pain continues and begins to assume a life of its own.

As this suggests, chronic pain is a symptom, not an illness. In some cases, it can be attributed to an injury or an illness, such as arthritis or cancer. In other cases, no specific cause can be identified; this is especially true for chronic headaches and back pain, the two most common types of chronic pain. In still other cases, some doctors diagnose **contested illnesses** (such as irritable bowel syndrome, fibromyalgia, or chronic fatigue syndrome) that other doctors question.

Living with Chronic Pain

Although the causes of chronic pain are often unclear, its consequences are obvious. Chronic pain is the most common underlying reason for disability among adults between the ages of 18 and 65 (American Pain Society, 2000). In addition to its physical toll (which includes sleep deprivation and exhaustion), chronic pain damages social relationships; increases depression, anxiety, and the risk of suicide; and costs the nation $600 billion yearly in medical costs and reduced productivity (Institute of Medicine, 2014). As Thernstrom (2010:5) writes:

> [As] the pain worsens, the body sensitizes, and other parts begin to hurt, too. She has trouble sleeping; she stumbles through her days. Her sense

> of her body as a source of pleasure changes to a sense of it as a source of pain. She feels haunted, persecuted by an unseen tormentor. Depression sets in. It feels wrong ... maddening ... delusional. She tries to describe her torment, but others respond with skepticism or contempt. She consults doctors, to no avail. Her original affliction—whatever it may have been—has been superseded by the new disease of pain (ellipses in original).

Treating those who live with chronic pain is notoriously difficult, and doctors don't agree on how to do so (American Pain Society, 2000). Physical therapy, strength training, meditation, and psychological treatments that help people think differently about their pain can often help but are *underused* (Foreman, 2014). Marijuana may well be useful, but little money is available for research on it, and it remains illegal in many states (and under federal law). Drugs derived from opiates such as OxyContin are often used. Ironically, although they are sometimes overused, they are also underused both because doctors share popular American beliefs about opiates and addiction and because they fear arrest under strict US drug-trafficking laws (Foreman, 2014). To make matters worse, few American doctors are trained in pain management (Institute of Medicine, 2014).

Gender, Ethnicity, Class, and Chronic Pain

Chronic pain is most common among women, poorer persons, minorities, and the elderly. Unfortunately, obtaining appropriate treatment is especially difficult for members of all these groups (Hoffman and Tarzian, 2001; Thernstrom, 2010). (Obtaining proper treatment is also difficult for children because doctors especially fear giving them potentially dangerous medications.)

Women—the majority of those living with chronic pain—are significantly more likely than men to encounter doctors who ascribe their pain to psychiatric causes and prescribe sedatives or psychotherapy rather than effective pain medications (Barker, 2005; Hoffman and Tarzian, 2001; Werner and Malterud, 2003). To avoid this fate, women with chronic pain must tread a fine line, striving to appear neither too sick nor too well and neither too assertive nor too passive in order to receive proper treatment and avoid being labeled hysterical or pushy, malingerers, or whiners (Werner and Malterud, 2003; Thernstrom, 2010). Not surprisingly, women consumers have been at the forefront of movements to medicalize chronic fatigue syndrome, multiple chemical sensitivity, fibromyalgia, and other contested illnesses (Barker, 2005).

A different set of stereotypes makes it difficult for persons of color to receive proper treatment for pain (Chen et al., 2005; Thernstrom, 2010). Numerous studies have found that doctors routinely dismiss minority patients' reports of pain on the assumption that such patients are really seeking drugs for recreational purposes.

The same bias limits poorer persons' access to pain medication. In addition, poorer persons' pain more often stems from their work. In these cases, however, individuals often have access only to company-employed doctors who have a vested interest in dismissing or downplaying—and thus undertreating—their pain.

LIVING WITH DISABILITY AND CHRONIC ILLNESS

Living with disability or chronic illness, whether or not it results in chronic pain, is a long-term process that includes responding to initial symptoms, injuries, or diagnoses; dealing with one's evolving situation; seeking and managing health care; and continually reconceptualizing one's future. In this section, we examine this process and explore how illness, pain, and disability affect individuals' lives, relationships with others, and sense of self. In addition, we look at how the experience of illness has in some cases led individuals to start or join social movements based around illness.

Responding to Initial Symptoms

Becoming a chronically ill or disabled person begins with recognizing that something about the body is troubling. This recognition can be slow to arrive. Health problems (such as stiffening joints caused by arthritis or gradual vision loss caused by cataracts) often build gradually, allowing individuals almost unconsciously to minimize and adapt to these symptoms. In addition, the signs of illness and disability often don't differ greatly from normal bodily variations. A child who doesn't walk by 12 months of age might have a disability or might simply be a slow developer. Similarly, children with epilepsy, for example, can for many years experience "headaches," "spaciness," and "dizzy spells" before they or their parents recognize these experiences as signs of epilepsy.

Social scientists refer to this process of defining, interpreting, and otherwise responding to symptoms as **illness behavior** (Mechanic, 1995). A review article by anthropologists Vuckovic and Nichter (1997), summarizing 20 years of research studies, concluded that US residents treat between 70 and 95 percent of all illness episodes without a doctor's assistance; it is likely that this percentage has grown considerably now that many turn to the Internet for health advice before seeking medical care (Fox and Duggan, 2013). Individuals typically begin by medicating themselves or those under their care with nonprescription medications recommended by friends, families, store clerks, or pharmacists or, more rarely, with prescription medicines left over from previous illnesses.

When and whether individuals seek formal, medical diagnoses for acute or chronic medical problems depends on a variety of factors. According to the **illness behavior model**, individuals are most likely to seek medical care if (1) their symptoms are frequent, persistent, visible, and severe enough to interfere with daily activities; (2) they lack alternative explanations for their symptoms; (3) their families and friends generally trust doctors and support seeking medical care for health problems; and (4) no psychological, economic, or practical barriers keep them from accessing health care (Mechanic, 1995). *Key Concepts: Predicting Illness Behavior* summarizes this model.

KEY CONCEPTS

Predicting Illness Behavior

Individuals Are Likely to Define Themselves as Ill and Seek Medical Care When	Individuals Are Unlikely to Define Themselves as Ill or Seek Medical Care When
Symptoms appear frequently or persistently (e.g., coughing blood once per day for a week).	Symptoms appear infrequently (e.g., coughing blood every few months).
Symptoms are very visible (e.g., rash on face).	Symptoms are not very visible (e.g., rash on lower back).
Symptoms are severe enough to disrupt normal activities (e.g., epileptic convulsions).	Symptoms are mild (e.g., annoying but tolerable headaches).
Illness is only likely explanation for physical problems (e.g., no recent changes in life circumstances that might explain headaches).	Alternative explanations for physical problems are available (e.g., recent stresses may explain headaches).
Access to health care is readily available (e.g., good health insurance).	Access to health care is poor (e.g., no health insurance).
Trust in doctors is high, and families and friends encourage seeking medical help.	Trust in doctors is low, and families and friends discourage seeking medical help.

Gender, Age, Class, Ethnicity, and Illness Behavior Illness behavior is significantly affected by gender, ethnicity, age, and social class. As Chapter 2 describes, for a variety of biological and cultural reasons, women are more likely than men to seek medical care when they experience bodily changes that might be symptoms of illness. Age has a more mixed effect on illness behavior. On the one hand, elderly persons experience more symptoms and more illness and so more often seek medical care. On the other hand, as the body declines, some elderly persons begin to expect a certain amount of physical discomfort, grow tired of constant visits to doctors, and so put off seeking diagnosis for new symptoms.

Similarly, working-class and poor individuals often accept physical pain as an unavoidable consequence of hard living and hard work, so they put off medical care until their symptoms interfere greatly with their daily lives. At any rate, even if they want medical care, many working-class and poor individuals can't afford to pay for care, can't get time off work to seek care, or lack transportation to go to a doctor or clinic. Moreover, those who can afford health care only at public clinics where long waits and rude treatment are common may put off seeking care as long as they can.

Ethnicity also affects illness behavior. Not surprisingly, members of poor ethnic groups are less likely to seek early diagnoses for all the reasons that poor persons are less likely to do so. In addition, some ethnic groups (especially those that include many recent immigrants) may feel more comfortable seeking care from traditional healers, at least initially. Similarly, among all ethnic groups, individuals are most likely to seek early diagnosis from a doctor if their friends and relatives trust doctors and encourage medical help seeking (Pescosolido, 1992; Pierret, 2003).

The Search for a Diagnosis Eventually, however, if symptoms persist—and especially if they progress—individuals and their families are likely to reach a point where they can't avoid recognizing that something is seriously wrong. As their previous interpretations of their symptoms crumble, individuals find themselves in an intolerable situation, torn by uncertainty regarding the changes in their bodies and their lives. At this point, any diagnosis can become preferable to uncertainty, so the incentive to seek diagnosis increases (Pierret, 2003).

Seeking a diagnosis, however, does not necessarily mean receiving one. Although some problems are relatively easy to diagnose—a 45-year-old white man who complains to his doctor of pains in the left side of his chest will probably quickly find himself getting tested for a heart attack—others are far less obvious. Persons with fibromyalgia, for example, often find that doctors initially dismiss their symptoms as psychosomatic or trivial (Barker, 2005). In addition, the same symptoms may more rapidly produce a diagnosis for some than for others. For example, as mentioned earlier, doctors more often dismiss women's complaints as merely emotional problems than they do men's complaints (Council on Ethical and Judicial Affairs, American Medical Association, 1991; Steingart, 1991).

Initially, both women and men can find these alternative diagnoses comforting and welcome—after all, it's far easier to hear that you are just responding to stress than that you have multiple sclerosis. When symptoms persist, however, individuals find themselves torn by ambiguity and uncertainty, experiencing anxiety about their failing health but often receiving little sympathy or help from relatives and colleagues (Barker, 2005; Charmaz and Rosenfeld, 2010; Stockl, 2007). As a result, most people eventually seek more accurate diagnoses, going from doctor to doctor until they find one who offers a plausible explanation for their symptoms.

When doctors cannot offer a convincing diagnosis, however, individuals may seek to diagnose themselves, a process known as **self-diagnosis**. The rise of the Internet and of new technologies has made self-diagnosis much easier in recent years (Copelton and Valle, 2009). Individuals can now go online and research their symptoms on a wide range of websites. They can also find online support groups that provide more details on others' experiences in obtaining diagnoses, including names of potentially helpful doctors or tests. Finally, in many cases individuals can now, on their own, pay for diagnostic tests offered direct to the public. For example, celiac disease is an autoimmune disorder that causes a wide range of gastrointestinal and other problems in affected individuals when they eat food containing gluten. Consequently, many who believe they have the disease stop eating gluten and, if they improve, feel even more confident of their self-diagnosis. Doctors, however, typically will not diagnose someone with celiac disease unless the diagnosis is confirmed by an established set of invasive and often highly unpleasant tests (which only work if individuals keep gluten in their diet for several months). Consequently, those who believe they have celiac disease increasingly are turning to laboratories that offer apparently scientific testing for celiac disease. Although doctors do not consider these tests definitive, the test results give individuals one more tool to convince doctors to confirm their self-diagnosis (Copelton and Valle, 2009). As this suggests, in the end any diagnosis—even one that requires a lifelong, highly restrictive diet—may be easier to live with than having no explanation for one's problems. In addition, having a diagnosis makes it more likely that others will believe that one's problems are real.

Managing Health Care and Treatment Regimens

Persons who live with chronic illness and disability can turn to both conventional and alternative health care for help. And increasingly, they use the Internet to help them in these decisions.

Using Conventional Health Care Living with chronic illness or disability often means living a life bound by health care regimens. That said, some individuals will strictly follow prescribed regimens of diet, exercise, or medication, whereas others won't. Researchers traditionally have framed this issue as a matter of *compliance*—that is, whether individuals do as instructed by health care workers.

The most commonly used framework for studying compliance is the **health belief model**. As we saw in Chapter 2, this model was developed to explain why healthy individuals adopt preventive health behaviors. The same model is also used to understand why people who have acute or chronic health problems comply with medical advice regarding treatment (see *Key Concepts: The Health Belief Model and Medical Compliance*). The model suggests that individuals will be most likely to comply if they (1) believe they are susceptible to a health problem that could have serious consequences, (2) believe compliance will help, and (3) perceive no significant barriers to compliance. For example, people who have diabetes will be most likely to comply with their prescribed diet if they believe that they face substantial risks of blindness due to diabetes-induced glaucoma, that blindness would substantially decrease their quality of life, that the prescribed diet would substantially reduce their risk of blindness, and that the diet is neither too costly nor too inconvenient.

The health belief model is a useful but limited one for understanding compliance with medical treatment because it largely reflects the medical model of illness and disability. First, the health belief model assumes that noncompliance with medical recommendations stems primarily from psychological processes internal to the patient. Although this is sometimes true, in other cases, patients don't comply because health care workers did not sufficiently explain either the mechanics of the treatment regimen or the benefits of following it (Conrad, 1985). Patients also might not comply because they lack the money, time, or other resources needed to do so.

Second, the health belief model implicitly assumes that compliance is always good (i.e., that health care workers always know better than patients what

KEY CONCEPTS

The Health Belief Model and Medical Compliance

People Are Most Likely to Comply with Medical Advice When They ...	Example: Compliance Likely	Example: Compliance Unlikely
Believe they are susceptible.	50-year-old man with hypertension who believes he is at risk for a heart attack	15-year-old boy diagnosed with epilepsy who has had only minor problems and does not believe he is at risk for convulsions
Believe risk is serious.	Believes that heart attack could be fatal	Believes that convulsions would not be physically dangerous
Believe compliance will reduce risk.	Believes he can reduce risk through taking medication regularly	Believes he doesn't really have a problem, so he doesn't see how medication could help
Have no significant barriers to compliance.	When medication is affordable and has no serious or highly unpleasant side effects	When medication makes the boy feel drowsy, dull, and set apart from his peers

patients should do). Yet, although health care workers often can help their patients considerably, this is not always the case. Especially with chronic conditions, the only available treatments may be disruptive to normal routines, experimental, ineffective, unpleasant, or potentially dangerous. As a result, many people who at first diligently follow prescribed regimens eventually abandon them (Conrad, 1985).

Contemporary Issues: Mobile Digital Health Devices discusses the skyrocketing increase in devices designed to increase compliance and health-protecting measures among both healthy individuals and those already diagnosed with illnesses.

Using Alternative Therapies As people's faith in mainstream medicine declines, some begin experimenting with their treatment regimens, learning through trial and error what works best for them not only physically but also socially, psychologically, and economically (Conrad, 1985). Others begin using **alternative** or **complementary therapies** (defined broadly as treatments not widely integrated into medical training or practice in the United States).

Interest in alternative therapies has grown rapidly in the United States, both among healthy persons interested in avoiding illness and among those with chronic or acute illnesses. According to data collected by federal researchers through a random national survey, almost 40 percent of US residents use some form of alternative therapy (Barnes, Bloom, and Nahin, 2008). The most commonly used therapies are herbal and other dietary supplements, deep breathing exercises, meditation, and chiropractic or osteopathic treatments.

CONTEMPORARY ISSUES

Mobile Digital Health Devices

The last few years have seen tremendous growth in the use of Internet-connected mobile devices that digitally monitor individuals' health. These include not only the Fitbit and the Apple HealthKit but also a cornucopia of other wearable digital devices: implanted sensors that monitor blood glucose or medications; watches, athletic shoes, and wristbands that measure exercise activity, heart rate, or sleep patterns; and thousands of smartphones apps that monitor mood, calorie intake, and much more. Users can upload their data to a computer or the Web and can share it with doctors, family members, or online groups of fellow sufferers.

The benefits of these technologies are obvious. Individuals who track their calorie intake or exercise routines, for example, are more likely than those who do not meet their personal goals. Those who track danger signs (such as persons with diabetes who track blood glucose levels) may be able to treat themselves before a problem develops and avoid the need to seek medical care. And those who share their data with their health care providers may benefit when their providers notice early signs of a problem developing.

Yet mobile health devices also have a downside (Lupton, 2013a). First, they encourage us to trust digital measures more than our sense of our own body. Yet our intuitive judgments are sometimes correct, and our digital devices sometimes fail us. Second, these devices encourage us to believe that our health is fully under our own control, potentially stigmatizing any individual who falls ill or who does not (or cannot afford to) rely on digital health devices. Finally, many of these devices frequently remind individuals (via beeps, texts, or pop-up messages) to monitor their body or moods, take a specified action, or upload data. As a result, they can add to the burdens of illness by increasing anxiety, resentment, and the sense that one's life has been overwhelmed by one's illness (Lupton, 2013b; Oudshoorn, 2011).

Users of alternative therapies are disproportionately likely to be female, younger than age 69, college educated, and suffering from chronic health problems, especially back pain, colds, neck pain, and joint pain (Barnes et al., 2008; Tindle et al., 2005). Most who use alternative therapies do so because conventional treatments have not helped them. That said, individuals typically use alternative therapies to *complement* rather than to replace mainstream medicine.

The popularity of alternative therapies rests on belief—or at least hope—in the efficacy of these treatments. These beliefs are supported both by personal experience and by recommendations from friends and acquaintances who have used alternative therapies. In some cases, the therapies no doubt did help, either because of the biological effects of the therapies or because consumers' belief in the therapy helped the body to heal itself, as happens in about 30 percent of all persons treated with **placebos** (drugs known to have no biological effect). In other cases, individuals attribute cures to alternative therapies when actually the problem went away on its own, as happens with 70 to 80 percent of all health problems (Lundberg, 2001:123). Finally, people sometimes convince themselves that therapies helped even though their health did not actually improve.

Use of alternative therapies also rests on the dangerous assumption that "natural" treatments are automatically safe. For example, a Chinese herb, ma huang, helps dieters but can cause heart attacks and strokes. Kava kava tea may

reduce anxiety but also can cause liver damage, and gingko biloba both stimulates circulation and increases bleeding during surgery (McNeil, 2002). Moreover, whereas the federal Food and Drug Administration is responsible for regulating the safety, potency, and effectiveness of prescription drugs, no governmental agency regulates herbal remedies or supplements. Although manufacturers can't legally claim that alternative herbs and supplements cure, they can claim that their products might help. Unfortunately, available research suggests that few if any of these treatments are useful, some contain dangerous contaminants such as lead or arsenic, and some don't even contain the herb or vitamin listed on the bottle (Offit and Erush, 2013; Guallar et al., 2013).

Finally, a study by Matthew Schneirov and Jonathan David Geczik (1996) suggests that use of alternative therapies sometimes stems from discontent with modern society's emphasis on science and rationality and from the poor match between doctors' concerns and patients' concerns. Whereas doctors typically are concerned with solving the puzzle of diagnosis and identifying a specific body part that requires treatment, patients are primarily concerned with reducing the impact of illness on their lives (Mechanic, 1995). This mismatch can leave patients feeling depersonalized and deeply dissatisfied with the care they receive, even if it is technically competent.

Schneirov and Geczik (1996) found that individuals typically turned to alternative care when confronted by a chronic illness and dissatisfied with the treatment they received from mainstream health care providers or when dealing with some other sort of life crisis. Such individuals shared several beliefs: that modern medicine focuses too much on treating symptoms through surgery and medication rather than on preventing illness through lifestyle changes; that government regulation of health care endangers both personal freedom and health;

that individuals should take responsibility for their own health; and that doing so means adopting stringent behavior regimens, such as restrictive diets and regular use of laxatives. Through these shared beliefs, users of alternative healing constructed not only a philosophy of health care but also a shared sense of identity and community. Thus, Schneirov and Geczik conclude, "the alternative health movement may be seen as part of a larger wave of discontent with the bureaucratic-administrative state, its reliance on expert systems, and the way it coordinates people's health care practices 'behind their backs'—without their knowledge and participation" (1996:642).

Seeking Information on the Internet Whether individuals rely primarily on mainstream or alternative therapies, many seek information about their conditions on their own rather than relying solely on information provided by health care professionals. In the past decade, public access to information has exploded due to the exponential growth of Internet use. One recent national survey found that 59 percent of Americans used the Internet to seek health information in the preceding year and 35 percent had used the Internet to diagnose themselves or others at some point in the past (Fox and Duggan, 2013). That said, Internet use is not evenly distributed across the population: Whites, women, and middle- to upper-class individuals are more likely than others to use it.

Unfortunately, there are no controls on the quality of materials posted on the Internet, and its vast size makes it impossible to police for fraudulent information, such as claims that herbs can cure cancer or AIDS. Moreover, more often than not, popular search engines take readers seeking health-related information to websites run by individuals or corporations that have vested economic interests in selling certain drugs or treatments (Green, Kazanjian, and Helmer, 2004). Partly in response to concerns about misleading websites, the US Department of Health and Human Services now runs its own website (www.healthfinder.gov) to link consumers to reliable online sources of health information.

Despite limitations in most people's ability to effectively search the Internet or evaluate the information they find there, the Internet has proven enormously beneficial to those living with chronic health problems. The Internet has allowed individuals to find online forums designed to help individuals who share similar health issues or concerns and to find information far beyond what they otherwise could access (Barker, 2008; Fox, 2012). This is especially useful for those with rare conditions, those confined to their homes by severe illness or disability, and those with stigmatized illnesses who might shy away even from doctors (Conrad and Stults, 2010; Vanderminden and Potter, 2010). Consequently, the Internet can help individuals to negotiate with health care providers regarding treatment and to navigate the daily difficulties of living with illness or disability.

The Internet has also given individuals access to options that their own doctors might reject as unethical. *Ethical Debate: The Sale of Human Organs* discusses one of these options.

Managing Social Relationships and Social Standing

For better or worse, chronic illness and disability necessarily alter relationships with friends, relatives, and others. Illness and disability can strengthen social relationships, as friends and families pull together to face health problems, old wounds are healed or put aside, and individuals realize how much they mean to each other. Illness and disability, however, can also strain relationships. Friends and family often help each other willingly during acute illnesses or the first few months of a chronic illness or traumatic injury but pull away over time. This is especially true for male friends and family, who less often than women are socialized to take care of others (Cancian and Oliker, 2000). Moreover, the growing burden of gratitude can make those who have chronic illnesses or disabilities reluctant to ask for needed help. Problems are especially acute among elderly persons who have outlived close relatives and friends and must rely on more distant social connections for help and support.

Relationships also suffer if individuals no longer can participate in previous activities. How do you maintain a relationship with a tennis partner if you no longer can hold a racket? How do you maintain a relationship with a friend when transportation barriers keep you from going to movies or restaurants? And how do you maintain a relationship with a spouse or lover when your sexual abilities and interests have changed dramatically—or when your partner no longer finds you sexually attractive?

Declines in financial standing also strain relationships. An individual might, for example, have the physical ability to go to a movie with a friend but lack the price of admission. Women and minorities are especially hard hit because they typically earn lower wages and have more erratic work histories before becoming ill or disabled, so they often qualify for lower Social Security benefits. At the same time, financial stresses can damage relationships with children, lovers, and spouses.

Managing Stigma Illness and disability affect not only relationships with friends and family but also less intimate relationships. Most basically—and despite the predictions of the **sick role** model—living with illness or disability means living with stigma. **Stigma** refers to the social disgrace of having a deeply discrediting attribute, whether a criminal record, a gay lifestyle, or a socially unacceptable illness. The term *stigma* does not imply that a condition is immoral or bad, only that it is commonly viewed that way.

Some illnesses and disabilities produce relatively little stigma, but others are so stigmatized that they can affect even relationships with health care providers. Illnesses and disabilities are most likely to result in stigma when they are believed to be the fault of the affected individual and when they cause fear or dread; visible disfigurement; loss of bowel, urinary, or other bodily functions; cognitive problems; or behavior that mimics cognitive problems (Charmaz and Rosenfeld, 2010). HIV/AIDS, for example, elicits particularly high stigma because it is often

ETHICAL DEBATE

The Sale of Human Organs

One of the most extreme situations an ill individual can face is the failure of a major organ, be it the heart, lung, kidney, or liver. Such situations are death sentences unless the organ can be replaced either with a mechanical substitute or with a donated human organ. But mechanical replacements are often poor substitutes for bodily organs. In addition, some mechanical replacements severely restrict individuals' lives by tethering them to machines. Human organs, on the other hand, can be difficult or even impossible to obtain legally. More than 100,000 Americans are now on organ waiting lists, and 18 die each day (US Department of Health and Human Services, 2010). As a result, an illegal, multimillion-dollar, international market in human organs has emerged (Bilefsky, 2010; Rohter, 2004).

Most commonly, the organs sold through this market are kidneys, although livers, lungs, corneas, and other organs also are sold. Because almost everyone is born with two kidneys and only one is needed to live, an individual can sell one kidney and still hope to live a normal and healthy life.

Selling an organ carries great risks but can seem worth it if an individual is poor enough. In Brazil, for example, a person can earn $80 per month working at minimum wage—if work is available—or can sell his or her kidney for $3,000. Such sales are illegal in many countries, but those laws are rarely enforced.

To some observers, the trade in human organs is a natural and reasonable market response in which supply (organs for sale) develops to fill an obvious need (organs wanted). These observers see no difference between selling organs and selling any other valued commodity, such as drugs, cars, or food. Similarly, they argue, people should have at least as much right to buy an organ that will save their life as they have to buy a television or a facelift and as much right to risk their health by selling an organ as they have to risk their life by selling their labor in dangerous occupations (Cherry, 2005).

Other observers, however, compare the trade in human *organs* to the trade in *humans* and consider selling organs no more ethical than selling slaves. They argue

interpreted as punishment for sin; is contagious (and thus frightening); and can cause wasting, facial sores, loss of bodily functions, and dementia.

Individuals with stigmatized illnesses and disabilities can use various strategies to manage that stigma. Many attempt to *avoid* stigma by hiding it or otherwise deflecting attention from it. For example, a man who bumps into furniture because of failing eyesight might try to convince others that he is merely clumsy, and a woman who has arthritis might choose not to go out with friends on days when her symptoms flare up.

Conversely, others manage stigma by *challenging* the very basis of that stigma. Some disabled men, for example, become star athletes in part to reject the assumption that a disabled man can't be "masculine." Others challenge stigma more directly by fighting for civil rights. Still others fight for acceptance of their bodies through displaying their own proud acceptance or even appreciation of their bodies. For example, one woman born without a hand, who, after a year of wearing a hot, uncomfortable, and functionally useless artificial hand, decided to switch to a much more useful metal hook, said:

> I never failed to get a reaction from people [who saw me], so I always looked too [whenever I passed a mirror or store window].

that no one truly sells their bodily organs freely but rather does so because they are coerced by poverty. They also argue that whenever a highly profitable commodity is for sale unregulated by laws, unscrupulous individuals will find ways to profit and vulnerable individuals—whether buyers or sellers—will be exploited. Individuals who purchase black-market organs have no guarantee that the donor was healthy or that the organ is a good match for them, and those who sell organs have no guarantee that the surgery will be conducted safely, that it won't harm their health, and that they will receive needed health care afterward. A study conducted in the Indian state of Tamil Nadu found that virtually all who (illegally) sell their kidneys do so to pay crippling debts. Yet because most (86%) were in worse health in the years after surgery, their average family incomes declined by one-third, even though average incomes in the state increased (Goyal et al., 2002). Despite these problems, though, the trade in organs is likely to continue so long as demand continues to outstrip supply.

Sociological Questions

1. What social views and values about medicine, society, and the body are reflected in policies that allow or forbid the selling of human organs? Whose views are these?
2. Which social groups are in conflict over this issue? Whose interests are served by laws forbidding the sale of human organs? By laws permitting it?
3. Which of these groups has more power to enforce its view? What kinds of power do they have?
4. What are the intended consequences of permitting the sale of human organs? What are the unintended social, economic, political, and health consequences of this policy?

> What the hell are they looking at? I looked and I saw a woman with a surprisingly short arm! But when I got the [cosmetic] hand, I looked and I thought, oh my God, that's what I would have looked like [if I had been born with a hand]! And I saw this person that I would have been. But maybe I would have been an asshole just like all the rest of them [the nondisabled].... And [now] when I see the hook, I say, boy, what a bad broad. And that's the look I like the best (Phillips, 1990:855).

In sum, this individual was able to reject the stigma others assigned to her by defining herself as feisty, independent, and rebellious and defining "normals" as voyeuristic "assholes."

Health Social Movements In addition to challenging stigma and discrimination individually, persons who live with or are at risk of illness or disability can also turn to collective political action to address their grievances. Like other social movements, **health social movements** are collective (rather than individual) efforts to change something about the world that movement members believe is wrong (Brown et al., 2004).

Health social movements have a variety of goals. Many focus on obtaining equal access to health care by, for example, fighting to loosen health insurers' restrictions on what treatments they will cover. Other health social movements focus on meeting the needs (including access to health care) of a particular group. For example, the feminist health movement has fought to gain women the same access to heart disease treatments that men have, to halt the unnecessary use of cesarean sections and hysterectomies, and to increase the number of women physicians. Finally, a growing number of health social movements focus on winning medical acknowledgement for contested illnesses. For example, few doctors believe in the existence of "multiple chemical sensitivity," which is theorized to make some individuals ill whenever they contact any of the many chemicals common to everyday modern life. Persons who believe they have this condition have organized to lobby for medical recognition of their condition and to sue insurance companies that refuse to cover their treatment.

The rise of health social movements reflects a variety of factors (Brown et al., 2004). The civil rights, women's rights, and gay rights movements set the stage for a broader discussion of rights and a broader acceptance of political action across American culture. Health social movements are partly a product of this changed cultural climate. In addition, the same cultural and technological forces that increased the use of the Internet and of alternative health care have fostered health social movements by reinforcing the idea that individuals have the right to challenge medical authority. Individuals are most likely to participate in health social movements when they come to believe that medical authorities have failed to protect them (or their loved ones) from diseases, to identify their diseases, or to treat their diseases appropriately. For example, the environmental breast cancer movement was organized primarily by women affected by the disease who questioned why medical research has focused almost exclusively on early diagnosis and treatment of breast cancer rather than on prevention. As this example suggests, people who live with illness and disability are not simply victims of their fate. Rather, they may actively work to better their situation and those of others like them.

The Body and the Self Regardless of a person's political stance toward his or her condition, all disabilities and chronic illnesses challenge the self (Brooks and Matson, 1987; Charmaz, 1991; Corbin and Strauss, 1987). Those whose bodies differ in some critical way from the norm must develop a self-concept in the context of a culture that interprets bodily differences as signs of moral as well as physical inferiority. The resulting stigma leads such individuals to feel set apart from others (Conrad, 1987; Kutner, 1987; Weitz, 1991).

Illness and disability threaten self-concept in various ways. People who become physically deformed or less attractive often find it difficult to maintain their self-images, as do those who lose their financial standing or their social roles as worker, student, spouse, or parent (Brooks and Matson, 1987; Weitz, 1991:97). In addition, the need to rely on others for assistance can shake individuals' images of themselves as competent adults.

Disability and illness create different problems for women than for men. American society expects men to be emotionally, physically, and financially independent, and the threat to self-esteem when men can't meet these expectations can be great. Conversely, American society expects women (except for African American women) to be dependent, so disability typically does not threaten women's self-esteem as much as it threatens men's self-esteem. For African American women, however, and for all other American women who cherish their independence, illness or disabilities can hamper the struggle to obtain that independence whenever prejudice and discrimination based on illness and disability compound prejudice and discrimination based on gender or ethnicity.

The sexual changes accompanying disability and illness also affect women and men differently. Social norms for both persons with and without disabilities expect men to be sexually active but regard women's sexual desires with suspicion. When men are unable to meet others' expectations for sexual performance, they can lose esteem in both their own eyes and those of their partners. And when disability leaves women unable to meet social norms for sexual attractiveness, they often find that others assume they have no sexual feelings at all (Lonsdale, 1990).

To cope with these threats to the self, individuals sometimes attempt intellectually to separate their essential selves from their recalcitrant bodies. They might mention how their leg is acting up today, as if they were talking about a neighbor rather than a part of their body, or talk about their lives and their selves with no mention of their bodies at all. This strategy succeeds best when symptoms follow a predictable course and the problem affects only one part of the body.

The impact of disability and illness on the self, however, is not solely negative. In fact, research consistently finds that even severe disability and illness have relatively little effect on overall life satisfaction or happiness, although they do affect satisfaction with health, income, and social relationships (Oswald and Powdthavee, 2008; Powdthavee, 2009). Whether individuals grow up with disability or become disabled over time, they may learn to devalue physical appearances, derive self-esteem from other sources, focus on the present rather than on an intangible future, and compare themselves to others who are ill or disabled rather than to the able-bodied (Powdthavee, 2009; Weitz, 1991:136–140). They may learn to set priorities in their lives and accomplish their most important goals rather than wasting precious energy on trivial concerns (Charmaz, 1991:134–166). Finally, they may come to define their condition simply as part of who they are, with good points and bad points, and to recognize that much of their personalities and accomplishments exist not *despite* their physical condition but *because* of it. As Mark Zupan, a husband, engineer, renowned wheelchair rugby athlete, and quadriplegic since age 18 writes:

> When I was growing up, sports meant everything to me. So you can imagine how I felt when it became clear that I was going to spend the rest of my life in a wheelchair. I thought I would never be able to play again—or drive a car, or have a job, a girlfriend, a house, a family of my own. [But] in truth, my accident has been the best thing that could ever

have happened to me. I'm not trying to be glib when I say this, or rationalize my mistakes, or offer you a steaming bowl of bullshit-flavored chicken soup for the soul. What I am saying is that it has been the single most defining event of my life. And without it, I wouldn't have seen the things that I've seen, done the things I've done, and met so many incredible people.... I wouldn't have come to understand and cherish my family and friends the way I do, and feel the kind of love they have for me and I have for them. In other words, I wouldn't be me, plain and simple.

Learning to live with limited function has forced me to take a good hard look at myself. When something catastrophic like this happens, the anger, frustration, and despair can become overwhelming.... There have been times that I have stared in the mirror and hated what I saw.... But here's the bottom line: At some point, life is going to give you a swift, hard kick to the nuts. You can't control everything that happens to you, but you can try to understand it. For me, this has been just one of the many things I've learned in this painful, beautiful, crappy, exhilarating, stupid, rewarding life that started the day I landed in this chair—which I thought was my cross to bear, but was actually my salvation (Zupan and Swanson, 2006:4–6).

IMPLICATIONS

Given the aging of the American population and the increasing ability of medical technology to keep ill and disabled individuals alive, many more of us can expect eventually to live with illness, chronic pain, and disability—whether our own, our parents', or our children's. Consequently, understanding what it means to live with these conditions has never been more important.

As both social constructions and social statuses, illness and disability affect all aspects of life. Most obviously, they force individuals to interact with health care providers and to manage health care regimens. But illness and disability also affect family relationships; friendships; work prospects; educational performance and opportunities; and, perhaps most importantly, sense of self and relationship with one's own body. Living with illness and disability also requires people to come to terms—or to refuse to come to terms—with uncomfortable questions and harsh realities regarding their past, present, and future.

Illness and disability can bring social disadvantages similar to those experienced by members of traditionally recognized minority groups. Yet the impact of illness and disability is not always negative, for illness and disability at times can provide individuals with the basis for increased self-esteem and enjoyment of life. Moreover, like other minorities, those who live with illness and disability have in recent years moved from pleas for tolerance to demands for rights. Those demands have produced significant changes in American architecture, education, transportation, and so on, and have laid the groundwork for the changes still needed.

SUMMARY

1. The medical model of disability defines disability as something located solely within the individual mind and body. The sociological model of disability, on the other hand, defines disability as restrictions or lack of ability to perform activities resulting largely or solely from either social responses to bodies that fail to meet social expectations or assumptions about the body reflected in the social or built environment.
2. Like members of minority groups, persons with disabilities experience prejudice and discrimination and increasingly share a sense of community.
3. The disability rights movement argues that persons with disabilities deserve the same rights as other members of society. This philosophy gained important legal support with passage of the Americans with Disabilities Act. However, until recently the impact of that Act was limited by restrictive court decisions.
4. Approximately 12 percent of noninstitutionalized persons living in the United States have a disability. African Americans, older persons, and poorer persons have higher rates of disabilities than others.
5. Chronic pain affects one of every six Americans and is most common among women, poorer persons, minorities, and elderly persons. Obtaining appropriate treatment for chronic pain is notoriously difficult, especially for members of these four groups.
6. Becoming a chronically ill or disabled person begins with recognizing that something about the body is troubling, a process that may develop slowly. The process of responding to symptoms and deciding whether to seek diagnosis and treatment is known as illness behavior.
7. Obtaining an accurate diagnosis is often difficult. According to the illness behavior model, individuals are most likely to seek medical care and diagnosis if (1) their symptoms are frequent, persistent, visible, and severe enough to interfere with daily activities; (2) they lack alternative explanations for their symptoms; (3) their families and friends generally trust doctors and support seeking medical care for health problems; and (4) no psychological, economic, or practical barriers keep them from accessing health care.
8. The health belief model predicts that individuals are most likely to comply with medical advice when they (1) believe they are susceptible to health dangers, (2) believe the risk is serious, (3) believe compliance will reduce their risk, and (4) have no significant barriers to compliance. Critical sociologists have noted that individuals sometimes have rational reasons for medical "noncompliance."
9. Users of alternative therapies are largely female, younger than age 69, college educated, and suffering from chronic health problems, especially back pain, colds, neck pain, and joint pain.
10. Alternative therapies are typically used in addition to mainstream medicine by individuals who find that conventional treatments have not helped them,

believe that alternative treatments are safe, or are discontent with modern society's emphasis on science and rationality.

11. Many Americans use the Internet to seek health information, although the quality of that information varies widely. The Internet is particularly useful for those living with chronic health problems, especially if the problems are rare or stigmatized.
12. Illness and disability can threaten social relationships in many ways, especially when they cause stigma. Individuals can manage stigma by hiding their illness or disability, deflecting attention from it, or challenging the norms that stigmatize them.
13. Health social movements are collective efforts to improve health and health care, sometimes by changing definitions of health and illness.
14. Disabilities and chronic illnesses can threaten one's sense of self and the body. However, they sometimes can improve individuals' self-concepts.

REVIEW QUESTIONS

1. How do the medical and sociological models of disability differ?
2. Are people with disabilities a minority group? Explain.
3. How common is disability, and which social groups are most at risk?
4. How common is chronic pain, and which social groups are most at risk?
5. What difficulties do individuals face in responding to initial symptoms of illness or disability and obtaining diagnoses?
6. What is illness behavior? Give an example.
7. Why do individuals sometimes ignore medical advice?
8. Why do individuals use alternative health care?
9. How can illness or disability affect social relationships and self-image?
10. How can individuals manage the stigma of illness or disability?
11. What is a health social movement, and why have they become more common?

CRITICAL THINKING QUESTIONS

1. Think of a recent experience you, a close friend, or a relative had with a chronic or acute illness. Explain which concepts from the sociological literature on the experience of illness applied to your experience. If *few* concepts applied, explain why these concepts generally did *not* apply.

2. To protect or improve their health, many individuals take actions that lack scientific proof of effectiveness, such as taking vitamin C to cure colds. Think of something that you, your friends, or your relatives do that falls into this category. Why did you or they decide to adopt this measure? Why have you or they continued? What beliefs or principles underlie these decisions? Why doesn't the lack of scientific proof affect these decisions?
3. What are some of the reasons why individuals seek *alternative* health care? What does the growing use of alternative health care tell us about modern *medical* care?

CHAPTER

7

The Sociology of Mental Illness

Lighttrace Studio/Alamy

LEARNING OBJECTIVES

After reading this chapter, students should be able to:

- Assess how ethnicity, gender, social class, social stress, and social capital affects mental illness.
- Compare the medical and sociological models of mental illness.
- Understand how medical and social ideas about mental illness have evolved over time.
- Offer a sociological description of the experience of living with mental illness.

Journalist Norah Vincent has struggled with depression for many years, although most of the time psychotherapy and psychotherapeutic drugs have helped her avoid serious problems. A few years ago, however, her depression landed her in a mental hospital. Dismayed by much of what she saw there, she decided, after her mental health stabilized, to fake symptoms of mental illness, have herself admitted to various mental hospitals, and write about her experiences. Summarizing her experiences, she writes:

> *There are few things more humiliating, more soul-destroying and depressing, than the process of being institutionalized. And the worst part is your own collusion in the process. You allow it to happen to you.... You become docile, subservient, frightened, dull, unthinking, susceptible to the mysterious self-fulfilling power of the rule. You loathe the tone of your own voice as you mewl and cower to the dingbat shoving you your meds or taking away your pen. You are demeaned by the routine as you regulate your life by mealtimes, loitering in the hall at eight, twelve, and six. You change as you acquiesce to rudeness, becoming less, becoming small....*
>
> *You do strange things. I tried, for example, to make shoelaces out of toilet paper [after the nurses removed them] so that I could walk like a normal person instead of limping like a gangster. The laces tore, of course, but it was a way to pass the time, rolling the long strands of tissue between my fingers as tight and string-like as they would go, and feeling, even though I failed to make the lashings tie or hold, the momentary elation of knowing that I could still exercise some form of creativity.*
>
> *I learned to flick on the light over my bed with the teeth of a comb or the tip of the forbidden ballpoint pen so that I could read late at night when I couldn't sleep and the dayroom was closed. The light switches were in the hall and recessed so that only the staff could access them ... and thus enforce lights out at eleven and lights on at eight. Controlling light is no small matter, as they well knew.*

Just one of many daily benefits you take for granted in the outside world (Vincent, 2009:41).

As Norah Vincent's story suggests, mental illness is a social as well as a psychiatric condition, and mental hospitalization has social as well as psychiatric consequences. We begin this chapter by considering the extent and distribution of mental illness. We then examine contrasts between the medical model of mental illness, which views mental illness as an objective reality, and the sociological model, which views mental illness as largely a social construction. Finally, we look at how social forces and values have affected both the history of treatment and the experiences of those who live with mental illness.

THE EPIDEMIOLOGY OF MENTAL ILLNESS

The importance of understanding mental illness becomes clearer when we realize how many people are affected. The following section discusses research on the extent, distribution, and causes of mental illness.

The Extent of Mental Illness

Since the 1920s, social scientists have tried to ascertain the extent of mental illness. These researchers essentially have adopted medical definitions of mental illness (which, as we will see later in this chapter, are problematic). However, whereas doctors and other clinicians have focused on how biological or psychological factors can foster mental illness, social scientists have focused on how *social* factors can do so.

Over the years, researchers using a variety of methods have reached two consistent conclusions regarding the extent of mental illness. First, all societies, from simple to complex, include some individuals who behave in ways considered unacceptable and incomprehensible (Horwitz, 1982:85–103). Second, symptoms of mental disorder are fairly common. According to the National Comorbidity Survey Replication, the largest national survey on the topic based on a **random sample** (Kessler et al., 2005a), during the course of a year approximately 31 percent of working-age adults experience a diagnosable mental illness, with 20 percent experiencing a moderate or severe disorder. The most common illnesses are major depression and problems with alcohol use, reported by 17 and 13 percent, respectively. These estimates, however, are probably high because they are based on reports of symptoms taken out of context (Horwitz, 2002, 2007). When an individual reports that he is extremely sad, survey researchers can't tell whether the sadness was caused by clinical depression or financial problems. Nor can researchers tell whether a woman who reports losing weight has done so because of depression or because she wanted to fit into her wedding dress.

Social Stress and Mental Illness

Although mental illness is common, it does not burden all social groups equally. So why do some social groups experience more mental illness than others do? For many sociologists, the answer lies in their different levels of social stress.

In the past, sociologists interested in the link between mental illness and stress largely focused on the **acute** stresses of **life events**, such as divorce, losing a job, or a death in the family. Researchers looked not only at the sheer number of life events individuals experienced but also at the *meaning* life events have for people and the *resources* individuals have for dealing with those life events. For example, an unplanned pregnancy means something quite different to an unmarried college student from a poor family than it does to a married, middle-class housewife.

Similarly, some individuals have resources that can reduce the stresses of life events (such as money, social support networks, and psychological coping skills), whereas others lack such resources (Ensel and Lin, 1991; Lennon and Limonic, 2010; Turner and Brown, 2010). For example, a person whose marriage fails but who has enough income to maintain his or her current lifestyle, close friends to provide companionship and social support, and good stress management skills will probably experience less stress than will someone whose economic standing plummets after divorce, who has few friends, and who responds to stress by drinking.

As we saw in Chapter 2, recent research finds that **chronic** stress is more important than acute stress for predicting poor physical health. Similarly, researchers have shown that chronic stresses affect mental health more than do acute stresses such as life events (Turner and Avison, 2003). Much research in this field now focuses on how exposure to chronic social stress may explain ethnic, gender, and social class differences in rates of mental illness.

Ethnicity, Gender, Social Class, and Rates of Mental Illness

Ethnicity, gender, and social class all affect rates of mental illness, as Table 7.1 summarizes. The rest of this section explains these effects.

TABLE 7.1 Sex, Ethnicity, and Social Class Groups with the *Highest* Lifetime Risks of Specific Mental Illnesses

	Mood Disorders	Impulse-Control Disorders	Substance Abuse Disorders	Schizophrenia
Sex	Females	Males	Males	Males
Ethnicity	Non-Hispanic whites	No ethnic differences	Hispanics and non-Hispanic whites	No ethnic differences
Social class	Lower class	Data unavailable	Lower class	Lower class

SOURCES: Aleman, Kahn, and Selten (2003); Kessler et al. (2005a).

The Impact of Ethnicity: Stress Effects Researchers have uncovered few significant ethnic differences in rates of schizophrenia or other major mental illnesses. However, for still unexplained reasons, African Americans seem *less* likely than do whites to develop anxiety or mood disorders. Nevertheless, African Americans are *more* likely—especially if they are poor—to report psychological distress, which overlaps with but is not the same as diagnosable mental illness (Kessler et al., 2005a). Researchers theorize that psychological distress among African Americans results from the chronic daily stresses of living with racism. This would explain why, for example, wealthier African Americans—who can use their income to shield themselves somewhat from the effects of racism—experience less stress than do lower income African Americans (Kessler and Neighbors, 1986; Turner and Avison, 2003).

Little recent research is available on psychological distress among other US minority groups. However, Hispanic Americans are less likely to develop anxiety disorders, mood disorders, or substance abuse problems (Kessler et al., 2005a). Importantly, the rate of mental disorders among new immigrants from Mexico is half that of US-born Mexican Americans, but those rates converge after immigrants have been in the United States for more than a decade (Vega et al., 1998). Researchers hypothesize that Mexican culture's strong emphasis on extended families protects immigrants from mental illness by offering social support and thus reducing chronic stress among persons who are single, childless, less educated, or employed in low-prestige jobs. As Mexicans integrate into American culture, they lose these protections.

The Impact of Gender: Socialization Effects The impact of gender on mental illness is at least as complex as the impact of ethnicity. Most mental illnesses are equally common among men and women. However, men have higher rates of schizophrenia, substance abuse, and impulse control disorders (such as compulsive gambling or chronic violence), whereas women have higher rates of anxiety disorders and of mood disorders (such as depression) (Aleman, Kahn, and Selten, 2003; Kessler et al., 2005a).

These differences in mental illness parallel differences in gender roles. Consistently, men display higher rates of disorders linked to violence. As a result, some researchers hypothesize that these forms of mental illness occur when men become "oversocialized" to their gender roles. For example, a young man who fails to plan ahead, shows "reckless disregard" for safety, and gets into fights often, and who before the age of 15 often bullied others, got into fights, or skipped school, would meet the criteria for diagnosis with "antisocial personality disorder." Yet these behaviors more or less parallel expectations within lower-class communities for how young men should act. Within these communities, men who meet these expectations are typically considered dangerous, but not mentally ill, because their behavior is comprehensible. Although they might be labeled criminal, they are unlikely to be labeled mentally ill unless they somehow come to the attention of doctors from outside their communities.

Similarly, many sociologists hypothesize that depression results when traditional female roles cause chronic stress by reducing women's control over their lives (Horwitz, 2002:173–179). Research has found that rates of depression are considerably higher among women with the least control over their lives: nonworking women and married mothers. By the same token, depression is especially common among men who have less power than their wives do, have little control over their work, or lose their jobs.

The Impact of Social Class: Social Stress or Social Drift? Of all the demographic variables researchers have investigated, social class shows the strongest and most consistent impact on mental illness. As social class goes up, the rate of both diagnosable mental illness and psychological distress goes down (Eaton and Muntaner, 1999; Kessler et al., 1994). But does lower social class status cause mental illness, or does mental illness cause lower social class? In other words, do the social stresses associated with lower-class life lead to greater mental disorder, or do those who suffer from mental disorder drift downward into the lower social classes? These two theories are referred to as **social stress** versus **social drift**, respectively.

Researchers interested in social class have focused primarily on schizophrenia, the disease that shows the most consistent relationship to social class; studies have found that schizophrenia and related disorders occur two to five times more often among those who have not graduated from college compared with those who have. Those who favor the social drift argument have shown that, for example, at first admission to a mental hospital, patients diagnosed with schizophrenia hold jobs lower in social class than one would expect given their family backgrounds. This suggests that mental problems caused these individuals to drift downward in social class (Eaton and Muntaner, 1999).

Most research, however, suggests that social stress better explains the link between social class and mental illness (Aneshensel, 2009; Schwartz and Meyer, 2010). For example, those diagnosed with schizophrenia are more likely than others to have grown up in lower-class homes and to have held stressful, noisy, hazardous, and physically uncomfortable jobs even *before* their first admissions to mental hospitals (Link, Lennon, and Dohrenwend, 1993; Muntaner et al., 2004). Recently, this theory has been vividly reinforced by reports of extremely high rates of mental illness among returning war veterans, most of whom come from poor or working-class families (see *Contemporary Issues: Invisible Wounds of War* for more details).

Social Capital and Mental Illness

As Chapter 3 explained, **social capital** refers to the resources available to an individual through his or her social network. Social capital is more common among those with higher social class but affects mental health across ethnic, class, and gender lines (Song, 2011). Not surprisingly, those with more social capital typically report less psychological distress. For example, people in their fifties and sixties often have excellent social capital: extensive connections to friends, neighbors, and relatives who have garnered a wide variety of resources over many years. This may partly explain why people in this age range report less

CONTEMPORARY ISSUES

Invisible Wounds of War

Over the past decade, almost 2 million US troops have served in Iraq or Afghanistan. Because of both military and medical advances, far fewer US soldiers have died there than in previous military engagements. However, longer, more frequent, and often involuntary deployments in a war characterized by suicide bombers and often-invisible explosive devices have taken a heavy psychological toll (Tanielian and Jaycox, 2008). Use of explosive devices also has caused a dramatic increase in traumatic brain injuries, which in turn can cause both psychological and cognitive damage. According to a major report by the nonprofit RAND Corporation (Tanielian and Jaycox, 2008), an extraordinary *30 percent* of returning troops experience disabling post-traumatic stress disorder (PTSD), major depression, or traumatic brain injury (TBI).

Individuals with these conditions are at substantially increased risk for suicide, divorce, unhealthy drug use and sexual activities, and other problems (Tanielian and Jaycox, 2008). Their children and spouses, too, may suffer, as veterans' ability to parent declines and rates of divorce, unemployment, domestic violence, and homelessness increase.

Unfortunately, fear of stigma and of treatment side effects has kept half of those who have neurological or psychiatric problems from seeking medical care. Moreover, only half of those who seek care receive even minimally effective treatment. Given that *not* treating these conditions costs the nation more than treatment would, the RAND report calls for a substantial investment in medical care for PTSD, TBI, and major depression among veterans. The report also calls for increased funding of programs designed to reduce the perceived and real career consequences of seeking such care.

psychological distress than do younger persons. Similarly, although women typically have more friendships than do men, women's friends often have relatively few resources. This may help explain why women are more likely than men to report psychological distress (Song, 2011).

DEFINING MENTAL ILLNESS

As with **disability** and physical illness, doctors and sociologists typically view mental illness in very different ways. In this section, we contrast the medical and sociological models of mental illness. Neither of these models is absolute, however, for both sociologists and doctors often blend elements from each in their work. Nevertheless, the contrast between these two "ideal types" provides a useful framework for understanding the broad differences between the two fields.

The Medical Model of Mental Illness

To doctors and most other clinicians in the field, mental illness is an illness essentially like any other. To understand what this means, it helps to understand the history of medical treatment for syphilis, the disease that first demonstrated the power of medicine to control mental illness and that in many ways established the frame through which doctors would understand all mental illnesses.

Since the fifteenth century, doctors had recognized syphilis as a discrete disease. Because of its mild initial symptoms, however, only in the late nineteenth century did doctors realize the full damage syphilis can inflict on the nervous system, including blindness, deformity, insanity, and death. Unfortunately, doctors could do little to help those with syphilis. The best available treatment consisted, essentially, of poisoning patients with arsenic and other heavy metals in the hopes that these poisons would kill whatever had caused the disease before they killed the patients.

In 1905, scientists first identified the bacterium *Treponema pallidum* as the cause of syphilis. Five years later, Paul Ehrlich discovered the drug Salvarsan as a cure for syphilis. Salvarsan, an arsenic derivative, was the first drug that successfully targeted a specific microorganism. As such, it opened the modern era of medical therapeutics. After this point, those who sought early treatment for syphilis could expect a complete cure, whereas those who put off treatment risked irreversible neurological damage and a horrible death.

The history of Salvarsan and syphilis provided ideological support for a **medical model of mental illness**. This medical model is composed of four assumptions about the nature of mental illness (Scheff, 1984):

1. Objectively measurable conditions define mental illness, in the same way that the presence of a specific bacterium defines syphilis.
2. Mental illness stems largely or solely from something within individual psychology or biology, even if doctors (such as those who studied syphilis before 1905) don't yet know its sources.
3. Mental illness, like syphilis, will worsen if left untreated but may diminish or disappear if treated promptly by a medical authority.
4. Treating mental illness, like treating syphilis, rarely harms patients, so it is safer to treat someone who might really be healthy than to refrain from treating someone who might really be ill.

The Sociological Model of Mental Illness

The sociological model of mental illness questions each of these assumptions (see *Key Concepts: Models of Mental Illness*). Perhaps most important, sociologists argue that definitions of mental illness, like the definitions of physical illness and disability discussed in Chapters 5 and 6, reflect subjective social judgments more than objective scientific measurements of biological problems.

What do we mean when we say someone is mentally ill? Why do we diagnose as mentally ill people as disparate as a teenager who uses drugs, a woman who hears voices, and a man who tries to kill himself? According to sociologist Allan Horwitz (1982), behavior becomes labeled mental illness when persons in positions of power consider that behavior both unacceptable and inherently

KEY CONCEPTS

Models of Mental Illness

The Medical Model	The Sociological Model
Mental illness is defined by objectively measurable conditions.	Mental illness is defined through subjective social judgments.
Mental illness stems largely or solely from something within individual psychology or biology.	Mental illness reflects a particular social setting as well as individual behavior or biology.
Mental illness will worsen if left untreated but may improve or disappear if treated promptly by a medical authority.	Persons labeled mentally ill may experience improvement regardless of treatment, and treatment may not help.
Medical treatment of mental illness can never harm patients.	Medical treatment for mental illness can sometimes harm patients.

incomprehensible. In contrast, we tend to define behavior as crime when we consider it unacceptable but comprehensible; we don't approve of theft, but we understand greed as a motive. (The judgment of "not guilty by reason of insanity" falls on the border between crime and mental illness.) Similarly, we might not understand why physicists do what they do, but we assume that those with appropriate training find their behavior comprehensible.

According to Peggy Thoits (1985), behavior leads to the label of mental illness when it violates **cognitive norms, performance norms**, or **feeling norms**. Someone who thinks he is Napoleon Bonaparte, for example, breaks cognitive norms (i.e., norms regarding how a person should think), and someone who can't hold a job breaks norms regarding proper role performance. Thoits argues that the last category—breaking feeling norms—accounts for most behavior labeled mental illness. Feeling norms refer to socially defined expectations regarding the "range, intensity, and duration of feelings that are appropriate to given situations" and regarding how people should express those feelings (Thoits, 1985:224). For example, laughing is highly inappropriate at a Methodist funeral but perfectly acceptable at an Irish wake, and feeling sad that your pet cat died is considered reasonable for a few days but unreasonable after a year.

Different social groups consider different behaviors comprehensible and acceptable. The friends of a drug-using teenager, for example, might consider drug use a reasonable way to reduce stress or have fun. Their views, however, have little impact on public definitions of drug use. Similarly, members of one church might consider a woman who reports talking to Jesus a saint, whereas members of another church would consider her mentally ill. The woman's fate will depend on how much power these opposing groups have over her life. The definition of mental illness, then, reflects not only socially accepted ideas regarding behavior but also the relative power of those who hold opposing ideas.

Researchers who use this sociological definition of mental illness don't mean to imply that emotional distress does not exist or that people don't feel real pain when they can't meet social expectations for thought, behavior, or emotions. Nor do these researchers mean to imply that biology has no effect on behavior or thought. They do, however, question the purpose and consequences of using medical language to describe such problems, and question why we label certain behaviors and individuals, but not others.

Not all sociologists raise these questions, however. Many, especially those working in health care settings and in **epidemiology**, use a **sociology *in* medicine** approach and use essentially medical definitions of mental illness in their research and writing. Nevertheless, sociologists are united in assuming that mental illness, like physical illness and disability, stems at least partially from social life rather than solely from individual psychology or biology.

The Problem of Diagnosis

The sociological model of mental illness gains credibility when we look at research on the problems with psychiatric diagnosis. These problems became a political embarrassment for psychiatrists (medical doctors who specialize in treating mental illness) after a famous experiment by psychologist David Rosenhan (1973). Rosenhan and seven of his assistants had presented themselves to 12 mental hospitals and complained of hearing voices but otherwise had acted normally. The hospitals diagnosed all eight "pseudopatients" as mentally ill and admitted them for treatment. After they were admitted, all behaved normally, leading 30 percent of the other *patients* to identify them as frauds. The *staff*, however, never noticed anything unusual about these pseudopatients. It took an average of 19 days for them to win their release, with their symptoms declared "in remission."

When these results were published, psychiatrists objected vociferously that the results were some sort of fluke. In response, Rosenhan agreed to send pseudopatients to another hospital and challenged the staff at that hospital to identify the pseudopatients. During the three months of the experiment, the staff identified 42 percent of their new patients as pseudopatients even though Rosenhan really had not sent any!

These two experiments vividly demonstrate the subjective nature of psychiatric diagnosis and its susceptibility to social expectations. Within the context of a mental hospital, staff members quite reasonably assume patients are ill and interpret everything patients do accordingly. When, for example, one bored pseudopatient began taking notes, a worker officially recorded this "note-taking behavior" as a symptom. Conversely, when staff members expected to find pseudopatients, they interpreted similar behaviors as signs of mental health.

The problems with diagnosis are particularly acute when a therapist and patient don't share the same culture. With the rise in immigration to the United States over the past generation, doctors increasingly must diagnose and treat patients whose symptoms don't appear in Western textbooks (Goleman, 1995). For example, a common symptom of psychological problems in the United

States is debilitating fear of being embarrassed by one's body, whereas a common symptom in Japan is debilitating fear (known as *taijin kyofusho*) that one's body will embarrass *someone else*. Similarly, a common symptom of mental illness in the United States is fear of abduction by aliens, whereas a common symptom in Malaysia is *koro*, the intense fear that one's penis and testicles will recede into the body and somehow cause one's death. Current guidelines of the American Psychiatric Association (APA) recommend that psychiatrists consider cultural and ethnic factors in their work and require training programs to cover cross-cultural issues.

The Politics of Diagnosis

Over the years, psychiatrists have worked to reduce problems with diagnosis by refining the definitions of illnesses in the ***Diagnostic and Statistical Manual of Mental Disorders (DSM)***, first published by the APA in 1952. Virtually all psychiatrists use this manual for assigning diagnoses, as do most other clinicians, because insurers usually require a *DSM* diagnosis before they will reimburse clinicians for care.

The *DSM* and the subsequent *DSM-II*, published in 1968, instructed clinicians to reach diagnoses based on the clinicians' inferences about such intrapsychic processes as defenses, repression, and transference. Because clinicians can't measure these processes, the same behavior often elicited quite different diagnoses from different clinicians (Helzer et al., 1977).

Partly because of these problems, in 1974, the APA announced its decision to revise the *DSM-II* (Spitzer, Williams, and Skodol, 1980). Ironically, although the resulting *DSM-III*, published in 1980, was designed to quiet questions about the ambiguities of psychiatric diagnosis, it instead illuminated those ambiguities because its writing became an overtly political battle, involving active lobbying by both professional and lay groups (Grob and Horwitz, 2009). This battle revealed wide differences among clinicians regarding what behaviors signified mental illness, what caused those behaviors, who should treat them, and how they should be treated.

To encourage support for the *DSM-III* and to avoid open political battles among psychiatrists, its authors decided to stress symptomatology and avoid discussing either causation or treatment (Kirk, 1992). In addition, to increase the odds that clinicians would use the *DSM-III*, the authors described the various diagnoses based not on available research, but rather on the consensus among practicing psychiatrists. These two strategies, they hoped, would produce a widely used and highly reliable document. **Reliability** refers to the likelihood that different people who use the same measure will reach the same conclusions—in this case, that different clinicians, seeing the same patient, would reach the same diagnosis. Yet even this modest goal was not achieved because studies continue to find high rates of disagreement over diagnosis (Kirk, 1992; Mirowsky and Ross, 1989). Moreover, reliability in the absence of validity is not particularly useful. **Validity** refers to the likelihood that a given measure accurately reflects what those who use the measure believe it reflects—in this case, that persons identified by the *DSM-III* as having a certain illness actually have that illness. As Phil Brown (1990:393) notes, "Anyone can achieve … reliability by teaching all people the 'wrong' material, and getting them to all agree on it…. The witch trials [of earlier centuries] showed a much higher degree of interrater reliability than any DSM category, yet we would not impute any validity to those social diagnoses."

Despite all of these problems, the *DSM-III* and the subsequent *DSM-IV* gained great support among clinicians because they served a variety of political needs (Horwitz, 2002). By stressing (even if inaccurately) the "objective" nature of diagnosis, clinicians were able to gain respect in the medical world, access to reimbursement from insurance companies, and funding from agencies that sponsor research. By assigning discrete diagnoses to all the different client groups and combinations of symptoms treated by different types of clinicians, they could gain widespread acceptance of the system from both clinicians and clients; the *DSM-IV* contains almost 400 different diagnoses. Finally, a system that emphasized diagnosis and symptoms, rather than underlying causes of illness, both stemmed from and was reinforced by the increasing reliance on drugs as the main treatment for mental illness.

The *DSM-V* was published in 2013. As with previous editions, critics argue that evidence for the validity of diagnostic categories remains limited and that the new edition will lead to further medicalization and overtreatment of everyday life struggles (Frances, 2012). For example, unlike *DSM-IV*, *DSM-V* defines ordinary forgetfulness among older adults, grieving for more than two months after a loved one dies, and what many would describe as "temper tantrums" as new forms of mental illness. Importantly, criticism of this new edition has come from within as well as outside of psychiatry.

A HISTORY OF TREATMENT

The history of treatment for mental illness further reveals the role social values play in medical responses to problematic behavior. In this section, we trace the treatment of mental illness from the prescientific era to the present.

Before the Scientific Era

Although the concept of mental illness is relatively new, all societies throughout history have had individuals whose behavior set them apart as unacceptably and incomprehensibly different. However, premodern societies more often could find informal ways of coping with such individuals (Horwitz, 1982). First, premodern societies could offer acceptable, low-level roles to those whose thought patterns and behaviors differed from the norm. Second, because work roles rarely required individuals to function in highly structured and regimented ways, many troubled individuals could perform at marginally acceptable levels. Third, in premodern societies, work occurred within the context of the family, whether at home or in fields or forests. As a result, families could watch over those whose emotional or cognitive problems interfered with their abilities to care for themselves. These three factors enabled families to **normalize** mental illness by explaining away problematic behavior as mere eccentricity. As a result, unless individuals behaved violently or caused problems for civil authorities, their families and communities could deal with them informally.

In some cases, however, individuals behaved too unacceptably or incomprehensibly for their communities to normalize. In these cases and as is true with all illnesses (as described in Chapter 5), communities needed to find explanations to help them understand why such problems struck some people and not others. Such explanations helped to make the world seem more predictable and safe by convincing the community that such bad things would never happen to "good people" like themselves.

Until the modern scientific age, societies typically viewed disturbing behavior as a punishment for sin or for violating a taboo; a sign that the afflicted individual was a witch; or a result of evildoing by devils, spirits, or witches. Therefore, they assigned treatment to religious authorities—whether shamans, witch doctors, or priests—who relied on prayer, exorcism, spells, and treatments such as bloodletting or trepanning (drilling a hole in the skull to let "bad spirits" out). Religious control of socially disturbing behavior reached a spectacular climax with the witchcraft trials of the fifteenth to seventeenth centuries, during which religious authorities brutally killed at least 100,000 people, including some we would now label mentally ill (Barstow, 1994).

As a capitalist economy began to develop, both religious control and informal **social control** began to decline (Horwitz, 1982; Scull, 1977). Under capitalism, work moved from home and farm to workshops and factories, making it more difficult for families to care informally for problematic relatives. In addition, a capitalist economy could less readily absorb those whose productivity could not be scheduled and regimented. At the same time, widespread migration from the countryside to cities weakened families and other social support systems, as did migration from Europe to the United States in subsequent centuries. Meanwhile, other changes in society weakened religious systems of social control.

These changes fostered a need for new, formal institutions to address mental illness. By the end of the eighteenth century, however, only a few hospitals devoted to treating people with mental illnesses existed along with a few private "madhouses" run by doctors for profit. Instead, most of those we would now label mentally ill were housed with poor people, people with disabilities, and criminals in the newly opened network of public **almshouses**, or poorhouses.

Conditions in both almshouses and madhouses were generally miserable, but they were especially bad for those considered mentally ill. Doctors and the public typically believed that persons with mental illness were incurable and were essentially animals. As a result, institutions treated people with mental illnesses like animals—chaining them for years to basement walls or cells, often without clothing or proper food, and beating them if they caused problems.

Bettmann/Corbis

Benjamin Rush, the "father of American psychiatry," invented this device to treat mental illness through removing distractions from the patient.

The Rise and Decline of Moral Treatment

By the late eighteenth century, however, attitudes toward persons with mental illness began to moderate (Scull, 1989:96–117). In place of punishment and warehousing, reformers proposed **moral treatment**: teaching individuals to live in society by showing them kindness, giving them opportunities to work and play, and in general treating mental illness more as a moral than a medical issue. The stunning successes that resulted convinced the public that mental illness was curable. The first American hospital designed to provide moral treatment, the Friends' (or Quakers') Asylum, was founded in 1817.

Despite this strong beginning, moral treatment in the end could not compete with medical models of mental illness (Scull, 1989:137–161). Because those who promoted moral treatment continued to use the language of medicine, talking of illnesses and cures, medical doctors could argue successfully that only they should control this field. In addition, because moral treatment required only kindness and sensitivity, which theoretically any professionals could offer, no professional group could claim greater expertise than doctors. As a result, by 1840, doctors largely had gained control over the field of mental illness both in the United States and Europe.

As care gradually shifted from laypersons to doctors, custodial care began to replace moral treatment. This shift reflected the growing belief that illness was genetic and untreatable, as well as the public's greater interest in *controlling* people with mental illnesses—especially if they were poor, nonwhite, or immigrant—than in *treating* them.

By the 1870s, moral treatment had been abandoned. Yet the number of mental hospitals continued to grow exponentially (Rothman, 1971). Historians refer to this change, and the similar but earlier developments in Europe, as the **Great Confinement**.

The Great Confinement drew energy from the well-meaning efforts of reformers to close down the brutal almshouses and to provide facilities specifically designed to care for people with mental illnesses instead of warehousing them with criminals, persons with disabilities, and poor people (Sutton, 1991). Because no agreed-upon definitions of mental illness existed, however, families and communities found it relatively easy to move the troublesome, poor, old, or sick into the newly established mental hospitals (Sutton, 1991). As a result, most of those labeled mentally ill continued to find themselves housed with others whom society had rejected. The only difference was that instead of residing in institutions filled with a varied group of marginalized individuals, they now lived in large institutions officially devoted to the "care" of people with mental illnesses.

Freud and Psychoanalysis

By the beginning of the twentieth century, then, doctors controlled the mental illness field. Yet doctors were deeply divided between those who assumed mental illness stemmed from psychological causes and those who assumed it had biological causes.

This split grew wider with the rise of Freudian psychiatry. According to Sigmund Freud, a Viennese doctor, to become a mentally healthy adult, one had to respond successfully to a series of developmental issues. Each issue occurred at a specific stage, with each stage linked to biological changes in the body and invested with sexual meanings. For example, Freud believed that during the phallic stage (between about ages 3 and 6) boys naturally begin noticing genitalia, experiencing sexual attraction toward their mothers, and therefore viewing their fathers as rivals. To become healthy adults, he argued, boys had to conclude that girls lack penises because their fathers castrated them after some wrongdoing. To avoid the same fate, boys must abandon their attraction for their mothers and instead pursue their fathers' love by adopting their fathers' values. Through this process, Freud argued, boys develop a strong internal sense of morality—something that girls, lacking penises, can never do.

Freud based this theory on his interpretations of the lives and dreams of his upper-middle-class patients; no scientific data underpin this theory. Looking back at this theory from the present, it is hard to comprehend how anyone could have believed in such notions. Yet the theory remained popular for decades, undoubtedly because it both reflected and supported popular ideas about men's superior bodies, intellect, and moral virtues.

For those who accepted Freud's theory, the only way to cure mental illness was to help patients resolve their developmental crises. To do so, Freud and his followers relied on psychoanalysis, a time-consuming and expensive form of psychotherapy geared to patients without major mental illnesses. In psychoanalysis, patients recounted their dreams and told a (usually silent) therapist whatever came to mind for the purpose of recovering hidden early memories and understanding their unconscious motivations.

Because psychoanalysis was so costly, most mental patients instead received far cheaper physical interventions such as electroconvulsive (shock) therapy or lobotomies (Valenstein, 1986). Neither therapy received scientific testing before becoming popular and both could cause brain damage (Valenstein, 1986). At any rate, therapy of any sort occupied only a minuscule proportion of patients' time in mental hospitals. Instead, patients spent their days locked in crowded wards with little other than radios or, later, televisions to ease their boredom.

The Antipsychiatry Critique

By the middle of the twentieth century, mental hospitals had become a huge and largely unsuccessful system (Mechanic, 1989). Patients with mental illnesses occupied half of all hospital beds in the United States. Virtually all (98%) were kept in public mental hospitals; insurance rarely covered mental health care, so private hospitals had no interest in the field. At their peak in 1955, public mental hospitals held 558,000 patients, for an average of eight years. Most were involuntarily confined and involuntarily treated, often with lobotomies as well as drugs that kept them highly sedated.

Beginning in the 1960s, however, many challenged this system, as the civil rights, antiwar, and feminist movements all promoted both individual rights and questioning authority. These ideas contributed to a growing critique of mental health treatment by sociologists, psychologists, and even some psychiatrists such as R. D. Laing (1967).

One of the most powerful critiques of large mental institutions appeared in a classic study by sociologist Erving Goffman (1961). Goffman's work fell within the tradition of **symbolic interactionism** theory. According to this theory, individual identity develops through an ongoing process in which individuals see themselves through the eyes of others and learn through social interactions to adopt the values of their community and to measure themselves against those values. In this way, a **self-fulfilling prophecy** is created, through which individuals become what they are already believed to be. So, for example, children who constantly hear that they are too stupid to succeed in school might conclude that it is senseless to attend classes or study. As a result, they fail in school, thus fulfilling the prophecies about them.

Goffman used symbolic interactionism theory to analyze mental hospitals and the experiences of mental patients. He pointed out that mental hospitals, like the military, prisons, and monasteries, were **total institutions**—institutions where a large number of individuals lead highly regimented lives segregated from the outside world. Goffman argued that these institutions necessarily produced

mortification of the self. Mortification refers to a process through which a person's self-image is damaged and is replaced by a personality adapted to institutional life.

Several aspects of institutional life foster mortification. Persons confined to mental hospitals lose the supports that usually give people a sense of self. Cut off from work and family, these individuals' only available role is that of patient. That role, meanwhile, is a **master status**—a status considered so central that it overwhelms all other aspects of individual identity. Within the mental hospital, a patient is viewed solely as a patient—not as a mother or father, husband or wife, worker or student, radical or conservative. According to Goffman's observations, and as in Rosenhan's (1973) experiment, all behavior becomes interpreted through the lens of illness. In addition, because each staff member must manage many patients, staff members lack the time to individualize care. In these circumstances, patients typically lose the right to choose what to wear, when to awaken or sleep, when and what to eat, and so on. Moreover, all of these activities occur in the company of many others. Individuals thus not only experience a sense of powerlessness but also can lose a sense of their identity—their desires, needs, and personalities—in the mass of others. As a result, patients experience **depersonalization**—a feeling that they no longer are fully human, or no longer are considered fully human by others. At the same time, the hierarchical nature of mental hospitals reinforces the distinctions between inmates and staff and constantly reminds all parties of the gulf between them. Consequently, patients can avoid punishment and eventually win release only by stifling their individuality and accepting the institution's beliefs and rules. These forces producing mortification are so strong that even Rosenhan's pseudopatients—knowing themselves sane and hospitalized only briefly—experienced depersonalization.

Implicit in Goffman's work is the idea that mental hospitals may be one of the worst environments for treating mental problems. Later research supports this conclusion. A review of ten **controlled** studies on alternatives to hospitalization, including halfway houses, day care, and supervised group apartments, found that all could boast equal or better results than those of traditional hospitalization, as measured by subsequent employment, reintegration into the community, life satisfaction, and extent of symptomatology (Kiesler and Sibulkin, 1987).

Deinstitutionalization

By the time the antipsychiatry critique appeared, the Great Confinement already had begun to wane. Beginning in 1955, the number of mental hospital inmates declined steadily as treatment shifted from **inpatient** care (in hospitals) to **outpatient** care. This process of moving mental health care away from large institutions, known as **deinstitutionalization**, gained further support during the 1970s, as mental patients successfully fought in the courts against involuntary treatment, against hospitals that provided custodial care rather than therapy, and for the right to treatment in the "least restrictive setting" appropriate for their care.

Those who adopt a medical model of illness typically assume that deinstitutionalization resulted from the introduction, beginning in 1954, of various drugs that were believed to reduce severe psychiatric symptoms. Yet the number of patients in public mental hospitals did not fall rapidly until more than a decade after these drugs were introduced. In fact, the sharp decline in institutionalization is better explained by changes in federal funding that enabled some chronically mentally ill persons to live on their own while encouraging **nursing homes** and private hospitals to seek other chronically mentally ill persons as residents (Mechanic, 1999; Mechanic and Rochefort, 1990).

Unfortunately, the promise that deinstitutionalization would herald a new era in which individuals would receive appropriate therapy in the community, avoiding the **stigma**, degradation, and mortification of mental hospitalization, has been met only partially. Few services are available outside of psychiatric hospitals to help individuals who have serious mental illnesses, and only 40 percent receive even minimally adequate treatment (Kessler et al., 2005b).

Meanwhile, as government funding for the nation's health care system declined, funding for the criminal justice system dramatically *increased* (Human Rights Watch, 2009). As a result, public mental hospitals now find that they can most easily pay their bills by accepting patients sent to them by the criminal justice system, whether mentally ill prison inmates, people found innocent by reason of insanity, or "sex predators," who under recent laws can be involuntarily confined even after finishing their sentences. These changes reflect the continuing stigma of mental illness. *Ethical Debate: Mental Illness and Gun Control* further illustrates the consequences of that stigma.

Simultaneously with these changes, the federal government also reduced funding for low-income housing. As a result, many mentally ill persons who can't afford treatment also can't find housing. Consequently, many persons with chronic mental illness now cycle between homelessness, jail time when they prove troublesome for local authorities, and brief stays in public mental hospitals (Earley, 2007); according to the US Department of Justice, more than half of all jail and prison inmates have a mental illness (James and Glaze, 2006).

Despite these severe gaps in our mental health system, however, observers generally agree that deinstitutionalization improved the quality of life for most seriously mentally ill persons (Grob, 1997).

The Remedicalization of Mental Illness

Since the 1980s, mental illness has undergone increasing **remedicalization** (Carlat, 2010; Kirsch, 2011; Leo, 2004). Psychiatrists have developed new techniques for diagnosis and treatment as well as new theories of mental illness that downplay any social causes and instead stress biochemical, neurological, or genetic abnormalities. Meanwhile, the mass media along with the medical and pharmaceutical establishments have "sold" the medicalization of mental illness to the public (Conrad, 2005); the majority of the public now believe that mental illness is a biological problem (Pescosolido, 2013).

ETHICAL DEBATE

Mental Illness and Gun Control

In May 2014, Elliot Rodger went on a killing spree in Isla Vista, California, a community largely populated by students at the nearby University of California–Santa Barbara. Within an hour, he fatally stabbed three people and shot another three to death before using his gun to kill himself.

Although Rodger had never been diagnosed with a mental illness, he had received psychological counseling on and off. Not surprisingly, after the fact many attributed his homicidal and suicidal actions to mental illness and many (including the National Rifle Association) responded by calling for stricter laws (or stricter enforcement of laws) to keep guns out of the hands of persons with mental illnesses.

In fact, federal law already forbids persons who have ever been declared mentally ill from owning guns. However, compliance with the law—which is based on voluntary reporting by mental health professionals to a national database—is low. But murder rates in the United States are far too high, and anything we can do to lower those rates would appear to be the ethical thing to do.

On the other hand, forbidding gun ownership based on a history of mental illness raises various ethical (and practical) issues. Most importantly, most persons with mental illness have never committed a violent act. In fact, mental illness is a very poor predictor of violence (Kangas and Calvert, 2014; McGinty et al., 2014). Instead, the best predictors of violence are behavioral factors such as arrests for domestic violence, assault, or dangerously inappropriate behavior while drunk (Kangas and Calvert, 2014; McGinty et al., 2014).

In addition, if mental health professionals are legally forced to report individuals to a national database, potential clients may shy away from treatment. Meanwhile,

Yet the data for the "biological revolution" in mental health is weak. Despite decades of research, scientists have failed to find evidence demonstrating any brain abnormalities that might explain mental illness (Brown, 1990; Carlat, 2010; Kirsch, 2011; Leo, 2004). Moreover, to the extent that brains of those labeled mentally ill differ from those of other people, the differences appear to be caused either by the drugs used to treat mental illness or by other factors such as poor nutrition.

Despite these weaknesses in the biological model of mental illness, most psychiatrists have adopted it (Whitaker, 2011). As a result, psychiatrists now present a more united front in their struggles for control against other mental health occupations such as psychology and social work. In addition, they have increased their political power relative to these other occupations because, having declared mental illness a biological problem, they now can argue that only persons trained in medicine can properly diagnose and treat it (Brown, 1990).

Reflecting this medical model, both psychiatrists and psychologists now often rely on psychoactive drugs not only to treat mental illness but also to diagnose it (Grob and Horwitz, 2009). For example, doctors now commonly diagnose patients with clinical depression whenever patients respond favorably to antidepressant drugs such as Prozac, even if the patients don't meet standard criteria for that diagnosis. Yet most people feel better whenever they take a mood-enhancing drug, whether it is Prozac or cocaine.

those already in treatment may feel betrayed if they learn that their doctors or counselors reported them, and may therefore break off treatment. Yet treatment can help individuals to recover from mental illness and reduce suicidal or homicidal thoughts (Kangas and Calvert, 2014).

Finally, laws that link mental illness with violence can increase the stigma experienced by those who live with it. Moreover, that stigma may be lifelong: It is hard to imagine how any national database could institute procedures for reliably identifying individuals who have recovered from mental illness and for removing those individuals' names from the database. Certainly, no such procedures exist currently. Thus legal requirements to report clients would in some cases require health care workers to break professional codes of ethics that require them to preserve confidentiality and to avoid harming their clients (Kangas and Calvert, 2014).

Sociological Questions

1. What social views and values about mental illness are reflected in laws and proposals aimed at keeping guns away from those ever diagnosed with it?
2. Who benefits from these laws and proposed laws? Why would the National Rifle Association favor them? Why aren't there more proposals to forbid gun ownership among those convicted of domestic violence or of drunken behavior?
3. What are the intended consequences of prohibiting gun ownership among individuals ever diagnosed with mental illness? What are the unintended social, economic, political, and health consequences of this policy?

Most of the drugs now used to treat persons with mental illness fall into one of three main categories: antipsychotics, mood stabilizers, and antidepressants. The use of antidepressants has grown particularly rapidly, especially among women, as Figure 7.1 shows (National Center for Health Statistics, 2014a).

Compared to older drugs for depression, the most common current antidepressants—known as for selective serotonin reuptake inhibitors (SSRIs)—carry fewer side effects and can't be taken to commit suicide, although they *increase* the odds of committing suicide by some other means. Moreover, statistical analyses based on *all* data from research on SSRIs—rather than only on data that pharmaceutical companies choose to publish—suggest that the drugs may offer no benefits at all: When patients are given placebos that carry side effects (and thus convince them that they are taking drugs), the placebos and drugs are both (slightly) effective (Kirsch, 2011). Moreover, in the long run the drugs may cause harm by changing the brain's ability to moderate emotions (Whitaker, 2011). Nevertheless, drug companies have proven highly successful at convincing both consumers and clinicians to redefine normal shyness as "social anxiety disorder" and to believe that it is best treated with SSRIs (Abramson, 2004; Lane, 2007).

Serious questions have also been raised about the increased use of antipsychotics. These drugs are now the top-selling drugs in the United States despite potential side effects that include depression, drooling, and severe weight gain.

FIGURE 7.1 Antidepressant Use in the past 30 days, United States*

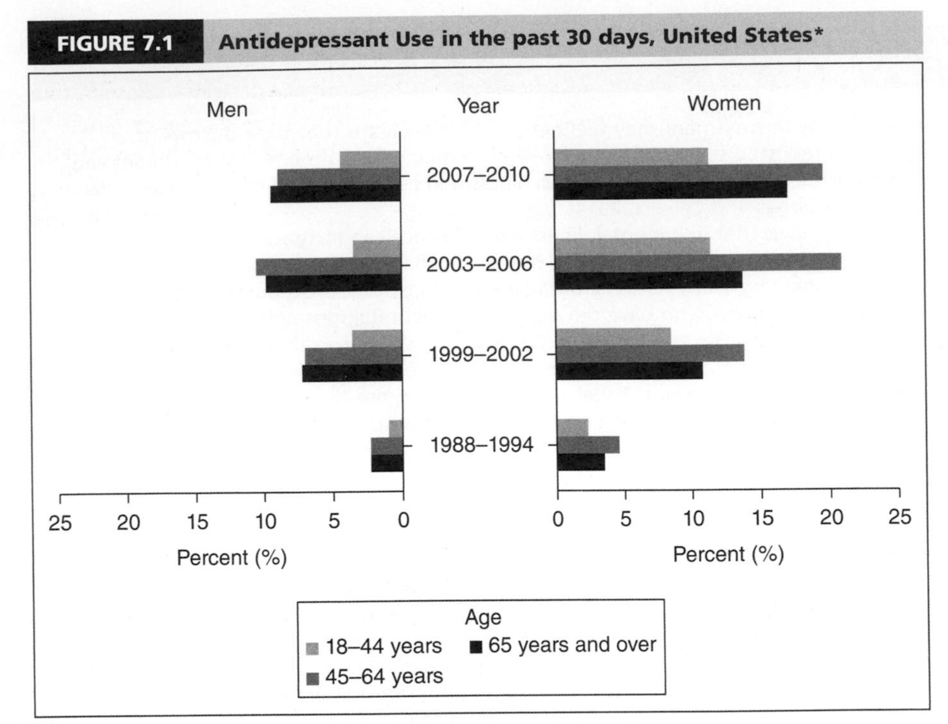

* Age-adjusted data, for civilian non-institutionalized population only.

SOURCE: National Center for Health Statistics (2014a).

Moreover, doctors increasingly are prescribing them to children, even though they have only been scientifically tested on adults. Nevertheless, between 2000 and 2007, usage of antipsychotics doubled among children under age five who had private insurance, although most experts believe that the diseases these drugs were designed to treat (such as schizophrenia) rarely begin before adolescence (Wilson, 2010a).

The Rise of Managed Care

Beginning in the 1990s and in response to consumer pressure, insurance coverage for mental illness became considerably more common. Coverage increased further with passage in 2010 of the Affordable Care Act (discussed further in Chapter 8), although it remains far less common than coverage for physical illness.

Most coverage for mental health care is offered through **managed care organizations (MCOs)**. Managed care is described more fully in Chapter 8 but essentially refers to any system that controls health care spending by closely monitoring where patients receive health care, what sorts of providers patients use, what treatments they receive, and with what consequences.

Research suggests that managed care can reduce the costs of mental health treatment, at least for less severe illnesses, by encouraging shorter rather than longer inpatient stays, outpatient rather than inpatient care, conservative rather than aggressive interventions, and use of lower-level clinicians (such as social workers) rather than psychologists or psychiatrists (Mechanic, 1995, 1999:160–162). According to David Mechanic, probably the most influential sociologist in the area of mental health care, it also may be able to improve the quality of care:

> By reducing inpatient admissions and length of stay, managed care programs potentially make available considerable resources for substitute services and other types of care. Managed care provides incentives to seek closer integration between inpatient and outpatient and primary and specialized services to achieve cost-effective substitutions.
>
> Managed care also offers the potential to bring ... science-based mental health care into the mental health system more quickly than traditional programs.... Many individual practitioners resist practice guidelines and scientific findings, preferring their own clinical experience, but managed care can put systems in place to measure performance and to enforce adherence to established standards (1997:45–46).

But managed care also carries risks. The emphasis on cost containment inherent in managed care has affected who offers mental health services, for how long, and of what type (Scheid, 2001). MCOs encourage the use of clinicians who charge less per hour, preferring those with master's degrees to those with doctorates and preferring those with doctorates to those with medical degrees. To further restrain costs, MCOs press clinicians to offer only short-term treatment of immediate problems rather than long-term treatment of underlying problems. As a result, therapists increasingly prescribe medications even if "talk therapies" might be more useful.

Managed care also has affected how mental disorders are diagnosed. One way managed care controls costs is by deciding in advance, based on data from past patients, how much and what type of care patients with specific diagnoses should receive. For this system to work, clinicians must assign a diagnosis to each patient. This in turn reinforces the medical model of mental illness and the idea that every person who seeks mental health services has a specific, diagnosable mental illness.

At the same time, to contain costs, MCOs are trying to curtail the breadth of the diagnostic system (Horwitz, 2002). Because each successive edition of the *DSM* has included more diagnoses, more individuals have become eligible for mental health care with each edition. For this reason, MCOs often oppose new diagnoses or any broadening of the criteria for existing diagnoses. For example, some MCOs deny treatment to individuals who have fewer than five symptoms on a depression checklist even if individuals' listed symptoms are severe and even if they have other, unlisted symptoms.

THE EXPERIENCE OF MENTAL ILLNESS

The previous sections described the nature, causes, distribution, and history of mental illness. Next, we look at the experience of mental illness.

Becoming a Mental Patient

As already noted, in any given year, 31 percent of working-age adults experience a diagnosable mental illness, but only 40 percent of these receive even basic treatment (Kessler et al., 2005a, 2005b). Ironically, as the stigma among the middle class against seeking counseling for minor problems has diminished and insurance coverage has increased, treatment has increased among basically well-functioning individuals who experience situational stress, sadness, or lowered self-esteem (Kessler et al., 2005b). Nearly half of those who receive outpatient treatment have no mental disorder that can be identified through surveys, although some of these might have disorders that could be identified by clinicians (Kessler et al., 2005b).

What explains this discrepancy between experiencing symptoms and receiving treatment? According to Allan Horwitz, "Symptoms of mental disorder are usually vague, ambiguous, and open to a number of varying interpretations.... Labels of 'mental illness,' 'madness,' or 'psychological disturbance' are applied only after alternative interpretations have failed to make sense of the behavior" (1982:31). The key question, then, is how and why does this happen?

Self-Labeling Regardless of how others define their situation, at least initially, individuals usually define themselves as mentally healthy. They downplay their situations as "problems" rather than illness; offer alternative explanations for their behavior and emotions ("anyone would be angry in these circumstances"); and reject any psychiatric diagnoses offered by friends, family members, or mental health workers as inaccurate or biased (Moses, 2009). For example, one 14-year-old boy who is currently receiving treatment told an interviewer:

> I have problems I need to work through, but other than that, I'm fine. Just regular family problems and daily personal problems. I don't consider those big issues where I need to take medication for it. Other people see that for me, but I think the medication is just making it worse (Moses, 2009:574).

Peggy Thoits (1985) offers a detailed model of how individuals avoid labeling themselves as mentally ill. According to Thoits, and as described earlier, most of the behavior that can lead to the label of mental illness involves inappropriate feelings or expressions of feelings. To avoid the label of mental illness, therefore, individuals can attempt to make their emotions match social expectations through what Arlie Hochschild (1983) refers to as **feeling work**.

Feeling work can take four forms. First, individuals can change or reinterpret the situation that is causing them to have feelings others consider inappropriate.

For example, a working woman distracted from her work by worries about how to care for an ill parent—and distracted while with her parent by worries about her work—can quit her job. Second, individuals can change their emotions physiologically through drugs, meditation, biofeedback, or other methods. The woman with the ill parent, for example, could drink alcohol or take Prozac to control anxiety. Third, individuals can change their behavior, acting as if they feel more appropriate emotions than they really do. Fourth, individuals can reinterpret their feelings, telling themselves, for example, that they only feel tired rather than anxious.

When feeling work succeeds, individuals can avoid labeling themselves mentally ill. This is most likely to happen when the situations causing the emotions are temporary and brief and when supportive others legitimize their emotions. If, for example, the woman with the ill parent has similarly situated friends who describe similar emotions, she might conclude that her emotions are understandable and acceptable.

Conversely, individuals are more likely to label themselves ill when their emotions repeatedly fail to meet social expectations and others repeatedly label them as mentally ill. Ironically, individuals also may label themselves mentally ill when they succeed too well at feeling work. For example, those who rely too heavily on drugs to manage their feelings can lose control of their lives, and those who consistently reinterpret their emotions—telling themselves that they are not angry, for example, even while punching a wall or a spouse—can find that others label them crazy when their emotions and behavior don't match. Finally, those who consistently engage in feeling work can lose the ability to interpret their feelings accurately and experience them fully. The resulting sense of numbness and alienation eventually can lead individuals to define themselves as mentally ill.

Labeling by Family, Friends, and the Public Like individuals, families only reluctantly label their members mentally ill (Horwitz, 1982). Instead, families can deny that a problem exists by convincing themselves that their relative's behavior does not depart greatly from the norm. If they do recognize that a problem exists, they can convince themselves that their relative is lazy, a drunkard, "nervous," responding normally to stress, or experiencing physical problems rather than mental illness. Finally, families might recognize that their relative is experiencing mental problems, but define those problems as temporary or unimportant.

Two factors explain how and why families can ignore for so long behavior that others would label mental illness. First, those who share cultural values, close personal relationships, and similar behavior patterns have a context for interpreting unusual behavior and therefore can interpret behavior as meaningful more easily than outsiders could. Second, families often hesitate to label one of their own for fear others will reject or devalue both the individual and the family. As a result, families have a strong motive to develop alternative and less stigmatizing explanations for problematic behavior.

Moreover, even when relatives and other intimates define an individual as mentally ill, they don't necessarily bring the individual to treatment. Instead, they can continue to protect the individual against social sanctions through a process Lynch (1983) refers to as **accommodation**. Accommodation refers to "interactional techniques that people use to manage persons they view as persistent sources of trouble" and to avoid conflict, such as humoring problematic individuals or minimizing contact with them (Lynch, 1983:152).

Nevertheless, despite these attempts to normalize and accommodate mental illness, families and friends may eventually conclude that an individual needs treatment. At that point, they must either get the individual to agree or coerce the individual into getting treatment despite his or her active resistance. One study of persons seeking care for a serious mental illness for the first time found that 42 percent had actively sought care and 23 percent had been coerced (Pescosolido, Gardner, and Lubell, 1998). Coercion was most common among those with bipolar disorder, who often enjoyed the "highs" of mania even though others regarded them as seriously disturbed, and among those with large, tight social networks. In another 31 percent of cases, families "muddled through": Either the individuals went along with treatment decisions made by others without accepting or rejecting those decisions, or no one in the family seemed to have been in charge of the decision-making process.

Labeling by the Psychiatric Establishment Once individuals enter treatment, a different set of rules applies, for whereas the public tends to normalize behavior, mental health professionals tend to assume illness. First, because the medical model of mental illness stresses that treatment usually helps and rarely harms, it encourages mental health workers to define mental illness broadly. Second, because mental health workers see prospective patients outside of any social context, behavior that might seem reasonable in context often seems incomprehensible. This is especially likely when mental health workers and prospective patients come from different social worlds, whether because they differ in gender, ethnicity, social class, or some other factor. Third, mental health workers assume that individuals would not have been brought to their attention if they did not need care. Finally, because normalization and accommodation are so common, mental health workers often don't see individuals until the situation has reached a crisis, making it relatively easy to conclude that the individuals are mentally ill.

The Post-Patient Experience

Research on the post-patient experience has focused on assessing the balance between the long-term positive benefits of treatment and the long-term harm caused by stigma. The latter is a critical issue because it challenges the medical model's assumption that psychiatric treatment is benign.

Studies consistently find that those considered mentally ill (including former mental patients) are often feared and rejected by others (Link et al., 1999; Moses, 2009; Pescosolido, 2013). Moreover, individuals who accept psychiatric diagnoses for themselves typically experience greater depression, lower self-esteem, and increased social isolation (Link et al., 1999; Moses, 2009). This does not, however, mean that treatment is not worth it: Substantial evidence suggests that both psychotherapy and drug treatment can reduce symptoms and prevent relapse, at least in the short term (Link et al., 1997). Nevertheless, it does seem that the negative effects of stigma partially cancel out the positive effects of treatment (Link et al., 1997; Rosenfield, 1997). These results led Bruce Link and his colleagues to conclude that health care providers need to "address stigma in its own right if they want to maximize the quality of life for those they treat and maintain the benefits of treatment beyond the short term" (1997:187).

IMPLICATIONS

In this chapter, we have compared the sociological and medical models of mental illness. As with the medical models of physical illness and disability discussed in Chapters 5 and 6, the medical model of mental illness asserts that mental illness is a scientifically measurable, objective reality, requiring prompt treatment by scientifically trained personnel. As such, this model downplays the role of social and moral values in the definition and treatment of mental illness and the effect of mortification and stigma on those who receive treatment.

In this second decade of the twenty-first century, we find ourselves facing a situation uncomfortably similar to that of past centuries. As in the years before the Great Confinement, thousands of mentally ill persons now live on the streets and support themselves at least partly by begging. Many more are confined in nursing homes, jails, or prisons in the same way that earlier societies confined persons with mental illness in almshouses. Although drugs largely have replaced shackles, society still allocates far too few resources to provide humanely for those with mental illnesses. Similarly, although the Affordable Care Act aims to expand access to care for mental illness, that Act will have no impact on the many people who continue to lack insurance coverage. Thus, we can only hope that in the future, with a greater understanding of the nature of mental illness and of the social response to it, we can develop more compassionate and effective means of coping with mental illness.

SUMMARY

1. All societies, from simple to complex, contain some individuals who behave in ways considered unacceptable and incomprehensible and who, in our society, might be labeled mentally ill.
2. During the course of a year, approximately one-third of working-age adults experience a diagnosable mental illness, with one-fifth experiencing a moderate or severe disorder.

3. Ethnicity has little effect on rates of *major* mental illnesses. However, African Americans are *less* likely than whites to develop anxiety or mood disorders but *more* likely to report psychological distress, perhaps because of the stresses imposed by racism. Hispanic Americans are less likely than whites to develop anxiety disorders, mood disorders, and substance abuse problems, perhaps because strong extended families protect against chronic stress.
4. Perhaps because of gender socialization, men consistently display higher rates of substance abuse and personality disorders, but women consistently display higher rates of depression and anxiety disorders.
5. Rates of both diagnosable mental illness and psychological distress increase as social class decreases. Research suggests that occasionally mental illness can cause individuals to drift into the lower classes, but much more often the chronic stresses of lower-class life *lead* to mental illness. Chronic social stress predicts mental illness considerably better than does acute stress, such as life events.
6. Psychological distress is less common among those with more social capital: resources available to individuals through their social network.
7. According to the medical model of mental illness, (1) objectively measurable conditions define mental illness; (2) mental illness stems largely or solely from something within individual psychology or biology; (3) mental illness will worsen if left untreated but is likely to lessen if treated promptly by a medical authority; and (4) treating mental illness rarely if ever harms patients.
8. The sociological model of mental illness argues that definitions of mental illness reflect subjective social judgments regarding whether behaviors are acceptable and understandable. Behaviors are labeled mental illness when they contravene cognitive norms, performance norms, or feeling norms.
9. Research suggests that psychiatric diagnoses are neither valid nor reliable and that the psychiatric diagnostic system has developed through an overtly political process.
10. Premodern societies often could find informal ways of coping with individuals we would consider mentally ill. When they could not do so, they typically blamed the problem on supernatural forces. The development of a capitalist economy fostered a need for new, formal, social institutions to address mental illness. The nineteenth century's "moral treatment" movement aimed to improve conditions at those institutions.
11. According to Sigmund Freud, mental illness occurred when children did not respond successfully to a series of early childhood developmental issues linked to the biological body. Freudian analysis was not based in scientific research and proved too costly to implement in large hospitals.
12. By the mid-twentieth century, most mental hospitals were huge, depersonalizing, "total institutions," which could worsen patients' mental health. The dramatic drop in inpatient censuses at these hospitals is referred to as deinstitutionalization. Deinstitutionalization stemmed primarily from

changes in federal funding rather than from improvements in medical treatment.

13. Mental health is currently undergoing remedicalization through new psychiatric techniques for diagnosis and treatment and new theories that blame mental illness on individual biological abnormalities.
14. Managed care organizations control health care spending by closely monitoring patient care. They can improve care by promoting the best, most cost-effective treatments but can worsen care by pressing clinicians to offer only short-term, drug-based treatment.
15. Rates of mental health treatment are highest among those who experience minor emotional problems or stress rather than significant mental illness. Persons with serious mental illness avoid seeking treatment when both they and their families define their behavior as comprehensible and can accommodate to that behavior. In contrast, mental health professionals tend to assume illness rather than health when they examine unusual individual behavior.
16. Although treatment can help, its benefits are reduced by the harm caused by the social stigma of mental illness.

REVIEW QUESTIONS

1. How and why do ethnicity, gender, and social class affect rates of mental illness?
2. What is the relationship between acute stress and mental illness? Between chronic stress and mental illness?
3. What are the differences between the medical and sociological models of mental illness?
4. What are the problems embedded in psychiatric diagnoses?
5. How did premodern societies respond to and cope with individuals we would now consider mentally ill?
6. What was moral treatment, and why did it fail?
7. What was the antipsychiatry critique?
8. What were the sources and consequences of deinstitutionalization?
9. What is the remedicalization of mental illness?
10. How is managed care affecting the treatment and experience of mental illness?
11. How do individuals become mental patients? How do they *avoid* becoming mental patients?
12. What are the consequences of labeling an individual mentally ill?

CRITICAL THINKING QUESTIONS

1. Explain how lower social class status can cause mental illness and how mental illness can cause lower social class status.
2. What similarities and what differences are there, in both causes and consequences, between the moral treatment movement of the nineteenth century and the deinstitutionalization movement of the twentieth century?
3. The APA's manual of mental illnesses includes an illness called Nicotine Dependence. It refers to persons who both want to stop smoking and who have tried unsuccessfully to stop smoking. Describe two possible harmful consequences and two possible beneficial consequences of medicalizing this situation.

PART III

Health Care Systems, Settings, and Technologies

In Part II, we looked at illness primarily from the perspective of the ill individual. In this part, we move to a *macro*sociological level to look at health care systems and settings. In Chapter 8, we consider the history and current nature of the US health care system. We examine why millions of Americans still find themselves uninsured or threatened with bankruptcy because of medical bills, as well as recent efforts to reform the US health care system. Chapter 9 begins by presenting a series of measures useful for evaluating any health care system and then uses these measures to explore health care systems in six other nations—Germany, Canada, Great Britain, the People's Republic of China, Mexico, and the Democratic Republic of Congo. Finally, in Chapter 10, we investigate the major settings in which health care is offered in the United States (other than individual doctors' offices) and the increasingly important role technology plays in those settings, as it simultaneously helps solve old problems and creates new ones.

CHAPTER

8

Health Care in the United States

Bloomberg/Getty Images

LEARNING OBJECTIVES

After reading this chapter, students should be able to:

- Discuss the history of health insurance in the United States.
- Understand the basic elements of the Affordable Care Act and its likely impact.
- Explain why health care in the United States is so costly.
- Describe the gaps in US health insurance coverage and the consequences of those gaps.

In March 2010, Congress passed the Patient Protection and Affordable Care Act, followed a few days later by the Health Care and Education Reconciliation Act of 2010. These acts are commonly referred to jointly as *Obamacare* or, more neutrally, as the **Affordable Care Act (ACA)**. Supporters argued that the acts would significantly reform the US health care system. Yet that system remains in crisis, as Peter Drier's story illustrates:

> *Before his three-hour neck surgery for herniated disks in December, Peter Drier, 37, signed a pile of consent forms. A bank technology manager who had researched his insurance coverage, Mr. Drier was prepared when the bills started arriving: $56,000 from Lenox Hill Hospital in Manhattan, $4,300 from the anesthesiologist and even $133,000 from his orthopedist, who he knew would accept a fraction of that fee.*
>
> *He was blindsided, though, by a bill of about $117,000 from an "assistant surgeon," a Queens-based neurosurgeon whom Mr. Drier did not recall meeting.... In Mr. Drier's case, the primary surgeon, Dr. Nathaniel L. Tindal, had said he would accept a negotiated fee determined through Mr. Drier's insurance company, which ended up being about $6,200. (Mr. Drier had to pay $3,000 of that to meet his deductible [the amount his insurance requires him to pay out of pocket].) But the assistant, Dr. Harrison T. Mu, was out of network and sent the $117,000 bill.*
>
> *"I thought I understood the risks," Mr. Drier, who lives in New York City, said later. "But this was just so wrong—I had no choice and no negotiating power" (Rosenthal, 2014a).*

The most basic element in any nation's health care system is how it provides and pays for health care. As Peter's story illustrates, however, the United States is the only **more developed nation** that does not guarantee affordable health care to its citizens. Nor, despite this chapter's title, does it really have a health care system. Instead, an agglomeration of public and private health care insurers (such as

Medicaid and Aetna), health care providers (such as doctors and physical therapists), and health care settings (such as hospitals and nursing homes) function autonomously in myriad and often-competing ways.

In this chapter, we first look at the origins of the US health insurance system. We then analyze two current crises in US health care: rising costs and lack of access. Finally, we explore the nature and the impact of the health care reforms passed in 2010.

A HISTORY OF US HEALTH INSURANCE

For most of US history, most Americans paid for their health care out of pocket. The upper class could buy any health care they wanted, the middle class could afford most needed health care, the poor mostly went without, and few questioned the system. But during the Great Depression of the 1930s, millions of Americans lost their jobs, savings, and the ability to pay for medical care. This financial crisis led to growing calls to adopt a national health care system such as those that had recently emerged in Western Europe.

Unlike in Europe, however, proposals for a national health system were stymied by **stakeholder mobilization**: organized political opposition by groups with vested interest in the outcome (Quadagno, 2005; Hoffman, 2012). This opposition came from numerous sources. For example, labor unions opposed national health care because it would eliminate one of the major benefits they offered: the ability to press employers to offer affordable health insurance to workers. Meanwhile, national health care also was opposed by politicians who considered it socialistic or who feared it would force racial integration in health care facilities.

The Birth of US Health Insurance

The most important source of opposition, however, was the American Medical Association (AMA), which feared that any sort of national health system would reduce doctors' incomes or autonomy. At the same time, however, the AMA knew that doctors' incomes were plunging because so many Americans could no longer afford to purchase health care. Consequently, the AMA and (for similar reasons) the American Hospital Association founded the nation's first major insurance programs: **Blue Shield** to cover medical bills and **Blue Cross** to cover hospital bills (Hoffman, 2012). These two plans (collectively known as "the Blues") continue to play an important role in the US health care system, currently insuring about one-third of all Americans (Blue Cross Blue Shield Association, 2014). Because these plans freed most middle-class Americans from worrying about paying their health care bills, they significantly cut popular support for any national health system (Quadagno, 2005; Rothman, 1997).

Given that the primary purpose of the Blues was to protect hospitals' and doctors' incomes, the plans had little incentive to control what kinds of care were given, to whom, or at what costs. Under Blue Cross/Blue Shield, doctors and hospitals were free to provide whatever treatments they thought were needed, at whatever price they thought was reasonable. Patients paid their bills up front, and then requested reimbursement from the Blues. Because patients were billed a fee for each office visit, test, or other service received, these plans were and are called **fee-for-service** insurance.

But although the primary goal of the Blues was protecting doctors' and hospitals' income, these plans still had to restrain costs in some way to stay financially solvent. To do so, the Blues sold their insurance only to people likely to be healthy (such as workers at major businesses) and covered members' expenses only until preset yearly or lifetime limits were reached. They also relied on **community rating**. Under community rating, each individual pays a "group rate" **insurance premium** (yearly fee) based on the average risk level of his or her community as a whole. Even if one individual in a community racked up high bills, those bills would be covered by the insurance premiums paid by the many healthy members of the same community.

The 1930s also saw the rise of a very different type of health insurance program, **health maintenance organizations (HMOs)**. Unlike the Blues, the early HMOs, such as Kaiser Permanente and the Group Health Cooperative of Puget Sound, were founded not to protect the incomes of doctors or hospitals, but to provide affordable health care. These plans also used community rating. But unlike the Blues, which reduced their costs by seeking only healthy individuals to enroll as members, HMOs reduced costs by keeping members healthy through preventive care, monitoring doctors' decisions to avoid unnecessary care, and requiring HMO members to use only salaried doctors who worked for HMOs rather than independent doctors paid fee-for-service.

The Government Steps In

Although the Blues, HMOs, and other insurance plans enabled most Americans to pay for health care, by the 1960s, many poor Americans, as well as many middle-class retirees, were finding it difficult to do so. Reflecting in part the rise of the civil rights movement and of the belief that government should use its power to improve Americans' lives, Congress in 1965 authorized two new health insurance programs: **Medicaid** to insure the poorest Americans and **Medicare** to insure Americans who were permanently disabled or over age 65 (Hoffman, 2012).

Importantly, *Medicaid* is funded jointly by state and federal governments, and is typically framed by politicians and citizens as a form of charity. Eligibility, coverage, and payments to providers vary considerably across the states, depending in part on how willing state residents are to offer such "charity." In contrast, *Medicare* is funded and organized by the federal government. Because most recipients are over age 65, the program is typically framed as an "entitlement" earned through a lifetime of working and paying taxes.

Both Medicaid and Medicare were established as fee-for-service insurance. Almost from the start, however, Medicaid offered relatively low reimbursement to health care providers, leading many to reject Medicaid patients. Medicare, however, broadened access to health care while allowing providers to set their own fees, at least initially. As a result, the incomes of doctors, hospitals, and others working in the health care field skyrocketed.

The Rise of Commercial Insurance

Recognition of the profits to be made in health care led commercial insurance companies to enter the field in large numbers. Whereas the early insurance programs were mostly nonprofits, **commercial insurance** programs by definition are organized on a for-profit basis and so must focus on earning a profit for their investors. To do so, they use **actuarial risk rating** rather than community rating. Under actuarial risk rating, insurers maximize their profits by doing whatever they can to avoid signing up individuals who are likely to have expensive medical bills. For example, until recently commercial insurers charged higher premiums to those who had back strain, kidney stones, or ulcers, and typically denied coverage to those who had diabetes or ulcerative colitis or who worked as airline pilots or in construction. (The ACA has changed this at least partly, as we will see later in this chapter.) Similarly, commercial insurers charged especially low rates to low-risk individuals. As a result, these insurers lured many low-risk individuals away from nonprofit insurers, leaving the nonprofits with a sicker clientele overall. To avoid having to raise their rates for all members to cover the bills of their sicker members, many nonprofit insurers have switched to actuarial risk rating or even become for-profit corporations.

The Rise (and Partial Fall) of Managed Care

By the 1980s, the amounts spent by government and insurers on health care had soared. This led to the explosive growth in **managed care** (Hoffman, 2012). Managed care refers to any system that controls costs through closely monitoring and controlling the decisions of health care providers; HMOs are one form of managed care organization (MCO). Most commonly, MCOs control costs in three ways. First, MCOs may negotiate prices with doctors and require consumers to use only doctors who accept their price schedule. Second, MCOs may offer bonuses to doctors who keep costs down and may require doctors to obtain approval before hospitalizing a patient, performing surgery, ordering an expensive diagnostic test, or referring to a specialist outside the MCO's "network." This system is known as **utilization review**. Finally, MCOs may rely on expert opinion to create lists (known as **formularies**) of the most cost-effective drugs for treating specific conditions. Doctors who work for an MCO must get permission before prescribing any drugs not on the MCO's formulary. Most insured Americans now belong to some form of managed care plan.

Despite evidence suggesting that managed care makes little difference in access to care, quality of care, or patient satisfaction, there has been a substantial backlash against the managed care revolution (Hoffman, 2012; Mechanic, 2004; Miller and Luft, 1997). A string of legislative and legal moves—often framed as "Patients' Bills of Rights"—have pressed insurers to drop some of the less popular aspects of managed care. For example, legislators have opposed the early release of women from hospitals soon after giving birth (labeled "drive-by deliveries" by the media), even though early release typically is safer because it reduces women's chances of contracting infections in the hospital. Similarly, legislators have fought to get patients access to experimental treatments, although patients are more likely to be harmed than helped by these treatments. In addition, even in the absence of legislative pressure, the need to keep both consumers and doctors happy has led insurers to scale back the use of formularies and utilization review and to increase consumers' access to doctors outside of the MCO's network (Bodenheimer, 1999; Hoffman, 2012).

Why has this backlash been so effective? Two important reasons can be found in American culture (Mechanic, 2004). First, a central theme in American culture is an emphasis on individual autonomy and independence. By its very nature, managed care reduces individual choices for both consumers and health care providers, which left it vulnerable to political attack. Second, Americans typically believe that more health care is always a good thing. Yet overtreatment can be both dangerous and costly. For example, mortality rates are *higher* in geographic regions where Americans receive more extensive medical care, apparently because the extra medical treatment often is more dangerous than helpful (Fisher et al., 2003; Wennberg, 2010). Because of this cultural belief in treatment, however, Americans less commonly fear the pressure to overtreat built into a fee-for-service system than the pressure to undertreat built into managed care. These cultural factors made managed care an easy target.

The Attempt at "Health Care Security"

Pressures for reform began simmering again in the early 1990s as more and more Americans found themselves uninsured or otherwise unable to pay their health care bills. These problems led US President William J. Clinton to propose his Health Care Security Act (HCSA) in 1993. The HCSA represented a liberal approach to health care reform. If adopted, the act would have broadened access to care without seriously threatening the basically entrepreneurial nature of the US health care system or the power of the "big players" in health care. Under the HCSA, Americans still would have received health insurance from many different insurers, retaining the complexity and costs of the current system. Wealthier Americans would have retained the right to purchase health care options unavailable to others, so health care would have remained a two-class system. And the proposal included no oversight mechanisms to restrain the costs (and profits) of hospital, drug, or medical care.

Nevertheless, opposition to the plan was fierce, especially from the insurance industry, which poured millions into fighting the bill (Quadagno, 2005; Hoffman, 2012). Moreover, the sheer complexity of the bill made it easier for opponents to raise fears among the American public, which since the 1980s had increasingly distrusted "big government" (Rothman, 1997; Skocpol, 1996). In the end, Congress rejected the bill. However, Congress did approve passage of the State Children's Health Insurance Program (SCHIP). That program has extended coverage (primarily through Medicaid) to many children under age 18 whose families earned too much to qualify for Medicaid but too little to pay for health care on their own. Still, millions of Americans were left without access to health care.

THE 2010 PATIENT PROTECTION AND AFFORDABLE CARE ACT

By 2008, with the election of US President Barack Obama, the time for larger-scale health care reform seemed to have arrived. The economy was spiraling into a recession, the costs of health care kept rising, and the ranks of the uninsured were growing rapidly, increasing public support for reform. Moreover, as the cost of insurance soared, many major employers who traditionally had paid most of their employees' insurance costs concluded that they could not compete in the global market unless those costs fell. As a result, the business community increasingly came to support health reform as well. Taken together, these factors led to passage in 2010 of the Patient Protection and Affordable Care Act.

Passing the Affordable Care Act

Stakeholder mobilization against the ACA, however, was strong among anti-tax and anti-government conservatives, older Americans who feared it would reduce their Medicare benefits, and parts of the health care industry. As a result, in designing the ACA, the Obama administration emphasized working within the existing health care system (Jacobs and Skocpol, 2010; Miller, 2010; Oberlander, 2010). To earn the support of hospitals, doctors, and insurance companies, the ACA included many millions in government subsidies for health care, all of which would eventually be paid to the health care industry. To assuage voters who opposed new taxes, the ACA would instead be funded by requiring individuals and employers to bear the costs of expanding coverage. To earn the vote of those who feared "creeping socialism," the government abandoned the idea of a government-run insurance system (such as an expanded version of Medicare). Finally, to earn the support of the major pharmaceutical manufacturers, the government promised new regulations that would reduce competition from foreign manufacturers and manufacturers of generic drugs (Jacobs and Skocpol, 2010; Miller, 2010; Oberlander, 2010). Thus, the Obama administration chose, in essence, health *insurance* reform over health *care* reform (Leonhardt, 2010). Nevertheless, opposition to the ACA remains strong. Numerous bills to alter or end it have been proposed in Congress, and numerous court challenges against it have been filed at the state and federal levels.

Understanding the Affordable Care Act

The ACA reflects the neoliberal premises underlying the US health care system. **Neoliberalism** is an economic and social philosophy that encourages free trade and private enterprise; disapproves of government involvement in education, health care, or other social services; and promotes the idea that each individual has both the freedom and the *responsibility* to make wise consumer choices in health care, as in all areas of life (Fisher, 2007; Fisher and Ronald, 2008). Although the government continues to play a role in health care under the ACA (especially in services for the poor), the law requires many individuals to obtain for-profit insurance coverage to purchase goods and services from for-profit pharmaceutical companies, hospitals, and doctors' offices. Moreover, the ACA holds individuals responsible for any bills not covered by their insurance.

The central goal of the ACA was to increase access to health care *within* the existing health care framework and *without* increasing costs. Creating universal access to health care was never stated as a goal (Hoffman, 2012). As a result, rather than requiring the *government* to provide health insurance or care to all citizens (as many nations do), the ACA established an **individual mandate**: that is, the requirement that each US citizen and legal resident obtain health insurance. To make that insurance affordable, the ACA proposed establishing both state-level "health exchanges" and a federal exchange through which individuals and small businesses could purchase coverage (helped by subsidies and tax credits for middle- and working-class individuals). In theory, the individual mandate would force healthy as well as unhealthy Americans to join, thus reducing the cost of insurance for each individual by spreading the bills across a large and mostly healthy population.

In addition, the ACA established an **employer mandate**: a legal requirement that employers with more than 50 employees are required to subsidize (for-profit) health insurance for their employees. (Small businesses will receive tax credits to encourage them to do the same.) The employer mandate was supposed to begin in 2014, but the date has been pushed back to at least 2016 in response to political pressure.

The ACA also called for Medicaid to be expanded to include all poor and near-poor Americans under age 65. This change was to play a major role in reducing the under-insured and uninsured population. However, in a landmark decision, the Supreme Court decided that the federal government could not require states to expand their Medicaid programs. As a result, about half of the states have decided against doing so, even though the federal government would have paid almost all the costs and about 8 million people would have gained insurance coverage (Dickman et al., 2014).

Finally, the ACA established various new restrictions on insurance companies. Among other things, companies are now prohibited from capping annual or lifetime benefits, refusing to cover those with preexisting health problems, or charging higher premiums to such individuals. Insurers also are now forbidden from charging more than $6,000 per individual per year (or $12,000 per family per year) for out-of-pocket expenses such as **deductibles** (required minimum amounts individuals must pay out of pocket before their insurance coverage kicks in) and **copayments** (unreimbursable fees paid out of pocket each time one sees a doctor). Insurers also must cover at least 60 percent of average medical costs and must allow young people to remain on their parents' insurance policies until they turn 26.

It will be some time, however, before the full impact of the ACA becomes known. Opposition to it remains fierce, and court battles over the laws will likely continue for years. Similarly, Congress will need to approve budgets annually for various aspects of the ACA's provisions, and these battles will likely be bloody. Finally, hundreds of new regulations will have to be written to implement the highly complex ACA, and this process, too, is likely to become a battlefront (Jacobs and Skocpol, 2010).

THE CONTINUING CRISIS IN HEALTH CARE COSTS

Unfortunately, even with adoption of the ACA, the cost of health care in the United States is perilously high. For example, in 1980, Americans spent on average about $1,000 per person (in current dollars) for medical care, drugs, supplies, and insurance. Those costs increased to more than $8,000 per person in 2014 and are expected to pass $12,000 per year by 2022, even with implementation of the ACA (Centers for Medicare & Medicaid Services, 2014).

Moreover, although costs have also risen in other nations, they remain by far the highest in the United States. Yet despite these costs, researchers consistently rank the US health care system below that of other more developed nations (Muennig and Glied, 2010; Schoen et al., 2010). Not surprisingly, compared to citizens in those other nations, Americans are considerably less likely to be able to afford needed health care and to believe their health care system works well (see Table 8.1).

The Myths of Health Care Costs

What accounts for the rising and unusually high costs of health care in the United States? If you ask the typical American—or member of Congress—he

TABLE 8.1 Citizens' Views on and Experiences with Health Care

Country	Percent Believing Their Country's Health Care System Works Well	Percent Who Could Not Visit Doctor or Afford Recommended Treatment in 2013
Canada	42%	13%
Germany	42	15
United Kingdom	63	4
United States	25	37

SOURCE: Commonwealth Fund (2014).

or she is likely to respond with one of four popular "myths" about US health care (Starr, 1994).

The first myth is that Americans receive more and better care than do citizens of other nations. Yet on average, the reverse is true. For example, despite our high health costs, Americans receive *fewer* days of inpatient hospital care and *fewer* doctor visits per capita, as Figures 8.1 and 8.2 show. And as Figure 8.3 shows, those higher health costs do not produce higher life expectancies.

The second myth attributes our high health care costs to our unique propensity for filing malpractice suits. Malpractice suits can raise prices both because doctors have to pay malpractice insurance premiums and because they may engage in **defensive medicine**—performing tests and procedures primarily to protect themselves against lawsuits. Federal researchers estimate, however, that defensive medicine accounts for no more than 2 percent of total US health care costs (Beider and Hagen, 2004). Moreover, their data suggest that changing the malpractice system would not significantly reduce the number of unnecessary tests and procedures.

The third myth attributes our rising health care costs to our aging population. Yet the population of the United States is no older than that of any of the other wealthy nations, and at any rate, economists have found no relationship

FIGURE 8.1 **Health Expenses and Inpatient Days in Acute Care Hospitals in 30 Nations***

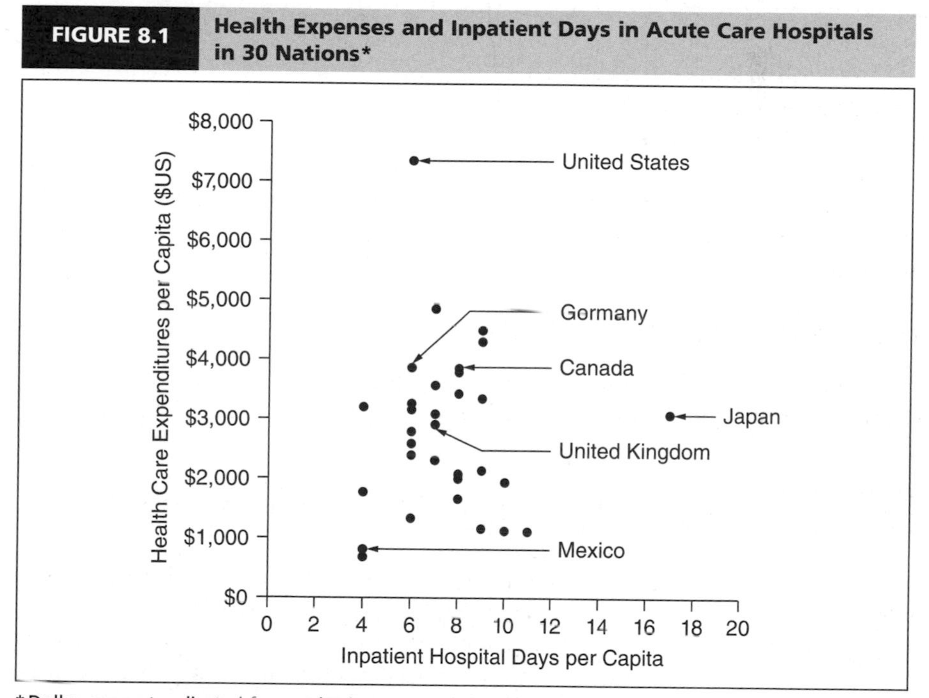

*Dollar amounts adjusted for purchasing power parity. This strategy controls for differences over time and across countries in the worth of a nation's currency by factoring in the number of units of a nation's currency required to buy the same amount of goods and services that $1 would buy in the United States.

SOURCE: Organization for Economic Cooperation and Development (OECD) (2014).

FIGURE 8.2 Health Expenses and Number of Doctor Visits in 30 Nations*

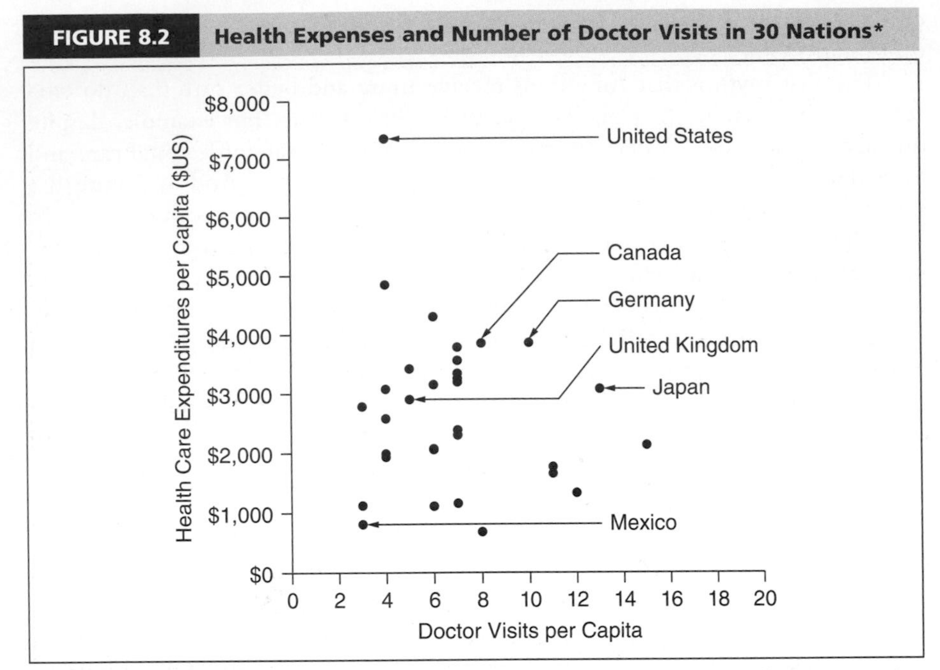

* Dollar amounts adjusted for purchasing power parity. This strategy controls for differences over time and across countries in the worth of a nation's currency by factoring in the number of units of a nation's currency required to buy the same amount of goods and services that $1 would buy in the United States.

SOURCE: OECD (2014).

between the age of a nation's population and its health care costs (Bodenheimer, 2005a).

The fourth myth is that health care costs are so high in the United States because of our advanced technologies. Although these technologies certainly play a role in health care costs, technologies (other than pharmaceutical drugs) account for only a small fraction of all health care costs. Moreover, the same technologies exist in the other wealthy nations without producing equally high health care costs. Thus, the mere existence of technology can't explain these costs.

Understanding Health Care Costs

If patient demand, malpractice costs, the aging population, and advanced technology don't explain the rising costs of health care, what does? Research points to three underlying factors: a fragmented system that multiplies administrative costs, the great power that health care providers (doctors, hospitals, pharmaceutical companies, etc.) hold relative to health care consumers (whether individuals, the government, or insurers), and the for-profit basis of the US health care system (Bodenheim, 2005a, 2005b, 2005c; Reinhardt, Hussey, and Anderson, 2004; Davis et al., 2014).

FIGURE 8.3 **Health Expenses and Life Expectancy in 30 Nations***

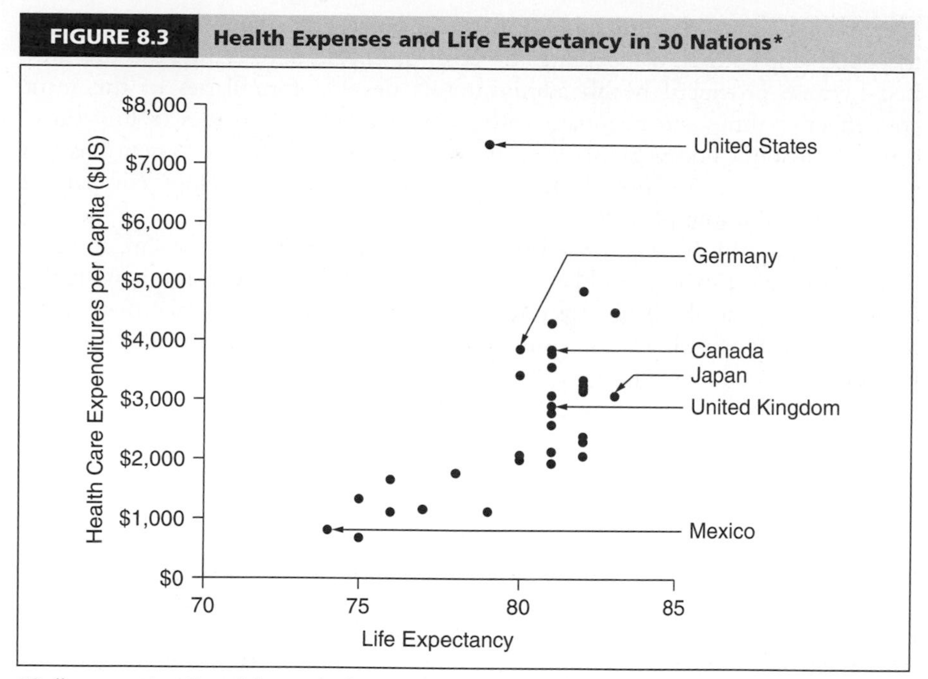

* Dollar amounts adjusted for purchasing power parity. This strategy controls for differences over time and across countries in the worth of a nation's currency by factoring in the number of units of a nation's currency required to buy the same amount of goods and services that $1 would buy in the United States.

SOURCE: OECD (2014).

Because Canadian society is probably the most similar to US society, comparing these two countries helps to illustrate why costs are so high in the United States. In the next chapter, we examine the Canadian health care system in detail. At this point, we need only note a few major points. Most important, Canadians receive their health insurance from a single payer: the government. For this reason, the Canadian system is referred to as a **single-payer system**. Similarly, hospitals receive an annual sum from the government to cover their costs. Those costs are restrained because, unlike in the United States, Canadian hospitals don't need an expensive administrative system to track patient expenses and to submit bills to multiple insurers. As a result, hospital costs per capita in Canada are almost 50 percent lower than in the United States (Himmelstein et al., 2014).

In Canada, costs are also restrained by government oversight on major capital development: If a Canadian hospital wants to add new beds or purchase new advanced technologies, it must first convince the government that such services are needed (Bodenheimer, 2005b). As a result, hospital costs are considerably lower in Canada than in the United States, even though admission rates are about equal and average stays are longer.

A unified rather than fragmented system also helps restrain Canada's medical and drug costs. Like hospitals, doctors must submit their bills only to the national insurance system rather than filing myriad different forms with different insurers.

Meanwhile, no one need spend money on advertising or selling insurance, trying to collect unpaid bills, or covering the costs of unpaid bills. Drug costs are limited because provincial health administrators develop formularies of the most cost-effective drugs and negotiate with pharmaceutical companies to buy those drugs at discount prices. Similarly, Canada's national health care system has the economic "muscle" to control the prices it pays doctors, technology companies, and other health care providers.

In addition to the fragmented nature of the US health care system, the fact that health care providers hold more power than health care consumers in the United States has also kept costs high. This results from the fact that profit making—by doctors, hospitals, insurers, pharmaceutical companies, and others—lies at the heart of the US health care system.

As the next section discusses further, in the United States, pharmaceutical companies largely control which drugs come to market, how they are advertised, and at what prices, with few constraints imposed by any national consumer or government forces. Similarly, US hospitals are free of the governmental oversight that constrains costs in Canada *and* are forced to compete for patients to pay their bills (and perhaps earn a profit). As a result, hospitals must create demand by adding beds, specialized units (such as heart transplant units), and expensive technologies (such as kidney dialysis machines), and then encouraging doctors and patients to use those services.

Similarly, because no national health care system controls the number or distribution of doctors in the United States, most of the country (other than poor and rural areas) has far too many doctors, especially specialists. To protect their incomes in the face of this competition, doctors may increase either the number of services they recommend to patients or their fees for those services (Aizenman, 2010; Bodenheimer, 2005c). This largely explains why US doctors are exceptionally likely to adopt new, expensive, and often unproven technologies, such as full-body scans and bone marrow transplants (Bodenheimer, 2005b). In addition, US doctors increasingly are trying to raise their incomes by purchasing surgical centers, CT scan machines, and other expensive technologies—actions that would likely not be permitted in a single-payer health care system. Not surprisingly, doctors who do so are considerably more likely to recommend those services to their patients (Ruggieri, 2014). For all these reasons, Americans living in areas with many doctors per capita receive more medical tests, surgeries, and other procedures; pay more for those services; and have *worse* health outcomes than those living in areas with fewer doctors (Bodenheimer, 2005b; Center for the Evaluative Clinical Sciences, 1996; Wennberg, 2010).

Finally, the for-profit basis of the US health care system, combined with its fragmented nature and the power it gives to health care providers, has made it difficult for reform efforts to succeed. For example, since the 1980s the US has tried to reduce Medicaid and Medicare costs through a system of **diagnosis-related groups (DRGs)**. Under this system, the government calculates the average cost of inpatient treatment for each possible DRG, and then reimburses hospitals for treatment based on those averages rather than on the actual costs per patient. If the hospital spends less than this amount, it earns money; if it spends more, it loses money. Theoretically, then, the DRG system should have limited the costs of providing care under Medicaid and Medicare. Instead, hospitals developed sophisticated computer software to identify the most remunerative, but still plausible, diagnosis for a given patient—a process known as "DRG creep." In addition, hospitals increasingly shifted services to outpatient units, where the DRG system does not apply. As a result, the DRG system only marginally reduced government costs for hospital care. Similarly, when the government restricted the fees it would pay health care providers for treating Medicare and, especially, Medicaid patients, many providers either stopped accepting such patients or increased the fees they charged patients who had other forms of insurance.

Health Care Costs and the ACA

Given the reasons why US health care costs are so high, it seems unlikely that the ACA can cut costs significantly. First, the ACA continues the nation's reliance on a vast web of insurers, thus guaranteeing huge administrative costs and inefficiencies. Second, health care providers (especially insurers) continue to have considerable control over the system. Most importantly, to appease health industry opponents, most proposals to incorporate well-established cost control mechanisms into the ACA were dropped from the bill before it was passed.

At the individual level, and as the story that opened this chapter illustrated, even insured Americans may continue to risk bankruptcy because of copayments, deductibles, and other services not covered by their insurance. In 2014, those who purchased the least expensive insurance plans available through state health exchanges were responsible for insurance deductibles averaging about $5,000 for individuals and $10,000 for families (Goodnough and Pear, 2014). In addition, individuals remain responsible for many costs not covered by their insurance, such as drugs not approved by their plans or emergency care at hospitals not included in their plan's network.

Finally, the ACA preserves the for-profit nature of our health care system. Within such a system, doctors, hospitals, and other health care providers will be pressured to find ways to generate profits, through their decisions regarding admissions, diagnoses, tests, treatment, and so on. For example, two-thirds of for-profit hospices for the dying will not accept patients whose pain needs to be managed through chemotherapy or other expensive forms of care (Rao and Hellander, 2014). And as we've seen, even those working in nonprofit environments will be pressured to do the same in order to survive. Similarly, we can expect that for-profit insurers will continue to seek ways to enroll members who are relatively healthy and to avoid potential members who might generate high medical bills. This will leave nonprofits and state health exchanges with a disproportionate number of members who have high medical bills, raising the cost of such plans in the end.

For all these reasons, the Centers for Medicaid and Medicare Services (2010a), a nonpartisan federal bureau that advises Congress and the president, estimates that the ACA will cost the federal government an additional $251 billion between 2010 and 2019. Those costs may make it impossible for the government to offer the insurance subsidies for poor and middle-class Americans that constitute the core of the ACA. If those subsidies are reduced, many will likely drop their insurance.

Health Care Costs and "Big Pharma"

Because the pharmaceutical industry, or "Big Pharma" as it is often known, has so quickly emerged as a major source of health care costs, it is worth exploring in more depth. This section looks at how the pharmaceutical industry affects doctors' and patients' ideas about illnesses and treatments and, as a result, affects health care costs.

Big Pharma Comes of Age The pharmaceutical industry is an enormous—and enormously profitable—enterprise. Indeed, it has been the most profitable industry in the United States since the early 1980s (Angell, 2004). Although the pharmaceutical industry routinely argues that their high profits merely reflect the high cost of researching and developing new drugs, such work accounts for only 14 percent of their budgets. In contrast, marketing accounts for about 50 percent (Angell, 2004). Largely because of this marketing, American citizens now spend a total of about $272 billion per year on prescription drugs, *not* including drugs purchased by doctors, nursing homes, hospitals, and other institutions (Centers for Medicare and Medicaid Services, 2014). Americans are buying *more* drugs, buying more *expensive* drugs, and seeing the *prices* of popular drugs rise more often than ever before. (The price of the popular antihistamine Claritin, for example, rose 13 times in five years.)

The pharmaceutical industry has not always been this profitable. Profits only began soaring in the early 1980s after a series of legal changes reflecting both the increasingly "business-friendly" atmosphere in the federal government and the increased influence of the pharmaceutical industry lobby—now the biggest spending lobby in Washington. First, new laws allowed researchers funded by federal agencies (including university professors and researchers working for small biotech companies) to patent their discoveries and then license those patents to pharmaceutical companies. This change dramatically reduced pharmaceutical companies' research costs—while giving these researchers a vested interest in emphasizing the benefits of new drugs.

Second, new laws doubled the life of drug patents. As long as a drug is under patent, only the company that owns the patent can sell the drug, allowing it to set its price as high as the market will bear. In addition, companies can now extend their patents by developing "me-too" drugs, which differ only slightly from existing drugs. For example, when the patent expired for Prilosec, a widely used treatment for common stomach troubles, its manufacturer released Nexium, an essentially identical new drug. Nexium now sells for $6 per pill and Prilosec for $1, whereas the chemically identical generic version, omeprazole, sells for 45 cents. Yet sales are highest for Nexium (Brawley, 2011).

Third, the pharmaceutical industry won the right to market drugs directly to consumers. Direct-to-consumer advertising has proven highly effective. According to a nationally representative survey conducted in 2008 for the nonprofit

Kaiser Family Foundation, almost one-third of American adults have asked their doctors about drugs they've seen advertised, and *82 percent* of those who asked for a prescription received one (Appleby, 2008).

Developing New Drugs Much of the recent rise in health care costs in the United States comes from the shift to new drugs. Whenever a new drug is developed, the crucial question for health care providers and patients is whether its benefits outweigh its dangers. For this reason, it is crucial that any new drug be extensively tested to determine whether it works better than already available drugs (which almost certainly are cheaper), whether it works differently in different populations, what dosages are appropriate, and what side effects are likely. But because pharmaceutical companies earn their profits by selling drugs, they have a vested interest in overstating benefits and understating dangers. And increasingly, these companies are both willing and able to manipulate the data available to outside researchers, doctors, federal regulators, and consumers (Abramson, 2004; Angell, 2004). For example, because scientific testing is typically designed to be accurate 95 percent of the time, manufacturers know that if they test a drug enough times, they will eventually hit the other 5 percent and obtain data that inaccurately suggest a drug works in some population. *Contemporary Issues: Race-Specific Medicine* describes one outcome of this process.

CONTEMPORARY ISSUES

Race-Specific Medicine

Is medicine a black or white matter? Increasingly, pharmaceutical manufacturers are acting as if it is. At least 30 drugs now on the market are claimed by manufacturers to be safer or more effective for African Americans than for whites (Epstein, 2007). Most commonly, these are drugs that proved ineffective in rigorous testing but that (perhaps accidentally) appeared to work in small studies of African Americans—some of which didn't even compare African Americans with whites. Yet as Chapter 3 discussed, there are no meaningful genetic differences between "races," so there are no biological explanations for these supposed differences in drug safety or efficacy. Indeed, one major review concluded that manufacturer's claims for "race-specific" drugs are "universally controversial" (Tate and Goldstein, 2004).

In addition to increasing drug costs as patients are shifted from older, less expensive drugs to newer and perhaps ineffective drugs, the rise of race-specific medicine reinforces the idea that racial differences are real and important (Epstein, 2007). Moreover, when drug companies focus on seeking racial differences, they may unintentionally hide more important causes of illness: Poor African Americans living in polluted neighborhoods in Mississippi, for example, may be no more susceptible to disease than their white neighbors, but this may be overlooked if researchers divide their subjects only by race and not by social class or living conditions. Similarly, the concept of race-specific medicine may lead doctors to quickly assign diagnoses and treatments based on race rather than on a holistic assessment of their patients as individuals. In fact, more than 80 percent of doctors responding in a national survey agreed that race should be used as a basis for diagnosis and treatment (Williams et al., 2010).

In the past, university-based drug researchers provided at least a partial check on the drug research process by bringing a more objective eye to their research. Since 1980, however, pharmaceutical industry funding for research by university-based scientists has skyrocketed (Lemmens, 2004). That funding comes in many forms, from research grants to stock options to all-expenses-paid conferences in Hawaii. Moreover, as other federal funding for universities declined over the past quarter century, university administrators came to expect their faculty to seek pharmaceutical funding. Importantly, when the pharmaceutical industry funds university-based research, it often retains the rights to the research results and so can keep university researchers from publishing any data suggesting that a particular drug is ineffective or dangerous (Angell, 2004; Lemmens, 2004).

At the same time that the pharmaceutical industry has increased its funding to university-based researchers, it has even more dramatically increased funding to *commercial* research organizations (Lemmens, 2004). These organizations are paid not only to conduct research but also to promote it. To keep on the good side of the companies that fund them, these research organizations must make drugs look as effective and safe as possible by, for example, selecting research subjects who are least likely to experience side effects, studying drugs' effects only briefly before side effects can appear, underestimating the severity of any side effects that do appear, and choosing not to publish any studies suggesting that a drug harms or doesn't help.

Doctors, medical researchers, sociologists, and others have raised concerns about the impact of bias on research publications (Bodenheimer, 2000). Researchers have found that medical journal articles written by individuals who received pharmaceutical industry funding are four to five times more likely to recommend the tested drug than are articles written by those without such funding (Abramson, 2004:97). Similarly, researchers have found that research studies suggesting a drug is effective are several times more likely to be submitted and accepted for publication than are those that suggest it is ineffective (Hadler, 2008; Turner et al., 2008). Concern about such biases led the *New England Journal of Medicine* (one of the top two medical journals in the United States) to forbid authors from publishing articles on drugs in which they had financial interests. The policy, however, was dropped quickly because it proved virtually impossible to find authors who did *not* have financial conflicts (Lemmens, 2004).

Even more astonishing than pharmaceutical industry funding of university-based researchers is the growing practice of paying such researchers to sign their names to articles written by industry employees (Elliott, 2004). For example, between 1988 and 2000, 96 articles were published in medical journals on the popular antidepressant Zoloft. Just over half of these were written by pharmaceutical industry employees but published under the names of university-based researchers. Moreover, these ghostwritten articles were *more* likely than other articles to be published in prestigious medical journals (Elliott, 2004).

Regulating Drugs In the United States, ensuring the safety of pharmaceutical drugs falls to the Food and Drug Administration (FDA). But during the same time period that the profits and power of the pharmaceutical industry grew, the FDA's power and funding declined as part of a broader public and political movement away from "big government." These two changes are not unrelated: The pharmaceutical industry now routinely provides funding of various sorts to staff members at government advisory agencies, doctors who serve on FDA advisory panels, and legislators who support reducing the FDA's powers (Lemmens, 2004).

Under current regulations, the FDA must make its decisions based primarily on data reported to it by the pharmaceutical industry. Yet the industry is required to report only a small fraction of the research it conducts. For example, the company that produced the antidepressant Paxil had considerable data indicating that among teenagers Paxil did *not* reduce depression but *could* lead to suicide. To avoid making this information public, the company submitted to the FDA only its data from studies on adults (Lemmens, 2004). Similarly, drug companies need only demonstrate that new drugs work better than **placebos**, not that they work better than existing (cheaper) drugs. For example, because of intensive marketing campaigns, new antipsychotic drugs such as Zyprexa have largely replaced older, cheaper drugs, even though the new drugs work little better than placebos and carry life-threatening risks (Wilson, 2010b).

Marketing Drugs Once the pharmaceutical industry develops a drug and gets FDA approval, the next step is to market the drug. One of the most important limitations to the FDA's power is that, once it approves a drug for a single use in a single population, doctors legally can prescribe it for *any* purpose to *any* population. For example, doctors increasingly are prescribing Botox injections to treat migraines even though the FDA has not approved its use for that purpose.

Drug marketing has two major audiences, doctors and the public. Marketing to doctors begins during medical school as students quickly learn that pharmaceutical companies provide a ready source not only of drug samples and information but also of pens, notepads, lunches, and all-expense-paid "educational" conferences at major resorts. After graduation, the pharmaceutical industry continues to serve as doctors' main source of information about drugs. The *Physicians' Desk Reference* (or *PDR*), the main reference doctors turn to for drug information, is solely composed of drug descriptions written by drug manufacturers. In addition, the pharmaceutical industry spends $6,000 to $11,000 (depending on medical specialty) per doctor per year to send salespeople to doctors' offices on top of the money it spends advertising drugs to doctors in other ways. Most doctors meet with pharmaceutical salespeople at least four times per month and believe their behavior is unaffected by these salespeople. Yet doctors who meet with drug salespeople prescribe promoted drugs more often than do other doctors, even when the promoted drugs are more costly and less effective than the alternatives (Angell, 2004; D. Shapiro, 2004). In addition, the pharmaceutical companies now surreptitiously provide much of the "continuing education courses" doctors must take each year by paying for-profit firms to teach the courses and to arrange with universities to accredit the courses (Angell, 2004).

In recent years, and as noted earlier, marketing directly to consumers has become as important as marketing to doctors. To the companies, such advertising is simply an extension of normal business practices, no different from any other form of advertising. Moreover, they argue, advertising to consumers is a public service because it can encourage consumers to seek medical care for problems they otherwise might have ignored. Finally, companies have argued that these advertisements pose no health risks because consumers still must get prescriptions before they can purchase drugs, thus leaving the final decisions in doctors' hands. Those who oppose such advertisements, on the other hand, argue that the advertisements are frequently misleading, encourage consumers to pressure their doctors into prescribing the drugs, and encourage both doctors and patients to treat normal human conditions (such as baldness) with pharmaceutical drugs (Angell, 2004; Hadler, 2008).

Marketing Diseases As this suggests, the pharmaceutical industry sells not only drugs but also diseases to doctors and the public alike. In some cases, drug companies have encouraged doctors and the public to define disease *risks* (such as high blood pressure) as *diseases* (such as hypertensive disease). In other cases (as Chapter 5 described), drug companies have defined symptoms into new diseases.

One example of this is the disease known as *pseudobulbar affect*, or PBA. PBA refers to uncontrollable laughing or crying unrelated to individuals' emotional state and can be caused by various disabling neurological conditions (such as head trauma, stroke, and Lou Gehrig's disease). The concept of PBA was developed by Avanir Pharmaceuticals, which markets the drug Neurodex as a treatment for it (Pollack, 2005). Although Neurodex seems to help some patients, its side effects are serious enough that at least one-quarter of users—all of whom already have serious health problems and must take numerous other medications—soon stop taking it.

To convince doctors that uncontrollable laughing and crying is a disease in itself, Avanir has advertised in medical journals and sponsored continuing education courses, conferences, and a PBA newsletter. Avanir also has marketed the concept of PBA directly to consumers through its PBA website and through educational grants it has given to advocacy groups for those living with stroke, multiple sclerosis, and other diseases (Pollack, 2005).

THE CONTINUING CRISIS IN HEALTH CARE ACCESS

The passage in 2010 of the Patient Protection and Affordable Care Act (ACA) reflected the growing consensus that health care in the United States is in crisis. But although the ACA has made a difference, shockingly high numbers of Americans nonetheless remain uninsured, underinsured, or precariously insured.

Uninsured Americans

According to the US Congressional Budget Office (2014), which provides nonpartisan analyses to Congress, 54 million Americans were uninsured in the months before the ACA began. The Office estimates that without the ACA, that number would have *risen* by 3 million over the next decade. In contrast, it estimates that the ACA will *reduce* the number of uninsured Americans by 26 million in its first three years alone. In fact, more than 8 million (most of them uninsured) purchased insurance through the ACA exchanges in the first months of the program (Kaiser Commission on Medicaid and the Uninsured, 2014), and many others gained insurance through expanded Medicaid coverage. This still leaves millions of Americans uninsured, however.

Young, childless adults—the population least likely to believe they might fall ill and least likely to be covered by government health care programs—are especially likely to be uninsured, as are African Americans, Hispanics, and poorer persons (Kaiser Commission on Medicaid and the Uninsured, 2014). Insurance coverage increased for all of these groups in the first months of the ACA. However, in part because most of the southern states opted out of the Medicaid expansion, southerners will remain especially likely to lack insurance (Garfield et al., 2014).

Surprisingly, given that insurance in the United States is typically linked to employment, most uninsured Americans live in families with one or more full-time workers (Kaiser Commission on Medicaid and the Uninsured, 2014). This reflects sharp reductions over the last two decades in the benefits employers offer their workers and sharp increases in the number of workers hired without benefits on a part-time or temporary basis. Ironically, because the ACA requires large employers to subsidize health insurance for employees who work 40 or more hours, many employers have cut workers' hours below that level (Rao and Hellander, 2014).

Finally, disabled and ill Americans remain disproportionately likely to be uninsured. In the past, most states allowed insurers to reject applicants for individual health insurance who showed any indications of health problems. The ACA now prohibits this practice, but experience suggests that insurers will continue to find ways to avoid enrolling individuals who seem likely to generate high medical bills.

Underinsured Americans

In addition to those who are *uninsured*, as of late 2014 more than 20 percent of all insured adults under age 65 are *underinsured* (Collins et al., 2014a). In other words, they have insurance but still can't afford to pay all their medical bills. Underinsurance is most common among poorer people and among those with chronic health problems (Collins et al., 2014a).

Underinsurance occurs when individuals can't afford to pay required insurance premiums, deductibles, or copayments. It can also occur when insurers either cap reimbursements per treatment or don't cover certain treatments, such as drugs or nursing home care. Since 2006, both the number of Americans who have to pay deductibles and copayments and the dollar amount of those payments have risen (Collins et al., 2014a). As a result, underinsured and uninsured individuals are equally likely to skip needed medical care (Collins et al., 2014a).

The ACA is expected to reduce underinsurance for those who buy insurance through the health exchanges or become eligible for Medicaid. However, the vast majority of Americans receive insurance through employers, and the ACA will not reduce underinsurance within this group (Collins et al., 2014b).

The Consequences of Underinsurance and Lack of Insurance

Uninsured and underinsured persons are considerably less likely than others to receive needed health care (Kaiser Commission on Medicaid and the Uninsured, 2010). As a result, they are also significantly more likely to suffer health problems and to die of potentially treatable conditions (Institute of Medicine, 2002).

This does not mean, however, that uninsured and underinsured persons have no access to health care. Federal, state, and some local governments provide clinics and public hospitals that offer low-cost or free care to such individuals. In addition, governments sometimes provide low-cost or free vaccination, cancer screening, and "well-child" programs. These facilities and programs, however, are not always geographically accessible to those who need them. In addition, these facilities are continually underfunded, so individuals may have to wait hours for emergency care and weeks or months for nonemergency care.

Uninsured and underinsured persons also sometimes can obtain health care through the private sector. First, some individuals can find private doctors who will reduce or waive their fees, and some live in communities where nonprofit hospitals offer inexpensive outpatient clinics. Second, uninsured persons can obtain care for both acute and chronic, emergency and nonemergency health problems from hospital emergency departments; although emergency departments legally can refuse care to anyone who is medically stable, many provide at least basic treatment to all who present themselves. Afterward, however, individuals can face stratospheric bills. Finally, uninsured persons increasingly have volunteered for experimental trials of new drugs to obtain at least sporadic treatment (Fisher, 2009). Yet in such experiments, some patients receive placebos, some receive drugs that prove ineffective, and some receive drugs that prove harmful. Moreover, even if the drugs work well, patients receive only temporary benefit because the drugs become unavailable after the experiments end.

THE PROSPECTS FOR STATE-LEVEL REFORM

Although the ACA mandates many elements of health care for the states, it also gives leeway for states to begin or continue their own reform efforts, some of which in the end may become models for national reform. So far, Vermont is the only state to have declared health care a right, and to have seriously considered adopting a single-payer system, operated under the ACA. Those plans are currently on hold, however. Vermont, though, is an unusual state, which leans heavily Democratic, and so few expect other states to follow its lead.

Hawaii's model is more likely to be adopted by other states. In 1974, Hawaii's legislators passed the Prepaid Health Care Act. Unlike the ACA, which is based on an individual mandate, Hawaii's program is based on an **employer mandate**—that is, on the requirement that employers offer health insurance to their workers and pay a specified percentage of the costs. Hawaii requires employers to pay at least 50 percent of the cost for any employees who work at least 20 hours per week for four consecutive weeks (Harris, 2009).

In addition, most employers voluntarily insure employees' families and pay more than their required 50 percent of costs.

The willingness of Hawaiian employers to care for their employees may reflect unusual aspects of Hawaii's history, geography, and culture. The state's geographic isolation makes it difficult or impossible for employers to move elsewhere, and decades of paternalistic control by pineapple plantation owners had established the idea that employers had some responsibilities to their employees. In addition, Hawaii's employers seem to share with many other Hawaiians the belief that all residents of these isolated islands should be treated like members of a family (Harris, 2009).

As in other states, elderly persons and very poor persons receive their health insurance from Medicaid or Medicare. Unemployed persons and part-time workers who earn too much to receive Medicaid but too little to purchase insurance on their own instead receive insurance through Hawaii's State Health Insurance Program (Harris, 2009). As a result, 90 percent of Hawaii residents are insured. Because such a high proportion of the state's population is insured, insurers can use community ratings rather than risk ratings—keeping rates affordable for all purchasers—and still remain financially viable. In fact, both insurance premiums and costs per Medicare enrollee are among the lowest in the nation (Harris, 2009).

In addition to ensuring a high level of coverage, the new system enabled Hawaii to achieve unusual success in restraining health care costs. First, because almost everyone has health insurance, residents can seek care early for illnesses and accidents. As a result, the system is protected from the tremendous medical costs that can accrue when illness or accidents are left untreated. Second, Hawaii benefited from the unintended development of monopolistic, nonprofit insurance plans. About 70 percent of Hawaiians receive their insurance from one of two nonprofit insurers, the Hawaii Medical Service Association or Kaiser Permanente. Because these two insurers control such a large share of the market, they can exert considerable control over medical costs. Doctors who refuse to accept their reimbursement schedules or salaries can attempt to seek patients elsewhere but will find few patients who don't belong to these plans. Finally, Hawaii restrained costs through reducing hospital use and costs. Unlike most US insurers, Hawaii's two major insurers pay only for hospital stays in multibed wards, not in semiprivate rooms. Meanwhile, Hawaii implemented a strict system for prospectively reviewing any hospital capital expenses. Hospitals can't purchase major equipment or construct new facilities unless they can demonstrate need for those services. Therefore, consumers need not pay the costs of maintaining unused hospital beds or duplicative technologies.

Conversely, the continued existence of Medicare and Medicaid has hampered Hawaii's ability to restrain health care costs. Because these plans don't reimburse hospitals at rates high enough to cover the actual costs of providing care, hospitals have shifted costs to patients with private health insurance. At the same time, Medicaid's and Medicare's low reimbursement schedules have hampered access to health care because many doctors won't accept patients who belong to these plans. These problems have been exacerbated by rising unemployment and by the (nationwide) shift toward replacing full-time workers with part-time workers, which means that more Hawaiians must turn to the state rather than employers for their insurance. As a result, costs have increased, and the state has had to reduce the benefits available through its insurance program. In addition, the costs of meeting various ACA requirements also have placed pressures on Hawaii's health insurance program.

In sum, the Hawaii experiment demonstrates both the advantages of moving toward a single-payer, nonprofit system with strong centralized control and the problems when multiple payers—in this case, public and private insurers—continue to function in the same economic sphere. It also demonstrates the benefits available from a reasonably unified managed care system and the difficulties of sustaining a strong system in the face of external economic pressures.

IMPLICATIONS

As we have seen, Americans obtain their health care through a wide range of funding mechanisms, from publicly subsidized health care programs to private fee-for-service insurance to nonprofit HMOs. Even with passage of the ACA, some Americans will continue to have nearly unlimited access to health care—including unneeded and potentially dangerous care—and others will lack access to even the most basic health care. Although millions will now gain insurance, millions will still face bankruptcy because of the limitations built into that insurance. Thus, the United States will continue to face economic and health problems caused by both overuse and underuse of health care services. Moreover, the ACA reforms won't change the underlying structure of the system and so may not reduce the nation's health care costs or other problems over the long run.

The failure to pass—or even seriously consider—any proposals for more dramatically changing the health care system reflects the political and cultural realities of the contemporary United States. American culture has always contained both liberal and conservative tendencies. The freedoms established in the Bill of Rights, the commitment to public education, and the establishment of programs such as Social Security reflect the widespread (liberal) belief that the government has a responsibility to protect and value all its citizens. At the same time, US culture has long linked belief in individual freedom with belief in individual responsibility: If the idea of an "American dream" suggests that anyone can succeed, it also suggests (as conservatives often emphasize) that those who do *not* succeed so have only themselves to blame. It remains to be seen whether changing US demographics, politics, or economic realities will shift the balance between these two tendencies and thus push either toward or away from further health care reform.

SUMMARY

1. The United States does not have a health care system. Rather, it has an agglomeration of public and private providers functioning autonomously in often-competing ways.

2. Stakeholder mobilization—organized political opposition by groups with vested interest in the outcome—has stood in the way of any true reform of the system.
3. The Blue Cross and Blue Shield insurance plans were established to protect the incomes of hospitals and doctors. Both plans were nonprofit, offered fee-for-service insurance (in which consumers are reimbursed for their medical and hospital bills), and were initially based on community rating (in which all members pay the same insurance premium based on the average risk level of their community as a whole).
4. HMOs also used community rating but were established to provide health care to all. HMOs reduced costs by encouraging preventive care, monitoring doctors' behavior to make sure it was cost effective, paying doctors on salary, and requiring HMO members to use only HMO doctors.
5. Medicare and Medicaid are government insurance programs that provide health care coverage to poor, disabled, and elderly persons. Because they initially were a form of fee-for-service insurance with the government paying all health care bills for members, these programs dramatically increased the profits available in health care.
6. Commercial insurers rely on actuarial risk rating in which insurance premiums are based on individual's health risks. Competition from commercial insurers has led Blue Cross, Blue Shield, HMOs, and other nonprofit insurers to begin operating more like each other and more like commercial insurers.
7. Managed care refers to any system that controls costs by monitoring and controlling health care providers' actions. Most US insurers now use managed care, but public backlash has substantially reduced its impact.
8. The ACA, passed in 2010, aims to reduce the number of uninsured Americans primarily through expanding Medicaid, requiring large employers to offer insurance and requiring other individuals to purchase health insurance (with the assistance of government subsidies and tax credits). The ACA includes only minimal efforts to control the costs of care and won't change the underlying structure of the health care system.
9. The cost of health care in the United States is perilously high for three reasons. First, a fragmented system multiplies administrative costs. Second, health care providers have considerably more power than health care consumers (whether individuals, the government, or insurers). Third, the for-profit basis of the US health care system makes it difficult to control costs.
10. Pharmaceutical companies are an important factor in rising health care costs because they largely control which drugs come to market, how they are advertised, and at what prices. Pharmaceutical companies market new diseases as well as new drugs.
11. Although the ACA has made a difference, shockingly high numbers of Americans nonetheless remain uninsured. Those who lack good insurance are significantly more likely than others to experience illness, disability, or death.

REVIEW QUESTIONS

1. How and why does commercial insurance differ from insurance offered on a nonprofit basis?
2. What is managed care? How can it restrain health care costs, and how can it harm individuals' health?
3. What are Medicaid and Medicare?
4. Why have health care costs in the United States risen?
5. Who are the uninsured?
6. Why do individuals who have health insurance still sometimes face financial difficulties in paying their health care bills?
7. How does underinsurance or the lack of insurance affect individuals' health and health care?
8. What are the benefits and limitations of the ACA?

CRITICAL THINKING QUESTIONS

1. Researchers believe they have identified a gene that increases women's risk of breast cancer. You are the chief administrator of a health insurance plan. One of your board members, whose mother died from breast cancer, argues that your plan should offer this test for free as a routine preventive procedure.
    a. Explain to the board member what information you would want before you could make this decision and why you would want that information. Be sure to think about the consequences for the plan as a whole as well as for individual patients.
    b. Would you want different information and reach a different decision if you were a doctor in private practice? If you were a patient?
2. How do we ration health care in our present system? What are the financial costs of this rationing? What are the social costs?
3. How are the costs of care distributed among US residents now? Be sure to think about not only costs paid out of pocket but also costs paid through taxes for government-provided care. How would those costs be distributed under a single-payer national health plan?

CHAPTER

9

Health Care around the Globe

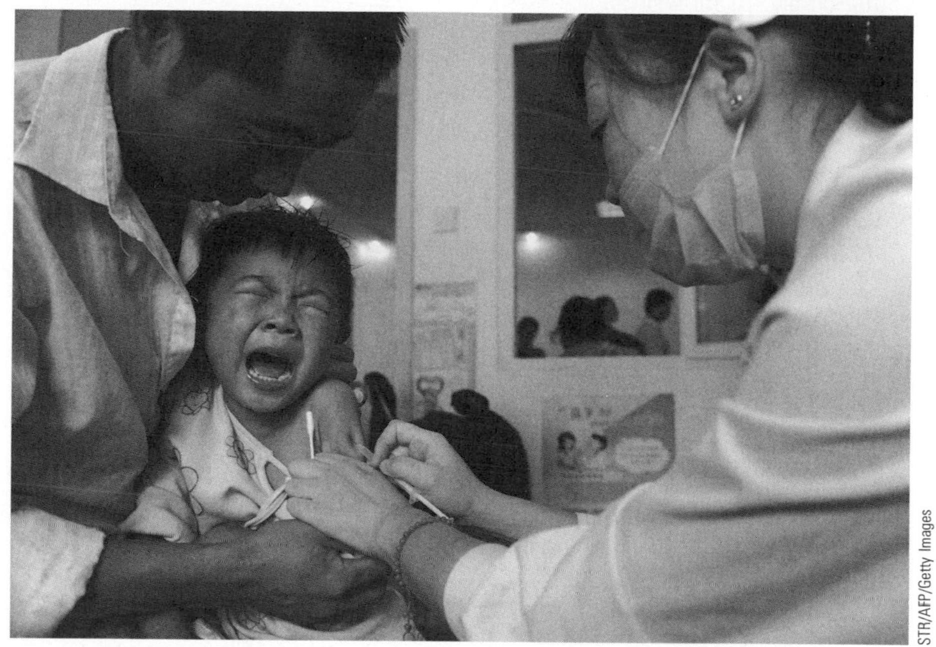

STR/AFP/Getty Images

LEARNING OBJECTIVES

After reading this chapter, students should be able to:

- Understand the most important measures for evaluating health care systems internationally.
- Compare the strengths and weaknesses of health care systems in the most developed nations.
- Identify the special problems faced by health care systems in the less developed nations.

A few years ago, American journalist T. C. Reid moved to London with his young family. As he writes:

> *Barely a week after we arrived in the U.K., our youngest daughter woke up with a painfully infected ear, bright red and swollen like a chestnut. We could guess the cause—it must have been that dubious ear piercing shop in one of the charming street markets—but had no idea how to fix the problem. We had barely unpacked our suitcases and certainly hadn't had time to find a local doctor. Feeling desperate, we piled into a roomy black cab and asked to go to the nearest hospital. Within minutes, we were in the emergency room (that is, "casualty" ward) at St. Mary's Hospital, an ancient, much-the-worse-for-wear institution.... St. Mary's on Praed Street was the place where Sir Alexander Fleming discovered penicillin in 1928; it looked as if no one had painted the walls since then.*
>
> *After a quarter-hour's wait there, a gentle nurse and an authoritative doctor took command of our daughter's case. They carefully removed the offending earring, reduced the swelling, treated the infection (with a form of penicillin), and offered a polite but firm instruction on the right way to care for the pierced ear. Our daughter—and her parents—felt an enormous sense of relief.*
>
> *I pulled out my check book and waited for the bill. I knew this treatment was going to be costly—emergency rooms always are—but frankly, I was perfectly willing to pay for the excellent and reassuring medical care we had received. The nurse, evidently accustomed to American patients, smiled at my mistake. "You can put away your checks," she said, crisply and proudly. "There won't be a bill to pay. We do it a bit differently here. In the National Health Service, we don't charge for medical treatment." With that, she sent us home.*
>
> *Had the same minor medical crisis occurred at home, we would have received the same level of professional treatment. But we would have received something else along with it: a pile of bills. Having had a similar experience with the emergency wards in the United States, I would expect that treatment ... would*

have brought in bills of about $200 from the hospital, $150 or so from the doctor, and $100 from some lab technician. And I would likely have faced a three-month battle with an insurance company trying to get the bills paid. In Britain, there was no need to argue with the insurance company over the bill, because there was no bill (Reid, 2009:117–118).

On television, in newspapers, and in public discussions, we often hear that the United States offers the best health care in the world. Yet other countries—both Western and non-Western, rich and not so rich—provide far better access to care for their citizenry at lower costs and with better health outcomes. In this chapter, we begin by looking at some basic measures for evaluating health care systems, before exploring the systems in six other countries—Germany, Canada, Great Britain, China, Mexico, and the Democratic Republic of Congo (DRC). The health care systems in Germany, Canada, and Great Britain have often been proposed as models for a revamped US system; all are ranked higher than the US system by the nonprofit Commonwealth Fund (Davis et al., 2014), based on overall health markers (such as life expectancy), equitable distribution of health care costs among citizens, good health outcomes relative to health expenditures, and responsiveness to consumer needs. The health care systems in China and Mexico are not useful as models for the United States, but they do help us understand how poorer countries have tried and sometimes succeeded in improving their nation's health despite limited resources. Finally, the DRC provides an example of what happens when health care systems collapse.

EVALUATING HEALTH CARE SYSTEMS

Universal Coverage

The most basic measure of any nation's health care system is whether it provides **universal coverage**, guaranteeing health care to all citizens and legal residents of a country. The United States is the only **more developed nation** that neither provides health care to all citizens nor recognizes a right to health care (a topic discussed in *Ethical Debate: Is There a Right to Health Care?*). Instead, the US government provides insurance to only a small percentage of the population, and even under the **Affordable Care Act (ACA)** allows private insurers considerable leeway in deciding who to insure and how much to charge for that insurance. In contrast, all legal residents of Great Britain or Canada, regardless of income, place of residence, employment status, age, or any other demographic characteristic, can obtain state-supported health care—although they are guaranteed neither immediate service nor every service they want.

ETHICAL DEBATE

Is There a Right to Health Care?

With the sole exception of the United States, every more developed nation in the world considers health care a basic right. In the United States, on the other hand—and as debate over the ACA in 2010 revealed—many question whether individuals have a right to health care, and no court has ever recognized such a right.

Those who argue against a right to health care draw on the language of individualism. **Individualism** refers to a set of cultural beliefs and practices that stresses the autonomy, equality, and dignity of individuals and therefore rejects the idea that society should *mandate* certain rights for all members of a society (Daniels and Roberts, 2008). Those who support individualism argue that in asserting individuals' rights to health care, we implicitly assert that health care workers have a duty to provide that care. In so doing, therefore, we restrict the rights of health care workers to control their time and resources. If we would not force a baker to give bread to the hungry, how can we force doctors to give their services away, or restrict what patients doctors see, what services they provide, and what charges they assess?

Similarly, in asserting a right to health care, we implicitly assert that all members of a society have a duty to pay the costs of that care. When we subsequently use tax dollars to pay for health care, we restrict the rights of individuals to spend their money as they please. Some individuals, both rich and poor, might consider this a good investment, but many others would prefer to choose for themselves how to spend their money.

Moreover, according to those who take this position, asserting a right to health care fails to differentiate between unfortunate circumstances and unfair ones (Daniels and Roberts, 2008). Although it is certainly unfortunate that some individuals experience pain, illness, and disability, it is not necessarily unfair. Society may have an obligation to intervene when an individual unfairly experiences disability because another acted negligently, but society can't be expected to take responsibility for correcting all inequities caused by biological or social differences in fortune.

Finally, if we assert that individuals have a right to demand certain social goods from a society, where do we draw the line? Do individuals have a right only to a minimum level of health care, or do they have a right to all forms of health care available in a given society? And if we grant individuals a right to health care, how can we deny them a right to decent housing, education, transportation, and so on?

Those who argue in favor of a right to health care, on the other hand, draw on the language of social justice (Daniels and Roberts, 2008). Believing each individual

Portability

A second important measure of health care systems is whether they offer portable benefits. As described in Chapter 8, most US citizens receive their health insurance through their jobs, their spouses' jobs, or their parents' jobs, so they may lose their insurance if their family or work situation changes. Similarly, individuals who receive **Medicaid** can lose this coverage if they move to another state or if their income rises above the legal maximum, and those who retire or go on disability often find that they can't move to another area because the health insurance they receive from their former employer won't cover them elsewhere, and obtaining insurance on their own would be too expensive. In contrast, in the other more developed nations, individuals need not worry about losing their insurance no matter what changes occur in their personal lives.

has inherent worth, they reject the distinction between unfortunate and unfair circumstances and the idea that health care is a privilege, dependent on charity or benevolence. Instead, they argue that each individual has a right to at least a minimum level of health care. Moreover, they argue that all members of a society are interdependent in ways that a rhetoric of individualism fails to recognize. For example, doctors who believe they should have full control over how and to whom they provide services fail to recognize the many ways they have benefited from social generosity. Medical training relies heavily on tax dollars, as do medical research projects, technological developments, hospitals, and other health care facilities. In accepting these benefits of tax support, therefore, doctors implicitly accept an obligation to repay society through the health care they provide.

Similarly, those who support a right to health care argue that to consider the decision to purchase health care as simply an individual choice misrepresents the nature of this decision because it hardly makes sense to define something as a choice when the alternative is death or disability. Nor does it make sense to talk about the purchase of health care as a choice when individuals can do so only by giving up other essentials such as housing or food.

Finally, those who support a right to health care recognize that society could never afford to provide all available health services to everyone but argue that this should not limit society's obligation to provide a decent minimum of care to all. Doing any less, they argue, denies the basic worth of all humans.

Sociological Questions

1. What social views and values about medicine, society, and the body are reflected in the debate over a right to health care? Whose views are these?
2. Which social groups are in conflict over this issue? Whose interests are served by offering universal health care? Whose interests are harmed?
3. Which of these groups has more power to enforce its view? What kinds of power do they have?
4. What are the intended consequences of our current system, which rejects the idea that individuals have a right to health care? What are the unintended social, economic, political, and health consequences of this system?

Geographic Accessibility

Even those who have health insurance can face obstacles to receiving care, depending on where they live. Both rural areas and poor inner-city neighborhoods in the United States typically have relatively few health care providers per capita. Meanwhile, other areas have an excess of doctors—a situation that can pressure doctors to increase their prices or perform perhaps unnecessary procedures to maintain their incomes despite competition for patients (Aizenman, 2010; Bodenheimer, 2005c). These problems suggest that for both economic and medical reasons, we should also evaluate health care systems according to whether they include mechanisms for encouraging an equitable distribution of doctors, such as providing low-cost loans to medical students who promise to work for a few years in underserved areas.

Comprehensive Benefits

Another important measure of health care systems is whether they offer all of the essential services individuals need. The difficulty lies in defining what is essential. Although all observers would agree that comprehensive health care must include coverage for **primary care**, agreement breaks down quickly when we begin discussing specialty care. Some individuals, for example, consider coronary bypass surgery an essential service, but others consider it an overpriced and overhyped luxury. Similarly, some favor offering only procedures necessary to keep patients alive, but others support offering procedures or technologies such as hip replacement surgery, home health care, hearing aids, or dental care, which improve quality of life but don't extend life.

Any system that does not provide comprehensive benefits runs the risk of devolving into a two-class system in which some individuals can buy more care than others can. To those who believe health care is a human right, such a system seems unethical. Others object to such systems on economic grounds, arguing that it costs less in the long run to plan on providing care for everyone than to haphazardly shift costs to the general public when individuals who can't afford care eventually seek care anyway.

Affordability

Guaranteeing *access* to health care does not help those who can't afford to *purchase* it. Consequently, we also must evaluate health care systems according to whether they make health care coverage affordable, restraining the costs not only of insurance premiums but also of **co-payments, deductibles**, and other crucial services such as prescription drugs and long-term care. Although the ACA offers some subsidies and tax credits to help people pay their premiums, it still leaves millions with many bills for these latter costs.

For health care to be affordable, individual costs must reflect individual incomes. As noted earlier, most insured Americans receive their insurance through employers. Typically, employers pay part of the cost for that insurance and deduct the rest from each employee's wages. Because low- and high-wage workers have their salaries reduced by the same dollar amount, low-wage workers are effectively hit harder: Paying $3,000 per year for health insurance might, for example, force a wealthier worker to scale back his vacation plans but force a poorer worker to put off fixing his roof. For this reason, the US system is considered **financially regressive** in that poorer people must pay a higher percentage of their income than do wealthier people. In contrast, in countries such as Great Britain and Canada, health coverage is paid for through graduated income taxes. Poorer persons pay a *lower* percentage of their income for taxes and therefore for health care than do wealthier persons, creating a **financially progressive** system. Either way—whether through taxes or lowered wages—the nation's citizens pay all the costs of health care. The only difference is who pays how much.

Financial Efficiency

Another critical measure of a health care system is whether it operates in a financially efficient manner. Currently, the multitude of private and public insurers in the United States substantially drives up the administrative costs of the health care system (Himmelstein et al., 2014). At the same time, the atomized and essentially entrepreneurial nature of our health care system makes it virtually impossible to impose effective cost controls. For example, doctors have responded to financial limits on Medicare payments by raising the fees they charge to non-Medicare patients, a process known as **cost shifting**. Neither of these problems was addressed by the ACA.

Consumer Choice

Finally, we need to evaluate health care systems based on whether they offer consumers a reasonable level of choice. Currently, wealthy Americans can purchase any care they want from any willing provider. In addition, Americans who have **fee-for-service insurance** can seek care from any provider as long as they can afford the copayments and deductibles, and if their plan uses managed care, as long as their insurer approves the care. Finally, those who have Medicaid or Medicare coverage can obtain care only from providers willing to accept the relatively low rates of reimbursement offered by these programs, and those who have no health insurance can obtain care only from the few places willing to provide care on a charity basis. The ACA, however, does seem likely to increase the options available for many Americans.

As we will see later in this chapter, in Mexico, China, and the DRC, some citizens have far greater choices in health care than do others, whereas in Germany, Great Britain, and Canada, all citizens have similar levels of health care choice.

HEALTH CARE IN OTHER COUNTRIES

With these measures in mind, we can now look at the health care systems in Germany, Canada, Great Britain, Mexico, China, and the DRC. Germany, Canada, and Great Britain are all considered to be more developed nations, and each guarantees portable, affordable, and universal health care coverage to its citizens. In contrast, both Mexico and China may soon join the ranks of the more developed nations. Mexico is gradually improving its health care system, but China's once exemplary record of providing health care to its citizenry has suffered recently. Finally, the DRC provides an example of the tremendous difficulties often faced by both the public and health care providers in the **least developed nations**. Despite their differences, however, most of the nations discussed in this chapter combine socialistic and entrepreneurial elements in their health care systems. In contrast, health care in the United States and the DRC is primarily organized as an **entrepreneurial system**, that is, a system based on private enterprise and the search for profit. Table 9.1 summarizes the characteristics of these six health care systems.

Not surprisingly, each of the systems described in this chapter has changed over time. More interestingly, the changes seem to have moved various nations at least somewhat toward **health care convergence**, that is, toward becoming more similar to each other (Stevens, 2010). For example, the United States, Great Britain, and Germany all now utilize **diagnosis-related groups** (DRGs) to restrain costs.

Two major causes of this convergence are **globalization** and economic pressures. Globalization has expanded access to medical and scientific knowledge. Increasingly, doctors use medical journals and Internet resources from around the world to learn about new treatments. Similarly, medical and pharmaceutical corporations now market new technologies internationally. Thus, doctors in many different countries are adopting the same technologies and placing similar economic pressures on their health care systems.

Second, whether a country's economy is booming or weakening and whether its health care system is largely capitalist or largely socialist, the cost of health care can press governments to reduce costs. Countries with largely capitalist health care systems may do so by *restricting* the role of the market in health care, whereas countries with largely socialistic health care systems may do so by *encouraging* the role of the market in health care. The latter situation, in which countries begin encouraging the private purchase of health care, the private practice of medicine for profit, and the operation of market forces in health care overall, is referred to as the **privatization of health care**. As in the United States, privatization reflects a **neoliberal** perspective.

Germany: Social Insurance for Health Care

Modern Germany is the product of a tumultuous twentieth-century history, including more than a decade of Nazism and the division of the country in two after its defeat in World War II. Yet despite the destruction wrought by two world wars and the economic stresses that accompanied the reunification of East and West Germany in 1990, the nation is a stable constitutional democracy and now enjoys one of the strongest economies in Europe. The gross national income (GNI) per capita is $44,500, about one-quarter lower than in the United States (Population Reference Bureau, 2014). (These figures are given in "international dollars," in which $1 equals the amount of goods and services one could buy for $1 in the United States.)

Structure of the Health Care System Health care in Germany is based on a system of social insurance (Commonwealth Fund, 2013). **Social insurance** refers to insurance provided by large social groups (such as regions, occupations, or industries) to their residents or members. This system was adopted in 1883 by politicians who hoped that offering workers accessible health care, as well as

TABLE 9.1 Characteristics of Health Care Systems in Seven Countries

Characteristics	United States	Germany	Canada	Great Britain	China	Mexico	Democratic Republic of Congo
Nature of system	Entrepreneurial	Social insurance	National health insurance	National health system	In flux from national health system to entrepreneurial	Multiple options providing increasingly equal access	System in tatters
Role of private enterprise	Very high	Moderate	Moderate	Low but rising	Moderate and rising	Moderate	Very high
Primary care doctors typically paid by:	Wide variety of payers and mechanisms (private, government, capitation salary, and so on)	Nongovernmental social insurance via capitation payments	Government via fee-for-service payments	Government via a form of capitation	Individuals on fee-for-service basis	Government as salaries	Individuals on fee-for-service basis
Universal coverage	No	Yes	Yes	Yes	No, but good access for urban residents	Yes, but variable quality and access	No
Payment mechanism for hospital doctors	Salaried and fee-for-service	Salaried	Salaried	Salaried	Salaried	Salaried	Theoretically salaries from government; actually fee-for-service plus international donations
Payment mechanism for hospital expenses	Varied	Lump sum from government and social insurance	Lump sum from government	Lump sum from government	Lump sum from government plus income from selling drugs and services	Lump sum from government	Theoretically salaries from government; actually funded by international donations

housing and unemployment and retirement benefits, would diffuse political tensions that might otherwise lead to a more radical redistribution of power and wealth in German society (Leith et al., 2009). Social insurance remains the center of the current German health care system, although about 10 percent of Germans now also or instead purchase private insurance.

Purchasing Care As in the United States, nongovernmental insurance forms the basis of the German health care system (Leith et al., 2009). But whereas in the United States, insurance providers must compete to survive in a profit-driven market, in Germany about 90 percent of health insurance is provided by nonprofit social insurance groups known as **sickness funds**. As of 2010, all Germans earning less than about $70,000 must join a sickness fund. The cost of belonging to a sickness fund is about 15 percent of income (about half paid by the individual for his or her entire family and the remainder paid by the employer). Because costs are based on income, the system, like Britain's, is financially progressive.

Paying Doctors and Hospitals German doctors are paid differently depending on the nature and location of their work. Those who work in hospitals receive annual salaries. Other doctors typically are paid on a fee-for-service basis. However, increasingly insurers are "bundling" payments, offering a set fee for doctors and other providers, both in and out of hospitals, who together care for patients with a specific condition such as diabetes or a hip joint that needs replacing. The hope is that integrating care across various providers will result in better health and lower costs. Hospitals receive their operating budgets from the sickness funds and receive their capital budgets (for items such as new magnetic resonance imaging machines) from the government.

Access to Care All Germans are required to have health insurance, and all German health insurance programs are required to provide a comprehensive package of health care benefits. With the exception of minimal copayments, insurance covers all costs of dental care, maternity care, hospitalization, outpatient care, prescription drugs, preventive measures such as vaccinations, and income lost because of illness. As a result, Germans have few incentives to put off obtaining needed care and see doctors an average of ten times per person per year, more than twice as often as do US citizens (OECD, 2014). Germans can see any doctors they like, although they must get referrals from primary care doctors before seeing hospital-based specialists.

Controlling the Costs of Care A major factor driving up costs of health care in Germany is the oversupply of doctors. To control this, Germany forbids doctors older than age 68 from working for the sickness funds, forbids doctors from opening practices in areas where many doctors already practice, and pays doctors by capitation rather than fee-for-service. Finally, to control drug costs, the sickness funds encourage doctors and consumers, through both education and economic incentives, to adopt more cost-effective drugs.

To control hospital costs, Germany now uses a system similar to the DRGs system in the United States. In addition, the government can restrain the purchase and use of unnecessary and expensive technologies because it determines hospitals' capital expense budgets.

Two factors still hamper efforts to constrain costs. First, hospitals have opposed policies designed to shift care when warranted to less expensive outpatient settings because the hospitals fear their incomes will fall. Second, the vast number of insurance providers in the German system has kept administrative costs high.

Health Outcomes Whether because of its health care system or because of its high standard of living and commitment to providing social services to its population, Germany enjoys a very high standard of health. Although conditions in the former East Germany remain poorer than in West Germany, those differences are rapidly disappearing. Life expectancy in Germany now averages 80, two years more than in the United States (and with far less variation among its citizenry). Infant mortality in Germany is among the lowest in the world: 3.5 per 1,000 live births compared with 6.5 in the United States (Population Reference Bureau, 2014).

Canada: National Health Insurance

Like the United States, Canada is a financially successful democracy made up of various provinces and territories more or less equivalent to US states. Although its GNI per capita of $42,600 is about one-quarter lower than in the United States, its economy is strong. In addition, because of steady immigration, Canada's population is younger on average than populations in the majority of more developed nations, which increases the likelihood of having a relatively healthy population.

Canada is also, however, a huge country, with vast social differences reflecting its vast geographic spaces. Its population is highly concentrated along its southern border, as are most health care personnel and facilities. Neither health status nor health care access is as good in rural areas or in its remote northern regions, where many of the residents are poor Native Americans (known in Canada as "First Nations").

Structure of the Health Care System The backbone of the Canadian health care system is the Canada Health Act of 1984, which stipulates that health care must be universal, must cover all medically necessary services, must be portable from province to province, must be publically administered on a nonprofit basis, and must be accessible to all regardless of ability to pay. The system is built around public insurance paid for primarily by each Canadian province, with assistance from the federal government (Duncan, Morris, and McCarey, 2009; Commonwealth Fund, 2013). For this reason, the Canadian system is referred to as **national health insurance**, or (as the previous chapter noted) a **single-payer system**. In fact, however, the Canadian system is a decentralized one, with each province retaining some autonomy and offering a somewhat different health care system. Underpinning the system are payments that the federal government gives the provinces yearly to run their health care systems. To receive these payments, provinces must offer comprehensive medical coverage to all residents through a public, nonprofit agency. Although the details of coverage vary across provinces, each province must charge residents only minimal fees and must allow residents to move to another province without losing their coverage.

Purchasing Care Through a combination of federal and provincial taxes, the public health insurance systems cover 70 percent of all health care costs, including most costs for hospital and medical care and some costs for prescription drugs, dental care, long-term care, and mental health services (Duncan et al., 2009). Because the system is based primarily on graduated income taxes, it is financially progressive: Wealthier persons pay a higher proportion of their income in taxes and therefore pay more toward health care than do others.

The remaining 30 percent of health costs are divided about equally between private insurers and private individuals paying out of pocket. Private insurance takes two forms. Most commonly, Canadians purchase private insurance to cover services not included in the national health insurance system. In addition, some provinces now allow residents to purchase private insurance that covers services that *are* included in the national health system. Such insurance enables individuals to buy these services immediately rather than having to wait their turn in the national health insurance system.

Paying Doctors and Hospitals Hospital doctors in Canada are paid on salary. Most non-hospital doctors work in private practices and are paid on a fee-for-service basis by the government insurance systems. Doctors submit their bills directly to the health insurance system using fee schedules negotiated annually between the provincial medical associations and provincial governments. Unlike in the United States, in Canada, doctors who consider these fees too low can't **balance bill** (or "extra bill," as it is known in Canada): billing patients for the difference between what the patients' insurance will pay and what the doctor wants to charge. In addition, some provinces control costs by setting annual caps on the total amounts they will reimburse doctors. In practice, this means reimbursing doctors less for each service rendered as the total number of services rises.

Canadian hospitals (almost all of which are nonprofit) annually receive an operating budget and a capital expenditure budget from their provincial insurance system. Hospitals can spend their budgets as they like as long as they provide care to anyone in their region who needs services.

Access to Care Canadians average eight doctor visits per person per year compared with four visits for US citizens (OECD, 2014). Waiting times for technologically complex care have been a problem in Canada, although rarely in life-threatening circumstances. However, recent data suggest considerable improvement (Canadian Institute for Health Information, 2010; Duncan et al.,

2009). Most importantly, Canadians are far less likely than US residents to go without needed health care for financial reasons or to risk bankruptcy if they do seek health care. Moreover, although Canadians are less likely to receive certain high-technology procedures, such as coronary artery bypass surgery, this may reflect *overuse* in the United States rather than *underuse* in Canada. For example, one large study found that after heart attacks, Americans were five times more likely than Canadians to receive coronary angiography and almost eight times more likely to receive coronary bypass surgery, but one-year survival rates for the two groups were identical (Tu et al., 1997).

Controlling the Costs of Care Costs of health care have risen rapidly in Canada, primarily because of population growth and increased prices for drugs and advanced technologies. Nevertheless, costs are far lower than in the United States (OECD, 2014). The United States currently spends 17 percent of its gross domestic product (GDP) on health care, whereas Canada spends 11 percent.

How does the Canadian system restrain health care costs? Most important, a single-payer system dramatically reduces administrative overhead (Himmelstein et al., 2014). In a single-payer, nonprofit system, no one need spend money selling or advertising insurance, paying profits to stockholders, sending bills to multiple insurers and individuals, or tracking down those who don't pay their bills. Nor is money spent collecting funds to run the system because those funds are already collected from the public through existing taxation systems.

The single-payer system also saves money by centralizing purchasing power. As the sole purchasers of drugs in Canada, the provinces have substantial leverage to negotiate with pharmaceutical companies regarding drug prices. Similarly, as the sole payer of doctors' bills, the provinces have considerable bargaining power when negotiating with doctors over how much to reimburse doctors per service. Finally, as the sole payer of hospital budgets, the government can implement efficient regional planning and avoid unnecessary duplication of expensive facilities and services.

Nevertheless, costs have risen substantially. Paying doctors on a fee-for-service basis makes it more difficult for Canada to control medical costs. When, for example, the provinces banned balance billing, doctors responded by increasing the number of services they performed (with the provinces responding by reducing the amount they reimbursed for each service). Finally, Canadian hospitals, like US hospitals, have reduced their costs by shifting toward outpatient services and shorter patient stays, thus moving some costs from the health care system to family caregivers.

Health Outcomes Despite continuing problems in access to health care, outcomes compare very favorably with those in the United States. Infant mortality in Canada is 5.1 per 1,000 births compared with 6.4 in the United States, and average life expectancy is three years longer in Canada. Of course, these health outcomes tell us more about social conditions than about the quality of health care. Nevertheless, these data suggest that the Canadian health care system, although certainly not perfect, is superior to the US system.

Great Britain: National Health Service

As the home of the Industrial Revolution, Britain for many decades was a leading industrial power. Along with its industrial strength came a strong labor movement as workers united to gain political power within Britain's parliamentary government. As a result, a commitment to protecting its citizens, including a commitment to universal health care coverage, has long been central to Britain's identity. Beginning in the 1980s, however, the nation's economy declined while health care costs rose. To restrain those costs, subsequent governments instituted a series of reforms designed to introduce market principles into the health care system while retaining universal health coverage (Lopes, Coppola, and Riste, 2009). Currently, GNI per capita in Britain is $35,800, about one-quarter lower than in the United States.

Structure of the Health Care System According to the esteemed Commonwealth Fund, Great Britain has the world's top health care system (Davis et al., 2014). Whereas Canada provides its citizens with national health *insurance*, Great Britain since 1948 has provided care through its **National Health Service (NHS)** (Commonwealth Fund, 2013). In Canada, the government provides insurance so individuals can purchase health care from private practitioners. In Great Britain, on the other hand, the government directly pays virtually all health care costs. As a result, the two systems look quite similar to health care consumers but differ substantially from the perspective of hospitals, health care workers, and the government. This section focuses on the structure of the NHS in England, one of the three countries that (along with Scotland and Wales) comprise Great Britain.

Purchasing Care Unlike US citizens, most English citizens rarely see a medical bill, an insurance form, or any other paperwork related to their health care. The NHS uses tax revenues to pay virtually all costs for a wide range of health care services, including medical care, visiting nurses for the homebound, homemakers for chronically ill persons, and some aspects of long-term care.

The NHS receives its funds almost solely through general taxation, with small supplements from employers and employees. As in Canada, because the health care system is paid for through graduated income taxes, it is financially progressive.

Paying Doctors and Hospitals As in Germany, almost all medical specialists work as salaried employees of the NHS at hospitals or other health care facilities, although they can earn extra income by seeing private patients. In contrast, most English general practitioners work as private contractors, increasingly in large group practices. General practitioners are paid by capitation. **Capitation** refers to a system in which doctors are paid a set fee per year for each patient in their practice, regardless of how many times they see their patients or what services the doctors provide. In such a system, doctors lose income when they provide more services. In addition, general practitioners receive financial supplements if they have low-income or elderly patients; practice in medically underserved areas; or meet government targets for preventive services, such as immunizing more than a certain percentage of children in their practices.

The vast majority of hospitals in England belong to the government (although some now include beds for private patients). The hospitals operate semi-autonomously, but regional NHS officials and hospital administrators work together to ensure that each hospital can offer quality care to patients.

Access to Care Under the NHS, individual financial difficulties no longer keep English citizens from receiving necessary medical care. Waits can be uncomfortably long for nonemergency care, but any case delayed more than 18 weeks is reported to national authorities for further action. In addition, the NHS has reduced substantially the geographic inequities that for generations made medical care inaccessible to many rural dwellers, although access to care remains a problem in poor, inner-city neighborhoods. Britons average five doctor visits per person per year compared with four visits for US citizens (OECD, 2014). Access to high-technology care and expensive new drugs, however, remains lower than in the United States. That said, in the United States access to treatments is limited only by the ability to pay, whereas in England a national panel of medical experts decide which services should be offered to citizens, based on their effectiveness, and then sets the prices for those services. Those prices are considerably lower than average prices in the United States and must be honored by drug manufacturers, private practice doctors, and anyone working under NHS auspices. For example, until it was taken off the market in 2011, many US health insurers (including **Medicare**) paid up to $100,000 per patient per year for the anticancer drug Avastin, even though strong evidence suggested it was ineffective (Kolata and Pollack, 2008). In contrast, the NHS decided against covering it from the start, arguing that NHS money would be better spent on less expensive drugs with better track records. In sum, both the US and UK systems limit access to care, but in very different ways with very different consequences.

Controlling the Costs of Care Great Britain spends about 9 percent of its GDP on health care, almost half the percentage spent by the United States (OECD, 2014). Like Canada, Britain has made its health funds go further than they otherwise would through national and regional planning and by keeping salaries relatively low. Because the government owns a large proportion of health care facilities and employs a large proportion of health care personnel, it can base decisions about developing, expanding, and locating high-technology facilities on a rational assessment of how best to use available resources and can avoid the unnecessary proliferation of expensive facilities. Similarly, because it is such a large buyer, the NHS can negotiate drug prices effectively with pharmaceutical firms.

In addition, England has attempted to restrain government health care expenditures by promoting privatization: Private companies can now run primary care practices funded by the NHS; private hospitals can compete for NHS contracts against public hospitals; and NHS hospitals can now offer a variety of services on a cash basis. In addition, officially commissioned groups of general practitioners in each region are now required to put services (such as nursing home or hospital care) up for bid rather than deciding based on their own judgment whether to use for-profit or NHS services. Finally, and as in Germany, both private and public hospitals have been pressured to control costs by a DRG sort of system.

Health Outcomes Despite some access problems in the NHS, health outcomes have remained good. Infant mortality is lower than in the United States (4.7 vs. 6.4 per 1,000), and life expectancy is two years longer. However, individuals' social class and ethnicity continue to affect both health and health care access, although considerably less than in the United States. Moreover, the current budget crisis may well bring substantial cuts to the NHS despite high popular support for the program.

China: Promises and Perils

Although many observers have proposed using the health care systems of Germany, Canada, and Great Britain as models for a restructured US health care system, few would seriously propose China (officially known as the People's Republic of China) as a viable model. China's culture differs greatly from that of the United States, so its citizenry has very different values regarding what constitutes an acceptable health care system. In addition, China's GNI per capita of only $11,900 (Population Reference Bureau, 2014) severely limits its options, and the remaining Communistic underpinnings of its economy make a different set of health care options feasible there than in the United States. (These data and this discussion don't include Hong Kong, which only became part of China in 1997 and operates under a separate political structure.) That said, China's story suggests *both* how poor countries can protect their citizens' health by committing to primary care and to public health improvements *and* how quickly those gains can be reversed if that commitment wavers (LaFraniere, 2010; Riley, 2007; Wang, Xu, and Xu, 2007).

China's health care system reflects its unique history and situation. When, after many years of civil war, the Communist Party in 1949 won control of mainland China, it found itself in charge of a vast, poverty-stricken, largely agricultural, and densely populated nation of about 1 billion persons. Most people lived in abject misery while a small percentage enjoyed great wealth. Malnutrition and famine occurred periodically, life expectancies for both men and women were low, and infant and maternal mortality were shockingly high. In urban areas, only the elite typically could afford medical care, whereas in rural areas, where most of the population lived, Western medical care barely existed.

Structure of the Health Care System In 1950, one year after winning control of mainland China, the Communist government announced four basic principles for the new nation's health care system (Anson and Sun, 2005:10). First, the primary goal of the health care system would be to improve the health of the masses rather than of the elite. Second, the health care system would emphasize prevention rather than cure. Third, to attain health for all, the country would rely heavily on mass campaigns. Fourth, the health care system would integrate Western medicine with traditional Chinese medicine.

These principles reflected both the political climate and the practical realities of the new People's Republic of China. The first goal—improving the health of the masses—stemmed directly from the Communist political philosophy underpinning the revolution. The years of bloodshed were to be justified by a new system that would more equitably distribute the nation's wealth and raise the living standards and health status of China's people. The second and third goals reflected unignorable facts about China's situation. Lacking both a developed technological base and an educated citizenry, China's greatest resource was the sheer labor power of its enormous population, which could be efficiently mobilized because of its now-centralized economy. Focusing on prevention through mass campaigns promised to deliver the quickest improvements in the nation's health. Finally, the decision to encourage both Western and traditional medicine similarly recognized the difficulties China would face in developing a Western health care system, as well as the benefits of including traditional medicine in any new system. By encouraging traditional as well as Western medicine, China could take advantage of its existing health care resources and gain the support of the peasantry, who remained skeptical of Western medicine. At the same time, incorporating traditional medicine into the new, modernized Chinese health care system offered a powerful statement to the world regarding the new nation's pride in its traditional culture. Simultaneously encouraging the growth of Western medicine, meanwhile, would help bring China into the scientific mainstream.

Given its large and poverty-stricken population and its lack of financial resources and medically trained personnel, China needed to adopt innovative strategies if it was to meet its goal of improving the health of the common people. Two of these strategies were the use of mass campaigns and the development of **physician extenders**—individuals (such as nurse practitioners and physician assistants in the United States) who can substitute for doctors in certain circumstances.

Mass Campaigns One of the more unusual aspects of China's health care policy has been its emphasis, especially in the early years of the People's Republic, on mass campaigns (Horn, 1969). For example, to combat syphilis, which was **endemic** in much of China when the Communists came to power, the government first closed all brothels, outlawed prostitution, and retrained former prostitutes for other work. Second, the government began the process of redistributing income and shifting to a socialist economy so that no young women would need to enter prostitution to survive. During the next decade, the government trained thousands of physician extenders to identify persons likely to have syphilis by asking ten simple questions, such as whether the person had ever had a genital sore. By so doing, the government made manageable the task of finding, in a population of 1 billion, the small percentage that needed to be tested and treated for syphilis.

To convince people to come to health centers for testing, these physician extenders held mandatory political meetings in villages, performed educational plays in marketplaces, and gave talks around the country, explaining the importance of eradicating syphilis and attempting to reduce the stigma of seeking treatment for syphilis by defining the disease as a product of the corrupt former regime rather than a matter of individual guilt. Those identified as likely to have syphilis were tested and treated if needed. These methods—coupled with testing, among others, persons applying for marriage licenses, newly drafted soldiers, and entire populations in areas where syphilis was especially common—dramatically reduced the **prevalence** of syphilis in China.

Physician Extenders The second innovative strategy for which China has won acclaim is its use of physician extenders in addition to medical doctors trained in Western and (rarely) traditional medicine. In urban areas, **street doctors** offer both primary care and basic emergency care, as well as health education, immunization, and assistance with birth control. Street doctors have little formal training and work in outpatient clinics under doctors' supervision.

In rural areas, **village doctors** played a similar role. Novice village doctors were selected for health care training by their fellow workers based on their aptitude for health work, personal qualities, and political "purity." After about three months of training (supplemented yearly by continuing education), village doctors returned to their rural communes, where they divided their time between agricultural labor and health care. Beginning in the late 1970s, however, village doctors were largely replaced by **assistant doctors**, who receive three years of postsecondary training in Western and traditional Chinese medicine and who can provide both primary care and minor surgery. Unfortunately, there are far too few assistant doctors, and so rural areas remain seriously underserved.

Purchasing Care As China's economy has changed from a largely socialized and centrally controlled system toward a more decentralized, economically heterogeneous model, so has its health care system (Chen, 2001; Wang et al., 2007). For the majority of urban residents, these shifts have brought few changes. As in the past, the government pays most costs of health insurance and health care for government employees, military personnel, and students. Public industries and urban industrial collectives also pay for care for their workers. The growing and now significant numbers of urban residents who work in private enterprises, however, often lack any health insurance.

For rural Chinese—about 60 percent of China's population—recent years have brought dramatically reduced access to health care (Wang et al., 2007). Before the 1980s, rural residents received their care at little or no cost through the agricultural communes where they lived and worked. Within these communes, members shared all profits and costs, including those for health care. Each commune had between 15,000 and 50,000 members, several village doctors, and a clinic staffed by assistant doctors.

Beginning in the early 1980s, most agricultural communes reverted to their original noncommunal village structures, with each family given land to farm by the village. Families now keep their profits but are responsible for their own welfare if costs exceed profits. Because of this shift in financing, the former communes no longer earn sufficient revenues to continue providing health care. Many village doctors returned to full-time agricultural work, and most assistant doctors moved to township or city clinics. Almost all rural residents now receive their primary health care on a fee-for-service basis, and financial difficulties have forced some to cut back on needed care. In addition, waning government support for large-scale public health activities has allowed previously conquered diseases to reemerge. For example, schistosomiasis, a debilitating and sometimes deadly disease once eradicated by mass campaigns that killed the snails that carry it, is again endemic in some rural areas (Yardley, 2005).

Paying Doctors and Hospitals Currently, non-hospital doctors in China work primarily on a fee-for-service basis, and hospital doctors work on salary. In addition, many townships (made up of six or more rural villages) have a clinic where doctors work on salary but are allowed to divide among themselves any profits that the clinic generates. As a result, doctors have an incentive to order unnecessary tests and procedures (Wang et al., 2007).

Unlike most medical care, hospital care has remained largely a public enterprise. Almost all hospitals receive their operating and capital budgets from federal or local governments. In recent years, however, budgets have been cut and great pressure has been placed on hospitals to generate income through selling drugs and services and by starting other enterprises. As a result, hospital patients run considerable risk of receiving unnecessary (and potentially dangerous) drugs, surgeries, and other treatments (LaFraniere, 2010).

Access to Care Because of the changes in China's health care system, prices for health care have risen and access has diminished, especially in rural areas, where fewer hospital beds and doctors are available per capita. Although primary care remains affordable, even for those who lack health insurance, there are very few primary care doctors per person. As a result, people often turn to hospitals—which can be prohibitively expensive—even for very basic care (LaFraniere, 2010).

To equalize access to care, the government has established a national fund to supplement the health care budgets of poorer regions and an insurance program for childhood immunizations. Those who, for a small premium, purchase this insurance receive free immunization for children up to age seven and free treatment if a child develops one of the infectious diseases the immunization program is supposed to prevent. More than half of all children in the country belong to this program. Finally, a similar insurance program offers prenatal and postnatal care to women and infants; it is not known how many are covered by this program.

Health Outcomes Although China's economy is developing rapidly, it still spends only about 5.2 percent of its GDP (about $400 per person) on health care, considerably less than that spent in the more developed nations (OECD, 2014). Nevertheless, China's commitment to equalizing both income and health care has allowed it to attain health outcomes far greater than its economic status or investment in health care might predict. Although median income in China remains similar to that in many other **less developed nations**, China boasts health outcomes only slightly below those of the more developed nations. Whereas in 1960 infant mortality was 150 deaths per 1,000 and life expectancy was 47 years, currently infant mortality is 21 per 1,000 and life expectancy is 74, only four years lower than in the United States (Population Reference Bureau, 2014). Although large and increasing differences in health status remain between rural and urban dwellers, China now stands on the cusp of the **epidemiological transition**, with chronic and degenerative diseases increasingly outpacing infectious diseases as the leading causes of death.

Nevertheless, some regions of China continue to face health problems that have long characterized the less developed nations, such as insufficient access to clean drinking water. The rise of a market economy has contributed to these problems as pressure to develop profitable industries has increased water and air pollution and decreased occupational safety, especially in rural areas (Chen, 2001). Similarly, pressures on the health care system to control costs and generate profits has led to a decreased emphasis on preventive care and increased emphasis on profit-generating treatments and diagnostic procedures.

Despite these problems, however, China does offer some lessons in how to improve health in the less developed nations. As Chapter 4 described, three factors seem to explain how China (like Sri Lanka, Costa Rica, Vietnam, and Cuba) achieved excellent health outcomes at low cost (Caldwell, 1993; Riley, 2007). Not surprisingly, health outcomes improved when access to medical care improved. But improved health outcomes depended even more on emphasizing family planning; raising education levels among men; and, especially, raising education levels among women. Once women's educational levels increased, their power in the family increased, giving them greater control over family planning. Women's lives thus were less often cut short by childbirth, and their babies were born healthier. In addition, as women's status rose, they and the children who depended on them more often received a fair share of the family's food, thus reducing malnutrition and increasing life expectancies.

Mexico: Moving toward Equitable Health Care

Understanding Mexico's health care system is particularly important for US citizens because Mexico shares a long and permeable border with the United States. People routinely travel across the border in both directions for work or pleasure, bringing their diseases with them. In addition, both Mexicans *and* US citizens sometimes cross the border to the other country to seek health care, although Mexicans more often travel north to seek medical care for life-threatening health conditions, and US citizens more often travel south to seek inexpensive cosmetic surgery, dental work, or medical drugs.

Mexico has only recently entered the ranks of the more developed nations and still has much in common with the *less* developed nations. As Mexican industry has developed, many have moved off the land, and now more than three-quarters of Mexico's population live in cities. Those cities contain both middle-class neighborhoods that enjoy health and living conditions similar to those found in the more developed nations and impoverished slums that lack such basic facilities as running water and sewer systems. These slums are inhabited primarily by migrants from rural areas. Rural areas, especially those inhabited primarily by Indians, generally are poor, and about 40 percent lack sewer systems (Pan American Health Organization, 2012). GNI per capita remains only $16,100—considerably higher than in China but far lower than in the United States or in the European nations discussed in this chapter (Population Reference Bureau, 2014).

Structure of the Health Care System Unlike any of the other countries described in this chapter, Mexico has a three-part system for health care: (1) private health care and health insurance for the wealthiest, (2) a government-provided insurance program for salaried workers (Social Security), and (3) a separate government-provided insurance program for everyone else (*Seguro Popular*) (Frenk et al., 2006). This three-tiered system is a product of Mexico's unique history, in which revolutionary fervor and conservative sentiments have always counterbalanced each other and in which the social and economic division between Indians (who now make up less than 10 percent of the population) and others (who are primarily a mix of Spanish and Indian) has remained important.

Over the centuries, Mexico has experienced several revolutions—some violent and some at the ballot box. Throughout the twentieth century, these revolutions resulted in gradual improvements in the health care available to Mexico's citizens. In 1917, Mexico's new constitution first gave the federal government responsibility for health care. The government soon began providing funds for rural clinics staffed by health aides and, by the 1930s, began requiring all new physicians to work for a year in a rural community.

The next major change in the health care system occurred in 1942, when the government established the Social Security program and opened a network of modern health clinics and hospitals around the country for Social Security members. However, that program covers only salaried workers—about half the population—leaving many others with no access to health care or with crushing debts if they seek such care. Consequently, in 2003, Mexico passed a law aimed at reforming this system (Knaul et al., 2012). Under the 2003 law, all Mexicans not eligible for Social Security can instead obtain membership in another government-run health insurance program, known as Seguro Popular (Public Insurance). Membership is free for the poorest 20 percent of Mexicans and available on a sliding scale to all others. Less than a decade later, more than 50 million people had enrolled in the program, giving Mexico essentially universal health insurance coverage (Knaul et al., 2012).

Nevertheless, inequities remain within the system (Knaul et al., 2012). Social Security provides a more comprehensive package of health benefits than does Seguro Popular, and urbanized and wealthier regions continue to have more and better health care providers and facilities than do rural and poorer regions under both Social Security and Seguro Popular. Although these inequities are diminishing, the system as a whole remains underfunded, so problems are likely to continue.

Purchasing Care Mexicans typically pay only small copayments or other fees for their health care. Fees are waived for the poorest Mexicans.

Paying Doctors and Hospitals Most Mexican doctors work as salaried government employees, although most also take private, fee-for-service patients on a part-time basis, and some work solely for private patients. Public hospitals receive their funds from the government out of tax revenues.

Access to Care Individuals who purchase health care in the private sector have, of course, a wide choice of doctors and hospitals. Most Mexicans, however, must use the doctor or the clinic to which they are assigned for primary care (although in theory they have some choice). To obtain specialty care, patients must first get referrals from their primary care doctors. Such referrals can be difficult to get, however, because of government cost controls that restrict the number of practicing specialists. For the same reason, patients who do get referrals typically have long waits before they can get appointments with specialists. As a result, many patients subvert the system by instead seeking specialty care at emergency clinics or, if they can afford to, from private doctors.

Mexicans average three doctor visits per person per year compared with four visits for US citizens (OECD, 2014). Their access to technologically intensive care, however, remains limited. In addition, these services are haphazardly distributed, with more services available in cities and wealthier regions (most in the north) than in rural areas and poorer regions (most in the south). Recent reforms, however, have reduced these differences (Knaul et al., 2012).

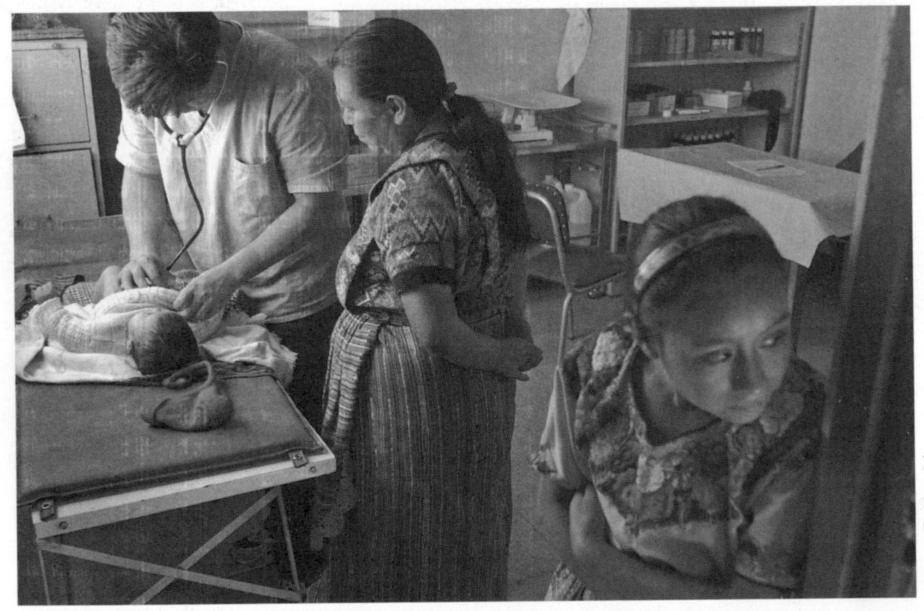

AP Images/Rodrigo abd

Although its health care system is far from perfect, Mexico is working hard to improve access to medical care across the country, including in rural areas inhabited largely by Indian minorities.

Health Outcomes Although Mexico remains rife with social and economic inequities and resulting inequities in health, it has nevertheless achieved notable improvements in health outcomes for much of its population. Consequently, by some measures, Mexico appears to have completed the epidemiological transition—cancer and heart disease now kill more Mexicans than do infectious diseases, and life expectancy is 76, only two years less than in the United States (Population Reference Bureau, 2014). Infant mortality, child mortality, and maternal mortality all decreased substantially between 2000 and 2010 (Knaul et al., 2012).

These health outcomes have been achieved at relatively little cost. Mexico spends less than $1,000 per person on health care—about 6 percent of its GDP, compared to 16 percent in the United States (OECD, 2014).

Democratic Republic of Congo: When Health Care Collapses

Current conditions in the DRC are the result of more than a century of corrupt and unstable governments. During the 1880s, King Leopold II of Belgium gained control over what was then known as the Congo Free State, holding it as his private property. Although Belgian administrators helped to develop basic infrastructure such as roads and hospitals, they also exploited natural resources for Belgium's benefit, used brutal force when it suited their purposes, kept virtually all power in Belgian hands, and flamed ethnic conflict among different Congolese ethnic groups to keep them from uniting against Belgian control.

Belgium relinquished control over the DRC in 1960. Since then, the country has been governed primarily by a series of corrupt, ruthless dictators (often funded by the United States for its own purposes). To make matters worse, beginning in 1998, the country was torn by a ferocious civil war that was fueled by inter-ethnic conflict and competition over valuable minerals. The numbers of dead from that war approach those from World War II, and the widespread use of mass rape as a weapon stunned the world.

Although the civil war officially ended in 2003, the violence continues, and the central government's control over the military, paramilitary groups, and the public remains fragile. Moreover, years of warfare led to environmental destruction, the abandonment or destruction of agricultural lands, and large-scale movement of citizens away from rural war zones into substandard, temporary housing in cities. All these factors have fed malnutrition and disease, including preventable outbreaks of cholera; measles; malaria; and, most ominously, Ebola virus disease (Doctors Without Borders, 2014). The DRC remains one of the world's poorest nations, with a GNI per capita of only $680 (Population Reference Bureau, 2014).

Structure of the Health Care System On paper, the health care system in the DRC consists of a network of hospitals, primary care clinics, and public health workers distributed around the country's numerous health districts (Inungu, 2010). These days, however, it is difficult to even talk about a health care system in the DRC. During the civil war, many doctors fled rural areas or fled the nation altogether, many hospitals were damaged or destroyed, and many pharmacists lost access to basic medications. Although conditions have improved since then in urban areas (primarily because of disease-specific health programs sponsored by the World Bank, World Health Organization, and other international nonprofit organizations), the situation remains dire. In areas where conflict continues, health care personnel have been harassed or threatened, and many clinics have been abandoned (Doctors Without Borders, 2014).

Purchasing Care Although treatment in state-run clinics and hospitals is supposed to be offered at low prices, in reality, patients and their families are often expected to pay for everything, including medicines, bandages, and other supplies. Moreover, even the lowest of fees are too high for many Congolese to pay. Others may decide against going to a doctor because they know they can't afford any medicines that the doctor might prescribe (Doctors Without Borders, 2014).

Because of both cultural traditions and a lack of access to Western medicine, many Congolese rely on homemade herbal remedies or seek care from traditional midwives or traditional healers called *ngangas* (Inungu, 2010). *Ngangas* are believed able to determine whether an illness was caused by natural or supernatural forces and to prescribe appropriate treatments, such as wearing a talisman to ward off evil or drinking an herbal potion. Some of the treatments used by traditional practitioners undoubtedly help (if only through a **placebo** effect), but others undoubtedly harm.

Paying Doctors and Hospitals In theory, doctors and hospitals receive regular salaries and budgets from the federal government. In practice, many doctors have been paid little or nothing for years, so they support themselves by charging fees to patients and their families (Michon, 2008). Most hospitals now receive most of their funding from international nonprofit organizations and donations from the more developed nations.

Access to Care As this discussion suggests, most citizens of the DRC—especially in rural areas—have extremely limited access to modern medical care. Many of the rest rely on temporary facilities staffed by international aid workers. Meanwhile, the wealthiest Congolese can travel to South Africa or elsewhere whenever they need care (Inungu, 2010).

Health Outcomes By all measures, health outcomes in the DRC are abysmal. Average life expectancy is only 48 years, far below that in most nations around the world. Similarly, 114 of every 1,000 babies die in infancy—*18 times* higher than in the United States (Population Reference Bureau, 2014).

IMPLICATIONS

A critical approach to health care reform suggests that for true structural changes to occur in the US health care system, we must be willing to challenge the power dynamics underlying the current system—something that did not happen with passage of the ACA. Anyone interested in promoting such change can benefit from the experiences of other countries that have successfully done so.

Germany's example suggests that it is possible to have an effective and cost-efficient system even with many different insurers—although in Germany, all the insurers are nonprofit. Similarly, Canada's history suggests that eliminating private insurers—major power holders in the current system—can reduce costs substantially by eliminating the costs of selling, advertising, and administering the various insurance plans. Eliminating private insurers also eliminates the costs that accrue when doctors, hospitals, and other health care providers must track and submit bills for each client to each insurance company.

Although it is unlikely that the United States would ever go so far as to establish a national health system as Britain has done, Britain's experience does illustrate the benefits of centralizing control. And all of the countries described in this chapter (except the DRC) illustrate how establishing government control over both operating and capital budgets for hospitals and other facilities can restrict the duplication of services and proliferation of technologies that have driven up the costs of the existing system. By the same token, establishing a national fee schedule for service providers, such as Canada uses, would enable the government to restrict the rise of those fees. Even more control is possible if the government, like Britain's, restricts doctors to salaried practices so they can't increase their incomes by increasing the number of procedures they perform. At the same time, mandating national health coverage, regardless of the nature of the system, would guarantee a large enough risk pool to make **community rating** feasible and affordable. Finally, using income taxes to pay for health care would more equitably distribute the costs of financing the system.

These issues may all come to the fore again as debate continues over repealing or revising the ACA and over the many regulations that will be needed to translate the law into practice.

SUMMARY

1. Health care systems can be evaluated according to whether they offer universal coverage, portable and comprehensive benefits, geographically accessible care, affordable coverage, financial efficiency, and consumer choice.
2. Globalization has combined with economic pressures to lead to health care convergence. Countries that are primarily capitalist have *restricted* market forces in health care; countries that are primarily socialist have *increased* market forces.
3. Health care in Germany is overwhelmingly obtained through nonprofit social insurance plans known as sickness funds. Social insurance refers to insurance provided on a nonprofit basis by social groups, such as cities, occupations, or industries.

4. Canada offers universal coverage through a single-payer (government-run) National Health Insurance program. Governments provide Canadian hospitals with their operating and capital expenditure budgets. Hospital doctors are paid on salary, and primary care doctors are paid fee-for-service, with fees negotiated between medical associations and the government.
5. Canadians have greater access to care than do US citizens, with better outcomes and lower costs overall. However, reductions in federal subsidies for health care have led to longer waiting lines and increased pressure for privatization.
6. Great Britain provides universal access to health care through its government-run National Health Service. Primary care doctors are paid by capitation; specialists are salaried government employees.
7. By centralizing purchasing power, Canada, Germany, and Great Britain have gained the ability to effectively negotiate prices with health care providers and thus keep down costs. By controlling hospital budgets, these governments have reined in unnecessary duplication of expensive services. By using single-payer systems, Canada and Great Britain have dramatically reduced administrative overhead.
8. To control costs, Canada, China, and Great Britain have made some moves toward privatization of health care.
9. By emphasizing mass campaigns and physician extenders, China was able to improve access to care and quality of care for millions of poor citizens. Life expectancy also rose because the government committed to family planning and to public education for men and, most importantly, women.

10. As China's economy has become more capitalistic and decentralized, it has moved toward a fee-for-service system. Access to care has declined, especially in rural areas and among urban residents who work for private businesses.
11. Mexico now provides essentially universal health care access. However, that access comes within a three-tiered system: (1) private health care for those who either have no health insurance or believe that purchased care is higher quality, (2) a government-provided insurance program for salaried workers, and (3) a separate government-provided insurance program for everyone else. Moreover, urbanized and wealthier regions continue to have more and better health care providers and facilities than do rural and poorer regions.
12. Conditions in the DRC illustrate how a health care system can collapse and the consequences when that happens.

REVIEW QUESTIONS

1. Define the eight measures of health care systems, and explain why each is important.
2. What is health care convergence? What evidence of convergence can be found in the histories of health care in Great Britain and China?

3. How are doctors and hospitals paid in Canada? In Great Britain?
4. What is the difference between national health insurance and a national health service?
5. How does access to primary and hospital care in Canada compare with access to care in the United States?
6. What aspects of the health care systems in Canada and Great Britain have helped them to restrain costs? What aspects have kept costs high?
7. How has the rise of market forces affected health care in Great Britain?
8. What aspects of its health care system have enabled China to provide good health at low cost to its people? What factors have led to a worsening of coverage?
9. What are the similarities and differences between health care in Mexico and the DRC?

CRITICAL THINKING QUESTIONS

1. Compare and contrast the health care system in the United States with the system in one other country. Explain which system you would prefer.
2. Do people have a right to health care? To primary care? To secondary care? To tertiary care? Do children? Adults? Elderly people? People with disabilities? Unemployed drug addicts? Illegal aliens?

CHAPTER

10

Health Care Settings and Technologies

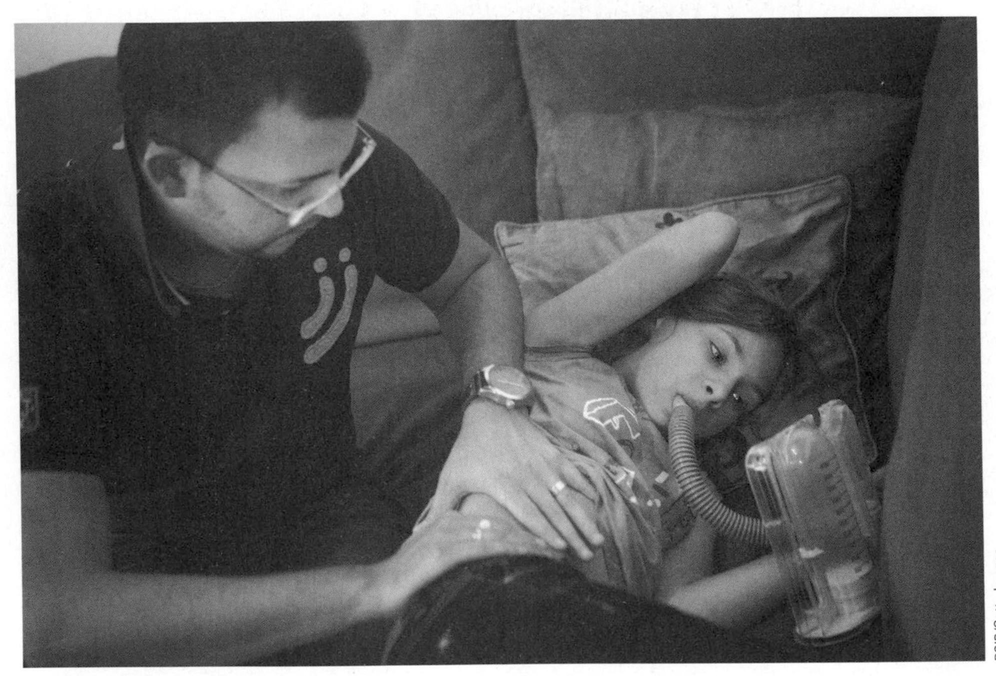

BSIP/Getty Images

LEARNING OBJECTIVES

After reading this chapter, students should be able to:

- Understand how hospitals, nursing homes, and hospices have evolved.
- Critique the ways that hospital, nursing home, and hospice care are now structured.
- Assess the benefits and burdens of caring for ill and disabled individuals at home.
- Analyze the nature and consequences of health care technologies.

Stacy Trebing yanked off the yellow paper hospital gown that covered her shorts and T-shirt, unhooked the surgical mask from behind her ears, and stuffed both items into the garbage pail in the entryway of her daughter's hospital room. She'd been at her three-year-old daughter's bedside practically every minute of the past ten days.

She needed a breather.

The next morning, Stacy's daughter would have a bone marrow transplant, a medical procedure that would either cure her or kill her. Every minute since Katie's birth had been leading to this day. Everything Stacy and her husband, Steve, had done, every decision they'd made, had propelled them here—including the most controversial of their choices: to [use in-vitro fertilization to] create a new human being they had selected as an embryo because he genetically matched a critical portion of his sister's DNA.

That one-year-old baby would be brought into the hospital the following morning to donate the life-changing bone marrow that was the only chance to heal his sister [who had an extremely rare genetic condition that would otherwise require painful, debilitating treatments and that would still likely kill her by her forties]. Christopher Trebing was born to be a member of the Trebing family [to parents who had wanted another child regardless], but he was also born with a job to do. He would be put under general anesthesia while a doctor inserted needles repeatedly into his hips and siphoned the tissue that could repair Katie's ailing body.

Katie and Christopher wouldn't see each other on what the doctors called Day Zero. Katie would stay in isolation in her room, and Christopher's marrow would be transported in an IV bag and dripped into her. Doctors told Stacy that because it had been so difficult to get an IV into Christopher's veins during his preoperative blood testing, they might have to go through a more dangerous route, a vein in his leg, to administer anesthesia. Stacy feared for both children.

As she sat, Stacy wasn't dwelling on the many ethical issues that troubled the bioethicists and critics who thought no baby should be conceived with a purpose: Who would protect the medical interests of what was referred to as a "savior sibling" when his parents were so focused on curing the older child? How would such a baby feel when he grew up and learned he had been brought into the family with a responsibility? Who would object if the child was later called upon to donate something more radical than bone marrow to help the sibling—a kidney perhaps?

As his mom, Stacy had more personal concerns: How would she feel if Christopher's much-anticipated bone marrow donation didn't work? What if Katie's body rejected Christopher's marrow and Katie died? Would it change how Stacy felt about Christopher? Would it make it hard to be his mother? If anything ever happened to Katie, Stacy asked herself uneasily, would I be resentful toward him? (Whitehouse, 2010: ix–xi)

In the end, the Trebing's gamble paid off, and both children emerged happy and healthy. Their story vividly illustrates the nature of modern health care technologies and the many questions those technologies raise, including questions about which technologies should be developed, who does or should receive them, and what are the financial and other costs of these technologies to both individuals and society in general. This chapter offers a sociological analysis of health technologies and of four key settings in which those technologies are offered: hospitals, nursing homes, hospices, and family homes. We begin with a discussion of hospitals.

THE HOSPITAL

The hospital as we know it is a modern invention. Before the twentieth century, almost all Americans, whether rich or poor, received their health care at home from friends, relatives, and assorted health care providers. Because these providers used only a few small and portable tools, hospitals were unnecessary.

Some form of institution, however, was needed for those Americans who lacked friends or relatives to provide care at home or the means to purchase such care. For these individuals, the only option was the **almshouse**. Here they—along with orphans, criminals, people with disabilities, people with mental illnesses, and other public wards—received essentially custodial care. Conditions in almshouses generally were appalling. Inmates often had to share beds or sleep on the floor, and rats often outnumbered humans. Hunger was common, and blankets and clothing scarce. These conditions, coupled with the lack of basic sanitation, made almshouses ideal breeding grounds for disease (Rosenberg, 1987:31–32).

The Premodern Hospital

Wealthy Americans considered almshouse conditions quite acceptable for those they regarded as lazy, insolent, alcoholic, promiscuous, or incurable (categories they believed included all nonwhites). By the end of the eighteenth century, however, wealthy Americans began to view these conditions as unacceptable for those they considered the "deserving" poor—the respectable widow, the worker disabled by an accident, or the sailor struck by illness far from home. With such individuals in mind, philanthropists decided to develop a new form of institution, the hospital, devoted to **inpatient** care of the "deserving" sick—so long as they didn't suffer from chronic, contagious, or mental illnesses (Rosenberg, 1987: 19–20). These hospitals were nonprofit and became known as **voluntary hospitals** because they relied heavily on unpaid (volunteer) charity work.

Although hospitals offered better conditions than did almshouses, they remained chaotic and dirty places. According to historian Charles Rosenberg:

> Nurses were often absent from assigned wards and servants insolent or evasive. Chamber pots [used for urinating and defecating] remained unemptied for hours under wooden bedsteads, and mattresses were still made of coarse straw packed tightly inside rough ticking. Vermin continued to be almost a condition of life among the poor and working people who populated the hospital's beds, and lice, bedbugs, flies, and even rats were tenacious realities of hospital life (1987:287).

These conditions, plus the severe limitations of contemporary medicine, kept **mortality** rates high and taught the public to associate hospitals with death rather than treatment. To make matters even worse, hospitals functioned as **total institutions** (described in Chapter 7) in which patients traded individual rights for health care (Rosenberg, 1987:34–46). Hospital rules regulated patients' every hour, even mandating work schedules for those who were physically capable. Patients who didn't follow the rules could find themselves thrown into punishment cells or frigid showers.

Beginnings of the Modern Hospital

Given the rigors of hospital life, the **stigma** of charity that accompanied hospital care, and the popular association of hospitals with death, early nineteenth-century Americans entered hospitals only as a last resort (Rosenberg, 1987:98–99). During the Civil War, however, the need to care for sick and wounded soldiers led to significant improvements in hospital organization and care, at least for the better-financed Union Army. These changes demonstrated that hospitals need not be either deadly or dehumanizing.

After the war, widespread adoption of new ideas about the dangers of germs and the importance of cleanliness helped to make hospitals safer and more pleasant. So, too, did technological changes such as the development of disposable gauze and cheaper linens, which made cleanliness feasible (Rosenberg, 1987:122–41). Concurrently, population increases (through births and immigration), the movement from farms to overcrowded cities, and the rise of dangerous factories led to a rise in contagious diseases and in serious accidents. Both these changes increased patient demand for hospitals. Meanwhile, doctors' also began pushing for hospital construction to gain access to new medical technologies and to sterile surgical theaters (Rosenberg, 1987:149).

Yet affluent Americans remained generally unwilling to tolerate the conditions on even the cleanest hospital wards. As a result, and to compete with the **for-profit, private hospitals** that began appearing during the second half of the nineteenth century, voluntary hospitals developed a class-based system of services (Rosenberg, 1987:293–94). Those who could pay for private accommodations received better heating and furnishings, exemption from many hospital rules, and privileges such as more anesthesia during operations. Thus voluntary hospitals began to lose their ethos of service and became increasingly similar to their for-profit competitors (Stevens, 1989:112).

The Rise of the Modern Hospital

By the early twentieth century, the hospital as we now know it had become an important American institution and a major site for medical education and research. Between 1873 and 1923, the number of hospitals increased from 178 to almost 5,000 (Rosenberg, 1987:341). These new hospitals also included **public hospitals**, established to provide services to those groups—people with mental illnesses, people with chronic illnesses, and the "undeserving poor"—that voluntary hospitals considered unworthy, and for-profit hospitals considered money losers. However, African Americans still could obtain care only in a few segregated, poorly staffed, and poorly funded wards and hospitals; in municipal hospitals where medical students and residents could learn skills by practicing on African American patients; and sometimes in other hospitals for emergency care (Stevens, 1989:137).

By this time, surgical admissions to hospitals far surpassed medical admissions (Rosenberg, 1987:150). Most patients went to a hospital to have their tonsils, adenoids, or appendixes removed; their babies delivered; or their injuries treated (Stevens, 1989: 106). The emphasis on technology as a defining aspect of modern hospitals further reinforced hospitals' tendency to focus on the care of **acute** rather than **chronic illness**.

Hospitals Today

The next major change in hospital care in the United States came in 1965, when the federal government implemented **Medicaid** and **Medicare**. These health insurance plans dramatically increased the profits available to hospitals. This in turn led both to a rise in for-profit hospitals and to increasing mergers of hospitals into ever-larger for-profit and voluntary hospital chains (such as Humana and Sisters of Charity, respectively).

As hospital profits grew, so did costs to the federal government via Medicaid and Medicare. As a result, the government for the first time developed a vested interest in controlling hospital costs. Ironically, the resulting price-control programs (described in Chapter 8) such as **diagnosis-related groups (DRGs)** have pressured voluntary hospitals (which remain the core of the hospital system) to focus more on the bottom line and thus to act more like for-profit hospitals (Stevens, 1989:305).

More recent cost-containment programs at the state and federal level have especially squeezed funding for public hospitals. Under any circumstances, it is difficult for these hospitals to make ends meet because so many of their patients go to public hospitals only because they can't afford to pay. To cover these costs, public hospitals rely on funding from state and local governments. This funding, however, has declined substantially in recent years, forcing hospitals to cut staff, reduce services, or close altogether.

Concern about costs and profits also has affected the mix of services offered by hospitals, especially voluntary and for-profit hospitals. Rather than fighting **managed care** insurers over which treatments and services the insurers will cover, these hospitals now increasingly offer extra services that patients are willing to pay for out of pocket, such as yoga, meditation, and massage (Abelson and Brown, 2002). Meanwhile, the search for profits has encouraged these hospitals to offer new, technologically intensive treatments and tests even if evidence of their benefits is weak and other nearby hospitals already offer them. The result has been a proliferation of expensive technologies such as magnetic resonance imaging machines, intensive care units, and open-heart surgical suites. Conversely, an increasing number of voluntary and for-profit hospitals have closed or shrunk money-losing units such as obstetrics wards, emergency departments, and psychiatric units. Because of these changes, these hospitals now treat an older and sicker mix of patients, most of whom suffer from the acute complications of chronic illnesses.

At the same time, as voluntary and for-profit hospitals have shifted toward providing more intensive care for middle-class Americans, public hospitals have increasingly become **primary care** providers for the poor. Patients who have neither health insurance nor money to pay for care sometimes turn to hospital **outpatient** clinics and emergency departments not only for treatment of acute problems such as gunshot wounds but also for chronic problems such as backaches.

The Hospital–Patient Experience

Although hospitals no longer terrify and endanger patients as they did in the nineteenth century, a hospital stay still can be alienating and frightening. The bureaucratic nature and large size of modern hospitals, coupled with the highly technological nature of hospital care, often mean that the patient as an individual person, rather than just a diseased body, gets lost.

The reasons behind this are obvious and to some extent unavoidable. First, increasingly, patients enter hospitals needing emergency care. Often, health care workers must respond immediately to their needs and have no time to talk with them to ascertain their preferences—which many are physically incapable of expressing in any case. Second, the highly technical nature of hospital care encourages staff to focus on the machines and the data these machines produce rather than on the patient as a whole person. In the modern obstetric ward, for example, workers often focus much of their attention on the electronic fetal monitor rather than on the laboring woman (Simonds, Rothman, and Norman, 2007). Third, and as we will see in Chapter 11, medical training encourages doctors to focus on biological issues much more than on patients' psychological or social needs. Fourth, as large institutions necessarily concerned with costs and profits, hospitals must rely on routines and schedules, with little leeway for individual needs or desires. Hence the common stories of nurses awakening patients from needed sleep to take their temperature or blood pressure.

The Shift away from Hospitals

Increasingly, changes in financing have led care to move away from hospitals. Because insurers (including Medicare under the DRG system) typically pay hospitals only preset amounts for inpatient surgery but will negotiate with hospitals over payment for outpatient surgery (i.e., surgery given without formally admitting the patient to the hospital), many hospitals have opened outpatient surgical clinics, which they use whenever feasible.

Similarly, insurers increasingly are reducing their costs by raising reimbursement for outpatient care and lowering it for (less expensive) inpatient care, a trend that the **ACA** further encourages. As a result, many hospitals have added outpatient medical centers as well as surgical centers. For example, between 2000 and 2014, San Diego's Scripps Health system closed one hospital and added 20 outpatient care sites (Vesely, 2014).

NURSING HOMES

From the start, American hospitals focused on caring for acutely ill persons and assumed that families would care for chronically ill persons. During the course of the twentieth century, however, average life expectancy increased; families grew smaller, more geographically dispersed, and less stable; and women less often worked at home. As a result, more and more Americans needed to seek long-term care from strangers, and **nursing homes**—facilities that primarily provide nursing and custodial care to groups of individuals over a long period of time—became part of the American landscape.

The number of nursing homes has more than tripled since 1980. Currently, there are about 16,000 skilled nursing homes in the United States, the majority run for profit (US Bureau of the Census, 2013).

Gender, Age, Ethnicity, Class, and Nursing Home Usage

About 1.4 million Americans currently live in nursing homes (National Center for Health Statistics, 2014a). Some groups, however, are more likely than others to find themselves in a nursing home.

Most strikingly, women are far more likely than men to become nursing home residents. Indeed, women now comprise about three-quarters of residents, partly because women more often live long enough to become enfeebled by age and partly because they more often survive their spouses, leaving no one to care for them if they need help.

Not surprisingly, older people are far more likely than others to live in nursing homes, and residents overwhelmingly are older than age 75. However, the numbers of nursing home residents younger than age 65—indeed, younger than age 30—has grown significantly in the past decade because of the rise in diabetes and in gang violence, among other factors (Persson and Ostwald, 2009). Many young people stay in nursing homes only temporarily while recuperating from a serious illness or accident, but others stay for years. Still others find themselves in nursing homes because they have a mental illness or disability and have no one who can help them with basic daily tasks (such as preparing food or dressing themselves).

Regardless of why young people find themselves in nursing homes, life for them can be grim and isolating: Few have interests in common with their elderly co-residents, and many are in nursing homes in part because they have limited contact with their families (Persson and Ostwald, 2009).

Historically, nursing homes residents overwhelmingly were white, but this is shifting as the US population shifts. Usage of nursing homes by African Americans and Hispanics increased significantly from 2000 to 2007, whereas usage by white Americans decreased, probably because whites increasingly can afford alternatives that allow them to stay in their homes or in other, less restrictive environments (Feng et al., 2010). In addition, African Americans and Hispanics far more often find themselves in lower-quality nursing homes, primarily because of their lower incomes (Fennell et al., 2010; Smith et al., 2007). The same is undoubtedly true for poorer white Americans, although data on this is unavailable.

Importantly, nursing home residents these days are considerably sicker on average than were residents 20 years ago. This change stems from the economic incentives built into DRGs, which have encouraged hospitals to discharge patients "sicker and quicker"—physically stable but still ill—once their bills and lengths of stay exceed the limits set by Medicare. Those who can't care for themselves at home often are discharged directly to nursing homes.

Financing Nursing Home Care

Currently, nursing home care costs about $74,000 per year for those who pay out of pocket (LongtermCare.gov, 2014). Few Americans can afford these costs. Nor can most afford insurance coverage for nursing home care, which typically is very expensive. As a result, at least initially, most residents rely on Medicare to pay their bills. However, Medicare pays only for skilled nursing care and at most for the first 150 days. As a result, most individuals who need only custodial care or need more than 150 days of care must turn to *Medicaid* to pay their bills. To be eligible for Medicaid, however, they must first sell all of their assets (minus their houses if they are married) and spend all of their savings (minus the cost of burial expenses and minimum living expenses for their spouses). Thus, long-term nursing homes residents (and their spouses) usually end up impoverished. Moreover, Medicaid reimbursement for nursing homes is often less than the homes' usual charges. As a result, nursing homes actively work to solicit Medicare patients and to avoid Medicaid patients. In addition, to free up beds for Medicare patients, nursing homes often move Medicaid patients to less desirable areas (such as dementia wards) or discharge them to lower-quality nursing homes (Rodriguez, 2014).

Working in Nursing Homes

Nursing home care is extremely labor intensive. To provide this care, nursing homes rely almost solely on **nursing assistants** (who often have no training) augmented by **licensed practical nurses** (who have completed approximately one year of classroom and clinical training).

Nationally, nursing assistants (half of whom work in nursing homes and one-quarter in hospitals) form one of the largest and fastest growing health care occupations (Bureau of Labor Statistics, 2014). Almost all are women, and most are nonwhite. Many come from Africa, Asia, or Latin America and are not native English speakers. These characteristics leave them particularly vulnerable in today's shrinking job market, with little ability to fight for better working conditions. Indeed, in August 2010, the federal government announced a new crackdown on employers that pressured nurses at all levels to work unpaid overtime and noted that the pressures on nursing assistants were especially severe (Pear, 2010). Currently, nursing assistants earn a median income of $24,000 (Bureau of Labor Statistics, 2014).

To understand the life of nursing home residents and the nursing assistants who care for them, sociologist Timothy Diamond (1992) became certified as a nursing assistant and worked for several years in a variety of nursing homes. He soon concluded that the core of working as a nursing assistant is caregiving but that those who train nursing assistants don't recognize this basic fact. Instead, his instructors taught him to recite biological and anatomical terms, measure vital signs, and perform simple medical procedures. Instructors divorced these skills from any social context or any sense that their patients were people rather than inanimate objects. Moreover, the skills Diamond most needed were never taught, such as exactly how do you clean an adult who has soiled a diaper in a manner that preserves the individual's sense of dignity? Only by labeling this caregiving as mere physical labor could those who hire nursing assistants label them "unskilled" and treat them so poorly.

Life in Nursing Homes

Diamond's research underlines how the fates of nursing assistants and nursing home residents intertwine and how even in the best nursing homes, the economics of a profit-driven system produce difficult conditions for both. According to Diamond, within nursing homes

> caregiving becomes something that is bought and sold. This process involves both ownership and the construction of goods and services that

> can be measured and priced so that a bottom line can be brought into being. It entails the enforcement of certain power relations and means of production so that those who live in nursing homes and those who tend to them can be made into commodities and cost accountable units (1992:172).

In this process of **commodification**, or turning people into commodities, "Mrs. Walsh in Bed 3" becomes simply "Bed 3." To keep down the price of this "commodity," only the most expensive homes provide private rooms or separate areas for residents who are dying, smelly, or psychologically disturbed. Privacy, then, also becomes a commodity that few residents can afford.

Nursing assistants, meanwhile, become budgeted expenses, which homes try to keep to an absolute minimum. This is not hard to do because federal standards require far less staff than is necessary to provide adequate care (Rodriguez, 2014). Yet almost no homes meet even those low standards (Pear, 2008). As a result, nursing home residents are unnecessarily placed at risk for numerous health problems. For example, residents can experience malnutrition when there are insufficient assistants to cut up their food, help them eat, and encourage them to take more than a few bites. Similarly, residents may be drugged, strapped to chairs, kept on a strictly regimented schedule, or left in a single central room during the day so they can be supervised by only a few assistants; one study found that 71 percent of new residents received psychiatric drugs even though most had neither been diagnosed with a psychiatric problem nor received such drugs before admission (Molinari et al., 2010).

Although all these problems also can occur in nonprofit nursing homes, a review of federal data from all US nursing homes found that both quality of life and quality of care were significantly worse in for-profit homes (Harrington et al., 2001). One reason for this is that within the profit-driven system, managers constantly stress to staff that *providing* care is less important than *documenting* care. As a sign proclaimed in one nursing home where Diamond worked, "If it's not charted, it didn't happen." For example, state regulations where Diamond worked required homes to serve residents certain "units of nutrition" each day. Consequently, each day nursing assistants collected cards placed on residents' food trays that named the foods and their nutritional content. Every few months, state regulators would inspect the cards and certify that the homes met state nutritional requirements. Yet these cards bore little relationship to reality because the appetizing-sounding names given to the foods rarely matched the actual appearance or taste of the food. Nor did the cards note if a resident refused to eat a food because it was cold, tasteless, or too hastily served. Similarly, sanitation regulations required homes to shower residents regularly but did not require that the showers be warm. Nor did they require the homes to hire enough nursing assistants so that residents who used diapers could be cleaned as soon as needed or so that residents could get the help they needed in using the toilet and thus avoid the indignity and discomfort of diapers. Unfortunately, more recent research suggests that low reimbursement rates by both Medicare and especially Medicaid now encourage nonprofit nursing homes to make similar choices (Rodriguez, 2014).

Ironically, in top tier nursing homes, the same process of commodification is now leading nursing home owners to *encourage* nursing assistants to emphasize caring and indeed to think of residents as their kin (Dodson and Zincavage, 2007). By so doing, nursing homes can both charge higher prices for their "family atmosphere" and get more work from assistants for the same low wages. But this "purchased intimacy"—similar to that offered by massage therapists, beauticians, and others—is a one-way transaction: Nursing assistants may treat residents like family, but residents still often sling racist slurs at nonwhite nursing assistants. And even in these "higher-quality" nursing homes, owners rarely allow nursing aides time off to deal with problems in their own families or to grieve when residents they cared about die.

HOSPICES

Origins of Hospice

Whereas nursing homes emerged to serve the needs for long-term care not met by hospitals, **hospices** emerged out of growing public recognition that neither of these options provided appropriate care for the dying.

Only in the past few decades has institutional care for the dying become a public issue. At the beginning of the twentieth century, few individuals experienced a long period during which they were known to be dying. Instead, most succumbed quickly to illnesses such as pneumonia, influenza, tuberculosis, or acute intestinal infections, dying at home and at relatively young ages. Now, however, most Americans live long enough to die from chronic rather than acute illnesses. In addition, as doctors and scientists have developed techniques for detecting illnesses in their earliest stages, they now more often identify individuals as terminally ill long before death occurs. Thus, dealing with the dying is to some extent a uniquely modern problem and certainly has taken on a uniquely modern aspect.

Although modern medical care has proved lifesaving for many, its ability to extend life can turn from a blessing to a curse for those who are dying. (*Ethical Debate: A Right to Die?* discusses this issue in more detail.) For various reasons, including legal concerns about restricting care, financial incentives for using highly invasive treatments, and a medical culture (described in the next chapter) that emphasizes technological interventions, thousands of Americans each year receive intensive, painful, and tremendously expensive medical care that offers little hope of restoring quality of life or extending lives (Byock, 2013; Gawande, 2014). In nursing homes, on the other hand, the emphasis on profit making and cost cutting often results in dying persons receiving only minimal and depersonalized custodial care.

This lack of appropriate care for the dying led to the development of the hospice movement. The first modern hospice, St. Christopher's, was founded in England in 1968 by Dr. Cicely Saunders to address the needs of the dying and to provide an alternative to the alienating and dehumanizing experience of hospital death. The hospice admitted only patients expected to die within six months and offered only palliative care (designed to reduce pain and discomfort) rather than treatment or mechanical life supports. The hospice provided care both in St. Christopher's and in patients' homes.

The first American hospice, which closely resembled St. Christopher's, opened in 1974 in New Haven, Connecticut. Other hospices soon followed, emerging from grassroots organizations of religious workers, health care workers, and community activists seeking alternatives to hospitals and nursing homes. Public support for hospices was so immediate and so great that in 1982, only eight years after the first American hospice opened, Congress approved covering hospice care under Medicare.

Modern Hospices

The US hospice movement has proved enormously successful, growing from one hospice in 1974 to about 5,500 hospices serving about 1.5 million clients annually (National Hospice and Palliative Care Organization, 2014). With that success, however, have come changes. Whereas the original hospices were independent, freestanding institutions, these days most hospice care is received in homes and nursing homes. In addition, the original hospices were nonprofit organizations, primarily staffed by volunteers, that emphasized individualized care and patient participation. Now that hospices are primarily funded by insurers, they have had to reconfigure their staffing and practices to meet standards for care based on hospital protocols. Finally, about two-thirds of all hospices are now run on a for-profit basis (National Hospice and Palliative Care Organization, 2014). For-profit hospices are less likely than other hospices to provide care at home and are more likely to declare clients ineligible for services even as they get closer to death, apparently to reduce their costs and increase their profits (Aldridge et al. 2014). For all of these reasons, hospice care is now more cost-oriented and less individualized or patient centered than it was originally.

Use of Hospice

About one-third of those who die in the United States use hospice services, for a median of 18 days (National Hospice and Palliative Care Organization, 2014). Women and whites are especially likely to use hospice care. In addition, most hospice clients are older than age 65. Not surprisingly given this age distribution, most hospice clients rely on Medicare to pay the costs (National Hospice and Palliative Care Organization, 2014).

Because Medicare only pays for hospice care for six months, hospices lose money if their clients survive beyond that time period. As a result, hospices disproportionately serve individuals with cancer, whose life expectancies can be fairly accurately predicted. However, use of hospice by individuals with other diagnoses (especially dementia) has been growing in recent years (National Hospice and Palliative Care Organization, 2014).

ETHICAL DEBATE

A Right to Die?

In 1983, 26-year-old Elizabeth Bouvia, who lived with almost total paralysis from cerebral palsy and near-constant pain from arthritis, presented herself for admission to Riverside General Hospital. In years past, and despite her physical problems, Bouvia had earned a degree in social work, married, and lived independently. However, after her efforts to have children failed, her husband left her, and the state stopped paying for her special transportation needs, she lost interest in living. Her purpose in coming to the hospital, she told the hospital staff soon after her admission, was to obtain basic nursing care and pain-killing medication while starving herself to death, cutting short what might otherwise have been a normal life span. The hospital's doctors took her case to court and won the right to force-feed her. The court concluded that although Bouvia did have the right to commit suicide, she did *not* have the right to force health care workers to engage in **passive euthanasia** by allowing her to die through their inaction.

In 1990, Janet Adkins, 54 years old and living with Alzheimer's disease, killed herself with the assistance of Dr. Jack Kevorkian—a process known as **physician-assisted death**. Kevorkian had designed a machine that allowed people with severe disabilities to give themselves a fatal dose of sodium pentothal and potassium in the privacy and freedom of their homes. Over the next decade, Kevorkian helped more than 100 people to kill themselves, without facing any legal penalties. In 1999, however, he was convicted of second-degree homicide and sentenced to 10 to 25 years in prison for committing **active voluntary euthanasia**: taking the steps needed to end the life of someone who has requested that he or she be killed. In this case, Kevorkian administered a lethal injection to a patient, rather than having the patient administer the dose himself. He was released from prison in 2007 on the condition that he not assist in any further deaths.

In the Netherlands, meanwhile, doctors legally can practice active voluntary euthanasia for patients who are mentally competent, incurably ill, suffering intolerable and unrelievable pain, and who authorize their doctors to do so in writing. Although so far only Oregon has legalized physician-assisted death, at least two-thirds of Americans believe that terminally ill people have both a right to die and a right to their doctors' assistance in so doing (*Contexts*, 2004).

Those who support a "right to die" argue that competent adults have the right to make decisions for themselves, including the ultimate decision of when to die. They argue that death sometimes can be a rational choice and that it's cruel to force individuals to suffer extreme physical or mental anguish (Seale, 2010).

If we accept that death can be a rational choice, then harder questions follow. Why is it rational only if one's condition is terminal? Doesn't it make even more

Outcomes of Hospice Care

Research suggests that hospice care saves Medicare more than $2,000 per person and that even more could be saved if individuals entered hospice care sooner (Taylor et al., 2007). To understand the full economic impact of hospice care, however, we must take into account that about 40 percent of hospice users now die in their homes (National Hospice and Palliative Care Organization, 2014). In these circumstances, family members provide most care. To do so, they often must take time off from work or drop out of the labor market altogether. Consequently, hospice care might not reduce the costs of caring as much as it shifts the costs from hospitals and insurers to families.

sense to end the life of someone like Elizabeth Bouvia, whose agonies may continue for another 50 years, than to end the life of someone who will die soon regardless? Why should this choice be forbidden to individuals simply because they can't, either physically or emotionally, carry it out themselves? And why should we allow individuals to choose death only through passive euthanasia, leaving them to languish in pain while awaiting death, if instead they could be killed quickly and painlessly?

Opponents of this view argue that the duty to preserve life overrides any other values and that euthanasia is merely a nice word for suicide or murder. They question whether Elizabeth Bouvia would have wanted to die if she still had the resources she needed to live independently, and they wonder whether euthanasia gives society a way to avoid responsibility for relieving the burdens imposed by illness and disability. Opponents who have studied the Netherlands suggest that Dutch doctors sometimes end patients' lives without consent and without first attempting to make patients' lives worth living (Hendlin, Rutenfrans, and Zylicz, 1997). In addition, opponents question whether Dutch acceptance of euthanasia explains why, compared with the rest of Europe, the Netherlands has fewer hospices and fewer doctors trained in pain relief.

In sum, the use of euthanasia, whether active or passive, raises numerous difficult questions: What are the consequences of, in effect, declaring it reasonable for disabled people to choose death? What pressures does this place on individuals to end their own lives rather than burdening others? What responsibilities does this remove from society to make these individuals' lives less burdensome? Finally, given that social factors, such as age, gender, and social class, affect our perceptions of individuals' worth, how do we ensure that society won't more willingly grant a right to die to women, minorities, or other socially disvalued groups?

Sociological Questions

1. What social views and values about medicine, society, and the body are reflected in the debate over a right to die? Whose views are these?
2. Which social groups are in conflict over this issue? Whose interests do the different sides of this issue serve?
3. Which of these groups has more power to enforce its view? What kinds of power do they have?
4. What are the intended consequences of recognizing a right to die? What are the potential *unintended* social, economic, political, and health consequences of doing so?

The health benefits of hospice care are clearer. One study using a large-scale **random sample** of terminally ill Medicare recipients found that hospice clients survive an average of one month *longer* than those who receive ordinary medical care instead (Connor et al., 2007). Moreover, another study, which randomly assigned patients recently diagnosed with a terminal illness to either hospice or regular medical care, found that the hospice patients reported a *higher* quality of life, experienced *less* depression, and survived almost *three months* longer than those who continued with medical care (Temel et al., 2010). These individuals may have benefitted both from the supportive care of hospice and from avoiding the traumatic surgeries and chemotherapies typically given to terminally ill patients.

HOME CARE

As the discussion of hospices suggested, most individuals who experience chronic or acute health problems—whether children, working-age adults, or elderly, and whether the problems are physical or mental—receive most of their care at home. This is even truer now than in the recent past because of technical, demographic, and policy changes. Because of technological advances, babies born prematurely or with birth defects and persons who have experienced severe trauma are increasingly likely to survive, although often with severe disabilities that require lifelong assistance. Much of this care is now given by family members in the home. In addition, technological advances also have made it possible for families to provide treatments at home that previously were available only in hospitals, such as chemotherapy or kidney dialysis.

At the same time, the rise in the numbers of frail elderly, many of whom have both multiple physical problems and cognitive impairments, has increased the number receiving care at home. In addition, the movement begun in the 1960s (as described in Chapters 6 and 7) to deinstitutionalize people with disabilities and mental illnesses, combined with the lack of community supports for such individuals after deinstitutionalization, have shifted much of the burden of care from state institutions to the home. Finally, as described earlier, policy changes now encourage hospitals to discharge patients to their homes "sicker and quicker," in essence replacing paid hospital workers with unpaid family caregivers (Glazer, 1993).

Because of limited public or private insurance coverage for home care, most who need long-term supportive care receive services only from family members and, less often, friends. Existing data suggest that home care has little impact on the overall costs of care or the mental or physical health of those receiving care but can produce small improvements in individuals' satisfaction with life (Arno, Bonuck, and Padgug, 1995; Weissert, 1991).

The Nature of Family Caregiving

The most recent statistics (as of 2014) indicate that 29 percent of US adults—66 million people—regularly provide care for elderly, ill, or disabled relatives or friends (National Alliance for Caregiving and AARP, 2009). About two-thirds of these caregivers are women. Ethnic minorities and poorer persons also are more likely to become caregivers, probably because these groups experience higher rates of illness and disability and have less access to formal services.

Those who care for the health needs of family members typically do so out of love and often reap substantial psychological rewards. Nevertheless, caregiving by family members should not be romanticized, nor should the financial, physical, social, or psychological costs of caregiving be underestimated (Arras and Dubler, 1995; National Alliance for Caregiving and AARP, 2009; Reinhard and Horwitz, 1996; Tessler and Gamache, 1994).

The financial costs of caregiving are substantial. The demands of caregiving force many to shift to part-time work or even abandon paid employment. In addition, caregivers must purchase, often out of pocket, both expensive drugs and technologies and many everyday items such as diapers and bandages. In addition, caregivers typically are responsible for purchasing a variety of services and therapies from a range of companies and health care workers.

The physical costs also can be high. Caregiving often includes strenuous tasks that can result in back injuries or exhaustion, such as lifting a physically disabled person into a bed. The time burdens of caregiving also can become physically draining. The typical caregiver spends 20 hours per week on caregiving, does so for more than four years, and holds at least a part-time job as well. Moreover, many are simultaneously responsible for more than one relative, such as a child with a disability and a parent with Alzheimer's disease. Not surprisingly, the more hours individuals spend each day in caregiving and the more months (or years) they spend in this way, the greater the toll on caregivers' health (National Alliance for Caregiving and AARP, 2009).

Taken together, the financial and physical burdens of caregiving often leave individuals with little time, energy, or money for social relationships. Caregivers often report that their relationships with both family and friends have suffered because of their responsibilities (National Alliance for Caregiving and AARP, 2009). For example, a mother who spends hours each day caring for an ill child might regret that she has so little time for her other children, and those children might resent the attention given to their ill sibling. Problems are particularly acute when the person receiving care is mentally ill and throws family routines into chaos, embarrasses other family members, or physically threatens their safety (Reinhard and Horwitz, 1996; Tessler and Gamache, 1994).

> Family life also can suffer when caregiving requires the use of high technology within the home. John D. Arras and Nancy Neveloff Dubler suggest that this invasion of the home by high-tech medical procedures, mechanisms, and supporting personnel exerts a cost in terms of important values associated with the notion of home. How can someone be truly "at home," truly at ease, for example, when his or her living room has been transformed into a miniature intensive care unit?… Rooms occupied by the paraphernalia of high-tech medicine may cease to be what they once were in the minds of their occupants; familiar and comforting family rituals, such as holiday meals, may lose their charm when centered around a mammoth Flexicare bed; and much of the privacy and intimacy of ordinary family life may be sacrificed to the institutional culture that trails in the wake of high-tech medicine (1995:3).

Finally, caregiving brings with it numerous psychological costs. Caregivers can easily become depressed when their efforts can't stop or even slow the disease process. This is especially true when caregivers must routinely inflict painful treatments on their charges or when the burdens of caregiving are unceasing, as when a parent must suction the lungs of a child with cystic fibrosis hour after hour, day after day, to keep the child from dying. Moreover, as this example suggests, caregivers also often bear the enormous psychological burden of being directly responsible for another person's life. In fact, family caregivers are now expected to manage in the home—often with little training or technical support—technology considered too complex for licensed practical nurses to manage in hospitals. Finally, caregivers of persons younger than themselves face anxieties about what will happen to their charges if the caregivers die first.

Easing the Burdens of Caregiving

The problems faced by family caregivers have led to the development of new organizations, new organizational structures, and a new occupation to ease the burdens of caregiving. Two major organizations, the National Alliance for the Mentally Ill (NAMI) and the National Alliance for Caregiving (NAC), are now devoted to family caregiving. Both organizations work to increase assistance to family caregivers and improve access to community-based care, and the NAMI additionally fights to decrease the stigma of severe mental illness.

Both **respite care** and **family leave programs** also were developed to ease the burdens of caregivers. Respite care refers to any system designed to give caregivers a break from their otherwise unrelenting responsibilities, including paid aides who provide care in the home for a few hours, day-care centers for elderly and disabled adults, and nursing homes that accept clients for brief stays. Unfortunately, only California and Pennsylvania fund respite care programs. In all other states, respite care is expensive and difficult to find; only 12 percent of those included in the NAC and AARP (2009) survey had ever used respite care. Minimal data are available on the quality of these services (Kitchener and Harrington, 2004).

Similarly, although federal law gives employees the right to as many as 12 weeks of unpaid leave from work yearly to care for family members, few can afford to take unpaid leaves. In addition, the law does not apply to part-time workers, temporary workers, or employees of small firms. The law is also problematic because it reinforces the idea that caring for ill and disabled persons is the responsibility of the family—which, in practice, usually means women relatives—rather than the responsibility of society as a whole (Abel, 2000).

Finally, those who provide care to relatives or friends may turn for assistance to paid caregivers. Each year about 1.7 million Americans receive paid home care, most commonly in the form of help with bathing, dressing, and light housework (National Center for Health Statistics, 2014a). Most paid home care is provided by **home health aides**, who typically have no formal training, or **registered nurses**, who have received at least two years of nursing training and passed national licensure requirements. Aides are overwhelmingly minorities and women, and they are highly likely to be immigrants. Few receive any job benefits, and most receive only minimum wage. Because the growth in paid home health care is so new, little more is known regarding these workers or their work.

HEALTH CARE TECHNOLOGIES

Doctors and other healers have always used technologies in their work. Two hundred years ago, doctors used knives to cut veins and "bleed" patients of their illnesses and used strips of cloth to bandage the wounds afterward. One hundred years ago, doctors used mercury compounds and electricity in attempting to cure patients of masturbation or syphilis. In modern medicine, health care technology includes everything from Band-Aids to computerized patient record systems, to heart–lung machines.

The Nature of Technology

Technology refers to any human-made object used to perform a task. In addition, the term is often used to describe processes that involve such objects. For example, the term *technology* can refer both to the overall process of kidney dialysis and to the equipment used in that process.

Although we often talk about technology as if it is inherently either good or bad—"technology has made our lives easier" or "technology has depersonalized medical care"—the reality is more complex (Heath, Luff, and Svensson, 2003; Timmermans and Berg, 2003b). The nature of a technology does determine the *range* of ways it might be used, but whether it is harmful, helpful, or both depends on who uses it in which ways. Electricity is helpful when used by doctors to stimulate muscle healing and harmful when used by doctors who are poorly educated or who work as torturers in dictatorships. Fetal monitors can depersonalize childbirth when nurses stare at the screens rather than pay attention to the pregnant woman. But ultrasound imaging of fetuses can *personalize* pregnancy when fathers literally visualize their future children as real for the first time. In addition, such technologies can create a setting in which fathers, mothers, and health care workers can discuss the emotional aspects of pregnancy and child rearing.

Similarly, we often talk about technology as if it is either a blank slate, lacking any inherent nature, or a force outside of human control. Again, the reality is more complex. For example, doctors and hospitals now face considerable pressure to collect data on patients, using computerized medical databases in hopes that doing so will help identify and thus reduce medical errors (Timmermans and Berg, 2003a). Because these databases prompt doctors to ask patients a specific set of questions in a specific sequence, they implicitly encourage doctors to focus only on certain questions and to organize the answers they receive in certain ways. At the same time, doctors quickly learn to regain some control over the databases through how they ask their questions and record the answers. Similarly, although doctors who use these databases may press patients to answer specific questions, patients can assert control by instead addressing a different set of issues (Timmermans and Berg, 2003a).

For these reasons, we need to understand not only the nature of a given technology but also the cultural system that determines how a technology will be used, by whom, and for what purposes. In addition, we must study not only how society and social actors shape the use of technology but also how technology shapes society and social actors.

In this section, we look at how technologies develop and gain acceptance. We also consider how different groups within the health care world interact with technology—and with each other.

The Social Construction of Technology

In the same way that we talk about the social construction of illness, we can talk about the **social construction** of technology: the process through which groups decide which potential technologies should be pursued and which should be adopted. This concept in turn leads us to question who promotes and who benefits from the social construction of any given technology.

Like the social construction of illness, the social construction of technology is a political process, reflecting the needs, desires, and relative power of various social groups. These groups can include manufacturers, doctors, the government, and consumers. As a result, harmful technologies are sometimes developed and adopted, and needed technologies sometimes are not.

The history of cardiopulmonary resuscitation (CPR) offers a fascinating example of the social construction of technology. (*Contemporary Issues: Newborn Screening* discusses another example.) The purpose of CPR is to restore life to those whose hearts and lungs have stopped working. In earlier times, the very notion of such resuscitation would not have made any sense to doctors or the public. Death was considered to be in God's hands, and dead was dead. But since the rise of modern medicine, doctors have struggled to find ways to restore life to those who die suddenly.

At the same time, doctors have grown increasingly able to understand the slow trajectory of dying associated with cancer. And with the rise of the hospice movement (described earlier in this chapter), both doctors and the public have come to hold as an ideal the "good death" in which an individual comes to terms with his or her dying, makes peace with family and friends, and receives appropriate terminal care to minimize physical and emotional suffering.

None of this, however, applies to the sudden—and common—deaths caused by stroke or heart disease. In his award-winning book *Sudden Death and the Myth of CPR*, sociologist Stefan Timmermans (1999) argues that CPR and associated resuscitation techniques have become part of American medical culture because they appear to offer a "good death" in these circumstances. Innumerable television dramas portray heroic doctors who save apparently dead patients through CPR, and millions of dollars have been spent teaching the general public to perform CPR and outfitting community emergency response teams and hospital emergency rooms with resuscitation equipment. Yet CPR almost never succeeds except when healthy individuals drown or are struck by lightning. The typical person who receives CPR has *at best* a 1 to 3 percent chance—and probably much less—of surviving, at an estimated cost of $500,000 per survivor. Moreover, "survival" may be brief and may be accompanied by severe neurological damage. As a result, according to Timmermans, emergency department doctors

CONTEMPORARY ISSUES

Newborn Screening

Over the last few years, newborn screening for genetic problems has exploded. All US states now mandate screening for at least 21 disorders, and most screen for considerably more. These numbers are likely to increase as new tests appear (Timmermans and Buchbinder, 2010).

Without question, such tests have proven lifesaving for the tiny percentage of infants with conditions that are both clearly identifiable and treatable. For example, infants who are correctly diagnosed with phenylketonuria (PKU) and put on appropriately restricted diets can avoid the mental retardation that PKU otherwise can cause. In contrast, there is little obvious benefit to learning that one's infant has an untreatable condition that will eventually prove fatal or severely disabling (Timmermans and Buchbinder, 2010).

In addition, for every infant who is correctly diagnosed with a genetic disease, many others are diagnosed incorrectly, at least initially. Further testing and observation resolve most of these cases, but the process can be agonizing for parents instructed to watch carefully for a range of potential problems, many of which (such as diarrhea or deep sleeping) are common among babies. Not surprisingly, most parents turn immediately to the Internet for information, which often focuses on worst-case scenarios and amplifies parents' fears. Moreover, even if doctors eventually conclude that the baby isn't at risk, the parents may find it difficult to stop watching and worrying.

Finally, as newborn testing spreads, growing numbers of infants are identified as having genetic anomalies with unknown effects or having genetic configurations that fall outside of the statically normal range. In these cases, parents' questions may never be answered and their fears may never fully dissipate, even as the infant grows into a healthy child. Such fears may lead parents to limit the children's activities, diets, exposure to other children, or whatever else they think might harm their children's health (Timmermans and Buchbinder, 2010).

and emergency medical technicians overwhelmingly regard resuscitation as futile, so they joke, complain, or simply go through the motions whenever they have to use it.

Why, then, has CPR become so widely adopted? Timmermans argues that the widespread use of CPR reflects modern Americans' discomfort with death. The real benefit of CPR, according to Timmermans, is that it "takes some of the suddenness of sudden death away" (1999:110). CPR allows families and friends to believe they have done everything possible by getting their loved ones to treatment as fast as possible. It also gives families and friends time to gather and to recognize that death may be imminent, and it gives medical personnel a sense of technical accomplishment as they fight to keep their patients' bodily organs functioning as long as possible. The use of CPR, then, is part of the broader project of **death brokering**: the process through which medical authorities make deaths explainable, culturally acceptable, and individually meaningful, such as through pain management, "death counseling," or the gradual removal of life supports from dying patients (Timmermans and Berg, 2005). For these reasons and despite all its emotional and financial costs, CPR has become a valued and expected ritual in American culture.

At the same time, adoption of CPR illustrates the economics and politics as well as the cultural forces that underlie the social construction of technology. CPR would not have been so widely adopted if corporations had not had a vested economic interest in promoting it. Nor is it likely that CPR would have become the norm if corporations had been required to demonstrate its effectiveness before selling it. In fact, however, there are almost no legal requirements for manufacturers to demonstrate the safety or effectiveness of technical devices, so they rarely fund such research. As a result, doctors must depend on promotional materials from manufacturers and on their own clinical experiences in deciding whether to use a technology, and patients must rely on doctors' judgments.

The Technological Imperative

Once a technology enters the mainstream, the technological imperative can make it difficult to avoid. The **technological imperative** refers to the belief—held by both doctors and consumers—that technology is almost always good, so it is almost always appropriate to use all existing technological interventions, regardless of their cost. This belief is deeply embedded in American culture (and to a lesser extent in Western culture more generally). The belief in intervention (including technological intervention) is also highly stressed in medical culture and training, as Chapter 11 discusses. In addition, and as the history of CPR illustrates, the technological imperative is often reinforced by corporations that have a vested economic interest in selling a particular technology and doctors with a vested interest in offering a technology. Finally, the technological imperative is cemented whenever insurance companies, government regulatory agencies, or medical associations identify use of a particular technology as the "standard of care" for treating or diagnosing a given illness. In these situations, doctors who don't use that technology may risk malpractice suits.

Prostate cancer testing offers an excellent example of the technological imperative. Among men, one almost inevitable consequence of aging is cancer of the prostate, a poorly understood bodily organ that produces chemicals believed necessary for reproduction. Most men develop prostate cancer by middle age, and virtually all do so if they live long enough (Kolata, 2005). Few, however, are killed by the disease because it typically grows so slowly that most who have prostate cancer instead die from heart disease, stroke, or some other condition before the cancer can become dangerous (Grob and Horwitz, 2009).

Currently, however, doctors have little ability to identify which men might *die* from prostate cancer. Instead, doctors can only hope to identify who *has* prostate cancer. To do so, most doctors routinely test middle-aged male patients at periodic intervals for prostate-specific antigen (PSA), a chemical produced by the prostate. If a patient's PSA level has increased significantly, doctors perform a biopsy—inserting a needle into the prostate to remove a few cells, which they then check for cancer. Unfortunately, PSA tests are highly inaccurate. About 30 percent of those who have cancer are *not* identified by the test and so derive no benefit from testing. In addition, about two-thirds of those who *are* told they have cancer based on test results in fact don't have it. These individuals too can't benefit from this inaccurate diagnosis and likely will experience unnecessary emotional trauma, financial costs, and painful procedures because of it. Others are correctly identified as having some cancer cells in their body but would likely have died from heart disease or something else long before the cancer would have caused any health problems; sociologists use the term **pseudodisease** in cases such as these to refer to harmless conditions that are diagnosed as diseases based on medical tests (Mechanic, 2006).

If a biopsy suggests cancer, doctors usually perform a prostatectomy (i.e., surgical removal of the prostate). The surgery succeeds in removing the cancer in about 80 percent of cases. Even in these cases, however, the risks of surgery can outweigh the benefits. Between 0.5 and 2 percent of patients die within a month of surgery, and another 5 percent experience serious and potentially deadly complications (Wilt et al., 2008). In addition, more than 30 percent experience serious difficulties in sustaining erections or controlling their urine. Perhaps most important, large studies using random samples and **controlling** for other variables have found no significant differences in survival rates between men who do and don't receive treatment (Wilt et al., 2008). In sum, as an editorial published by the *New England Journal of Medicine* declared, "PSA screening has at best a modest effect on prostate-cancer mortality ... and comes at the cost of substantial overdiagnosis and overtreatment" (Barry, 2009). The widespread use of PSA screening thus offers a perfect example of the technological imperative.

Technology and the Changing Nature of Health Care

In addition to making certain tests and treatments almost unavoidable, new technologies have dramatically changed the very nature of health care for both health care providers and consumers (Casper and Morrison, 2010; Clarke et al., 2010).

Just a few decades ago, health care was an intensely "hands-on" experience. The doctor would literally lay his (or, rarely, her) hands on the patient, feeling the belly to check for swelling or lumps, thumping the chest to listen for abnormal lung sounds, and so on. A thorough physical examination could take up to an hour of intense questioning and physical probing. As this suggests, medical care also relied heavily on listening to the patient's report of his or her symptoms and concerns. Surgery, too, was bloody, intimate, hands-on work.

These days, doctors give far less attention to patients' reports of their health and far more attention to results from medical tests. Those tests, meanwhile, are largely performed not by doctors, but by technicians who collect and analyze blood samples, ultrasound readings, CT scans, and other tests. Meanwhile, an increasing proportion of doctors work primarily *within* the body as surgeons, spending little time interacting with patients. Moreover, surgeons now often use computers to manipulate microscopic surgical tools, further distancing their own bodies from their patients' bodies.

As this suggests, technology also has led to the rise of a wide range of new health care occupations, shifting many tasks from doctors to these new providers and threatening the balance of power within health care (a topic discussed more fully in the next chapter). At the same time, new technologies (coupled with changes in the structure of insurance and hospital care) increasingly have shifted care onto patients and their families and away from health care providers altogether. As the earlier discussion of family caregiving suggested, many families now have both the opportunity and often the financial need to provide high-technology care in the home, from injecting insulin in children with diabetes to operating ventilator machines for those who can't breathe on their own.

IMPLICATIONS

In this chapter, we examined three difficulties inherent in the ways we provide care to individuals who have illnesses or disabilities. First, we looked at some of the inherent contradictions of trying simultaneously to provide care and to make a profit. Health care workers—from medical students to home health aides—who labor long hours in difficult conditions to keep their employers' costs low can't provide the quality of care they might like. Even institutions such as nonprofit hospices must contend with the demands of a wider system that emphasizes cutting costs and generating profits. Meanwhile, many other health care institutions have emerged specifically to make money, relegating caregiving to a secondary priority.

Second, we considered the difficulty of providing individualized care in institutional environments. To stay within their budgets, large institutions must provide standardized care, ignoring individual preferences and desires. Patients must follow rules, schedules, and regimens established for efficiency and cost cutting, regardless of the impact on their quality of life.

Third, we explored some of the inherent difficulties of treating health care as an individual or family responsibility rather than a social responsibility. As we have seen, the burdens of caregiving can be enormous. Yet the United States offers little support to those who take on this responsibility. In contrast, other wealthy nations provide far more assistance; both Sweden and Finland, for example, provide long-term paid leaves and free or inexpensive assistance with domestic chores to elderly persons who might otherwise need help from family members (Swedish Institute, 1997, 1999; Zimmerman, 1993).

In sum, the data presented in this chapter regarding the virtual social abandonment of ill and disabled individuals and of their caregivers suggest the low priority that American society places on caring for those who are weak or ill, especially if they also are poor. Technology won't cure these problems. Nor for that matter, is it inherently dehumanizing or otherwise problematic. Rather, technology is a tool, adopted for a combination of cultural, medical, emotional, and financial reasons, which can be used for good or ill. Only when our underlying social values and commitments change can we expect the lives of persons with illnesses and disabilities and their caregivers to improve significantly.

SUMMARY

1. Until about 1900, most Americans received all health care at home. Those who could not care for themselves or obtain care at home were relegated to almshouses—charity institutions with terrible conditions where orphans and criminals, as well as people with illnesses or disabilities, were "warehoused."
2. Voluntary (nonprofit private) hospitals first emerged in the late 1700s as a means of providing care to the "deserving sick." Early voluntary hospitals were run as total institutions. Hospital conditions improved dramatically during the Civil War and improved further as the new belief in germs made cleanliness desirable, and technological changes made it economically feasible.
3. For-profit, private hospitals emerged as a way of offering better conditions to more affluent consumers. Public hospitals were developed to provide services to individuals with chronic mental or physical illnesses as well as to those considered the "undeserving poor." By the 1920s, hospital care had become a major part of American life and a center of medical education and research.
4. The initiation of Medicare and Medicaid dramatically increased the profits available to hospitals. Skyrocketing costs led the federal government to implement a system of *diagnostic-related groups*, under which hospitals receive a prepaid fee for each patient with a given diagnosis, regardless of the actual cost of treatment. To maintain their profits, hospitals shifted toward remunerative outpatient services, technologies, and surgeries.
5. Hospitals now treat an older and more seriously ill mix of patients than in the past, primarily for the acute complications of chronic illnesses. Hospitals—especially public hospitals—have become primary care providers for the poor, which has increased hospitals' financial problems.
6. Nursing homes offer care to those who need nursing or custodial care, but not hospital care. Those who use nursing homes tend to be female and elderly, although young people increasingly live in nursing homes. Most residents and their spouses are bankrupted quickly by the cost of care.
7. Nursing homes are primarily staffed by nursing assistants, who are overwhelmingly female, nonwhite, and low paid. Within nursing homes, both nursing assistants and residents are commodified: Nursing assistants become commodities to purchase as cheaply as possible, and residents become expenses to control.
8. Hospices are institutions designed to serve the needs of the dying. To gain social and financial support, hospices over time have become more routinized, medicalized, and profit-oriented.
9. Hospice care saves insurers money, partly by shifting the cost of care to family members. Terminally ill patients who enter hospices typically live longer and experience a higher quality of life than do those who continue with medical care.

10. Most Americans still receive most of their health care at home, typically from female family members. The physical and emotional strains of caregiving have led to the development of respite care, paid home caregivers, and family leave programs.
11. *Technology* refers to any human-made object used to perform a task, from Band-Aids to kidney dialysis machines. Technology is never inherently good or bad. Its nature determines the *range* of ways it might be used, but whether it is harmful, helpful, or both depends on *who* uses it in *which* ways.
12. The *social construction of technology* refers to the *political* process through which groups decide which potential technologies should be pursued and which should be adopted. This process reflects the needs, desires, and relative power of various social groups, including manufacturers, doctors, and consumers.
13. CPR was designed to restore life to those whose hearts and lungs have stopped. It is almost never successful but has been widely adopted because it helps people come to terms with sudden death, because its manufacturers had a vested interest in promoting it, and because neither the government nor doctors required manufacturers to prove it was effective.
14. The technological imperative refers to the belief, among both doctors and consumers, that technology is almost always good and therefore it is almost always appropriate to use all existing technological interventions, regardless of their cost.
15. Because most men develop prostate cancer, but few experience any related health problems, increased testing for this disease has led to skyrocketing rates of medical procedures and increased health problems, but no increase in survival rates. The identification of men with very early stage prostate cancer is an example of pseudodisease, and the treatment of these men is an example of the technological imperative.
16. Partly because of new technologies, medical care is now far less "hands on," results from medical tests are now often given more credence than patient reports of their symptoms, many tasks have shifted from doctors to new health care providers, and many tasks have shifted from health care providers onto patients and their families.

REVIEW QUESTIONS

1. Why do sociologists consider nineteenth-century hospitals to have been total institutions?
2. What led to the development of voluntary hospitals? Public hospitals? The modern hospital as we know it?
3. Why did hospices emerge? What are their strengths and weaknesses?

4. Who uses nursing homes, how is this changing, and why?
5. How does the process of commodification affect nursing assistants? How does it affect nursing home residents?
6. Why has home care grown? What are the difficulties faced by family caregivers?
7. What is technology? What do sociologists mean when they say that technology is inherently neither good nor bad and neither a blank slate nor a force outside of human control?
8. What is the social construction of technology? What does it mean to say that this is a political process?
9. Why was CPR so widely adopted even though it is so ineffective?
10. What is the technological imperative? What social factors encourage it?

CRITICAL THINKING QUESTIONS

1. What are the burdens faced by family caregivers, and how have these changed over time? If you were on your governor's Task Force on Health Care, what policies might you want to implement to lighten those burdens?
2. If you could go back into the past and change one thing in the history of hospitals, what would it be? Why?
3. How can society shape technology? How can technology shape society?
4. Why might sociologists and other observers argue *against* early detection and treatment of prostate cancer?

PART
IV

Health Care, Health Research, and Bioethics

In this final section, we shift our perspective to health care providers and researchers. Chapter 11 provides an overview of the history of medicine as a profession and describes how the social position of doctors has changed over time. In this chapter, we also explore the process through which new doctors learn both medical skills and medical culture. Finally, we look at how medical education and medical culture, as well as broader social and cultural factors, affect relationships between doctors and patients.

Although doctors typically are the first persons who come to mind when we think of health care, they form only a small percentage of all health care providers. In Chapter 12, we consider some of these other providers both within and outside the mainstream health care system, including nurses, midwives, and acupuncturists.

The final chapter in this part—and in this book—provides a history of bioethics as well as a sociological account of how bioethics has become institutionalized and how it affects health care and health research. We will see how issues of power underlie ethical issues and why we need a sociological understanding of bioethics.

CHAPTER

11

The Profession of Medicine

2007 Fancy/Getty Images

LEARNING OBJECTIVES

After reading this chapter, students should be able to:

- Understand how the medical profession gained professional dominance.
- Assess the current threats to medical dominance.
- Describe medicine's core cultural values.
- Critique the consequences of medical values for patients.
- Provide a sociological analysis of doctor–patient relationships.

To become a doctor, students must spend long years studying biology, chemistry, physiology, and other subjects. In addition, students must learn the way of thinking about medicine, patients, and medical care that characterizes medical culture.

Michael J. Collins learned this culture during four years as a surgical resident at the Mayo Clinic. After a particularly brutal day of surgery in which he watched a teenager die, Dr. Collins found himself emotionally traumatized, questioning the meaning of his work and the effect it had on him. Although he wished he could discuss his feelings with B. J. Burke, the director of his residency program, Dr. Collins knew from experience how B. J. would respond. As he wrote in his memoir:

> *BJ Burke was not interested in what I thought or understood. He was interested in what I did.*
>
> *"If you want to learn to be sensitive and introspective," he would say, "do it on your own time."*
>
> *I imagined myself being called into his office. As I enter the room he is seated at his desk, reading the report in front of him. He makes certain I know I am being ignored.*
>
> *At length he looks at me over the top of his glasses.*
>
> *"Dr. Collins, what is your job?"*
>
> *"My job, sir?"*
>
> *"You have a job, don't you? You get a paycheck, don't you?"*
>
> *"Yes, sir."*
>
> *"Well, what do you do?"*
>
> *"I'm a second-year orthopedic resident at the Mayo Clinic."*
>
> *"Do you want to be a third-year resident someday, Dr. Collins?"*
>
> *"Yes, sir."*
>
> *"What is an orthopedic resident supposed to do?"*
>
> *Where was this going? "Follow orders?" I venture.*
>
> *"An orthopedic resident is supposed to practice orthopedics, Doctor. He is not supposed to go around asking patients if they have ever considered the ontological implications of their fragile, mortal state."*

"I didn't exactly—"

He jumps to his feet and points his finger at me. "We fix things. Do you understand that? We don't analyze things. We don't discuss things. We don't wring our hands and cry about things. We fix them! If somebody wants to be analyzed they can see a shrink. When they come to the Department of Orthopedics at the Mayo Clinic they want only one thing: they want to be fixed. Now get the hell out of here and go fix things. And I better not get any more reports of touchy-wouchy, hand-holding sessions in this department" (Collins, 2005:152–153).

Dr. Collins's story illustrates two basic elements of modern-day medical culture—emotional detachment and a belief in medical intervention. In this chapter, we look at how these and other aspects of medical culture and training evolved and at the consequences for both doctors and their patients. We begin by looking at how doctors became the dominant profession within health care and at the forces that now threaten their dominance.

AMERICAN MEDICINE IN THE NINETEENTH CENTURY

When confronted by disquieting illness, most modern-day Americans seek care from a doctor of medicine. Little more than a century ago, however, that would not have been the case. Instead, Americans received most of their health care from family members. If they required more complicated treatment, they could choose from an array of poorly paid and typically poorly respected health care practitioners (Starr, 1982:31–59). These included **regular doctors** (the forerunners of contemporary doctors) as well as **irregular practitioners** such as midwives; patent medicine makers, who sold drugs concocted from a wide variety of ingredients; botanic eclectics, who offered herbal remedies; and bonesetters, who fixed dislocated joints and fractured bones.

Regular doctors were also known as **allopathic doctors**, or allopaths (from the Greek for "cure by opposites") because they sometimes treated illnesses with drugs selected to produce symptoms *opposite* to those caused by the illnesses. For example, allopaths would treat patients suffering the fevers of malaria with quinine, a drug known to reduce fevers, and treat patients with failing hearts with digitalis, a drug that stimulates the heartbeat. Their main competitors were **homeopathic doctors**, or homeopaths (from the Greek for "cure by similars"). Homeopaths treated illnesses with drugs that produced symptoms *similar* to those caused by the illnesses—treating a fever with a fever-producing drug, for example. Although in retrospect the homeopathic model might seem odd, it drew on the same logic as smallpox inoculation, the one successful inoculation available at that time: People who developed a mild form of cowpox after inoculation with a few cowpox cells somehow became immune to the related but far more serious smallpox. Homeopaths, therefore, concluded that patients who received a miniscule amount of a drug that mimicked the symptoms of a given illness would become better able to resist that illness. Modern science now tells us that drugs given in such small quantities can't biologically affect patients. However, we also now know that belief in a drug's effectiveness will lead about 30 percent of patients to experience at least temporary benefit from the drug—even if in reality they only receive sugar pills. This process is known as the **placebo effect** (Evans, 2003). It seems likely, therefore, that homeopathic drugs did not harm patients and sometimes helped them, if only through the placebo effect.

That Americans before the twentieth century placed no greater trust in allopathic doctors than in any others who claimed knowledge of healing should not surprise us. Although by the nineteenth century, science—the careful testing of hypotheses in **controlled** experiments—had infiltrated the curricula of European medical schools, where many of the wealthiest Americans trained, it had gained barely a foothold in US medical schools. Moreover, the United States licensed neither doctors nor medical schools (Ludmerer, 1985). Instead, and until about 1850, most doctors trained through apprenticeships lasting only a few months. After that date, most trained at any of the multitude of uncertified medical schools that had sprouted around the country, almost all of which were private, for-profit institutions, unaffiliated with colleges or universities and lacking any entrance requirements beyond the ability to pay tuition (Ludmerer, 1985). Nor were standards stricter at the few university-based medical schools. For example, in 1871, Henry Jacob Bigelow, a Harvard University professor of surgery, protested against a proposal to require written graduation examinations on the grounds that more than half of Harvard's medical students were illiterate (Ludmerer, 1985:12). Training averaged far less than a year and depended almost entirely on lectures, so few students ever examined a patient, conducted an experiment, or dissected a cadaver. Any student who regularly attended the lectures received a diploma. This situation began to change significantly only in the 1890s and only in the better university schools.

Lacking scientific research or knowledge, allopathic doctors developed their ideas about health and illness either from their clinical experiences with patients or by extrapolating from abstract, untested theories. The most popular theory of illness, from the classical Greek era until the mid-1800s, traced illness to an imbalance of bodily "humors," or fluids. Doctors had learned through experience that ill persons often recovered after episodes of fever, vomiting, or diarrhea. From this, doctors deduced—in part correctly—that fever, vomiting, and diarrhea helped the body restore itself to health. Unfortunately, lacking methods for testing their theories, doctors carried these ideas too far, often inducing life-threatening fever, vomiting, purging, and bloodletting. Consider, for example, the following description of how Boston doctors in 1833 used what was known as **heroic medicine** to treat a pregnant woman who began having convulsions a month before her delivery date:

> The doctors bled her of 8 ounces and gave her a purgative. The next day she again had convulsions, and they took 22 ounces of blood. After

> 90 minutes she had a headache, and the doctors took 18 more ounces of blood, gave emetics to cause vomiting, and put ice on her head and mustard plasters on her feet. Nearly four hours later she had another convulsion, and they took 12 ounces, and soon after, 6 more. By then she had lapsed into a deep coma, so the doctors doused her with cold water but could not revive her. Soon her cervix began to dilate, so the doctors gave ergot to induce labor. Shortly before delivery she convulsed again, and they applied ice and mustard plasters again and also gave a vomiting agent and calomel to purge her bowels. In six hours she delivered a stillborn child. After two days she regained consciousness and recovered. The doctors considered this a conservative treatment, even though they had removed two-fifths of her blood in a two-day period, for they had not artificially dilated her womb or used instruments to expedite delivery (Wertz and Wertz, 1989:69).

As this example suggests, because of the body's amazing ability to heal itself, even when doctors used heroic medicine, many of their patients survived. Thus, doctors could convince themselves they had cured their patients when in reality they either had made no difference or had endangered their patients' lives.

By the second half of the nineteenth century, most doctors, responding to the public's support for irregular practitioners and fear of heroic medicine, had abandoned their most dangerous techniques. Yet medical treatment remained risky. Allopathic doctors' major advantage over their competitors was their ability to conduct surgery in life-threatening situations. Unfortunately, until the development of anesthesia in the 1860s, many patients died from the inherent physical trauma of surgery. In addition, many died unnecessarily from postsurgical infections. Dr. Ignaz Semmelweis had demonstrated in the 1850s that because midwives (whose tasks included washing floors and linens) had relatively clean hands, whereas doctors routinely went, without washing their hands, from autopsies to obstetrical examinations and from patient to patient, more childbearing women died on medical wards than on midwifery wards. Yet not until the 1880s would hand washing became standard medical practice.

Until well into the twentieth century, then, doctors could offer their patients little beyond morphine for pain relief; quinine for malarial and other fevers; digitalis for heart problems; and after 1910, Salvarsan for syphilis—each of which presented dangers as well as benefits. According to the 1975 edition of *Cecil's Textbook of Medicine*, one of the most widely used medical textbooks, only 3 percent of the treatments described in the 1927 edition of the textbook were fully effective, whereas 60 percent were harmful, of doubtful value, or offered only symptomatic relief (Beeson, 1980). Doctors' effective pharmacopeia did not grow significantly until the development of antibiotics in the 1940s. Similarly, surgery in the early nineteenth century relied on only a few basic technologies and remained rare and dangerous, if nowhere near as dangerous as it had been before the development of anesthesia and antiseptic techniques.

THE RISE OF MEDICAL DOMINANCE

Despite the few benefits and many dangers inherent in allopathic medical care, by 1900, doctors had eliminated most of their competitors and gained control over health care (Starr, 1982:79–112). In this section, we will see how this change came about.

From its inception in 1847, the **American Medical Association (AMA)** had worked to restrain the practices of other health care occupations. State by state, the AMA fought to pass laws outlawing their competitors or restricting them to working only under allopathic supervision or to performing only certain techniques, such as spinal manipulation.

Most of these efforts met with little success initially because nineteenth-century Americans considered health care an uncomplicated domestic matter, unrelated to science, and requiring no special training (Starr, 1982:90–92). By the beginning of the twentieth century, however, as improvements in public health and in living conditions ended scourges such as cholera and typhoid, and as Americans began reaping practical dividends from scientific advances such as electric lights and streetcars, public faith in science swelled. Increasingly, Americans defined health care as a complex matter requiring expert intervention, assumed the superiority of "scientific" medicine, and turned to allopathic doctors for care (Starr, 1982:127–142).

Like the public, homeopaths and botanic eclectics (allopathic doctors' two major groups of competitors) also came to recognize the benefits of science and therefore to realize that a lack of scientific foundation could doom their fields. However, they still received considerable popular support. Moreover, because, like allopaths, most were white men, homeopaths and botanic eclectics generally held social statuses similar to those of allopaths. Thus, homeopaths and botanic eclectics retained sufficient influence to pressure allopaths to accept them into medical schools and licensing programs, and their fields eventually faded away.

Other health care workers could bring far less power to their dealings with legislators and with allopathic doctors. Newly emerging occupations such as chiropractic (described in Chapter 12) lacked the long-standing history of popular support that had allowed homeopaths to push for incorporation with allopathy. Older occupations, meanwhile, such as midwives and herbalists, lacked the social status, power, and money needed to fight against doctors' lobbying. Because most of these practitioners were women or minorities, they were assumed to be incompetent by both legislators and doctors (Starr, 1982:117, 124).

The Flexner Report and Its Aftermath

These differences between allopathic doctors and other health care practitioners increased during the early twentieth century. Since the 1890s, the better medical schools had begun tightening entrance requirements and stressing academic standards, scientific research, and clinical experience. These changes increased the pressure on other medical schools to do the same. Those pressures increased following publication in 1910 of the **Flexner Report** on American medical education (Ludmerer, 1985:166–90). The report, which was written by Abraham Flexner and commissioned by the nonprofit Carnegie Foundation at the AMA's behest, shocked the nation with its descriptions of the lax requirements and poor facilities at many medical schools. The Flexner Report increased the pressures on all medical schools to improve their programs and accelerated the changes already underway. In the next few years, responding to pressure from both the public and the AMA, all US states began enforcing stringent licensing laws for medical schools (Ludmerer, 1985:234–249). These laws hastened the closure of all proprietary and most nonprofit schools, many of which were already suffering financially from the costs of trying to meet students' growing demand for scientific training. As a result, the number of medical schools fell from 162 in 1906 to 81 in 1922 (Starr, 1982:118, 121).

The Flexner Report, in conjunction with the changes already underway in medical education, substantially improved the quality of health care available to the American public and paved the way for later advances in health care. However, these changes in medical education also had some more problematic results. The closure of so many schools made medicine as a field even more homogeneous because only two medical schools for African Americans and one for women survived (Ludmerer, 1985:248; Starr, 1982: 124). In addition, few immigrants, minorities, and poorer whites could afford the tuition for university-based medical schools or meet their strict educational prerequisites. Moreover, many of these schools openly discriminated against women, African Americans, Jews, and Catholics. Thus, even though the technical quality of medical care increased, fewer doctors were available who would practice in minority communities and who understood the special concerns of minority or female patients. At the same time, simply because doctors were now more homogeneously white, male, and upper class, their status grew, encouraging more hierarchical relationships between doctors and patients.

Doctors and Professional Dominance

By the 1920s, doctors had become the premiere example of a **profession** (Parsons, 1951). Although definitions of a profession vary, sociologists generally define an occupation as a profession when it has three characteristics:

1. The autonomy to set its own educational and licensing standards and to police its members for incompetence or malfeasance
2. Technical, specialized knowledge, unique to the occupation and learned through extended, systematic training
3. Public confidence that its members follow a code of ethics and are motivated more by a desire to serve than a desire to earn a profit

During the first half of the twentieth century, doctors clearly met this definition of a profession (Timmermans and Oh, 2010). Most doctors worked in private practice (whether solo or group) and set their own hours, fees, and other conditions of work. Those who worked in hospitals or clinics were typically supervised by other doctors, not by nonmedical administrators. And even in these settings, only doctors had the authority to review other doctors' clinical decisions, and this authority was rarely exercised. Similarly, only doctors served on boards that evaluated medical schools and granted or revoked medical licenses. Finally, the public placed great trust in doctors, believed most doctors worked selflessly to serve their patients, and routinely ranked medicine as the most prestigious occupation. These expectations were confirmed by doctors' adoption of a professional code of ethics. *Ethical Debate: A Duty to Provide Care?* explores one aspect of that code.

As this suggests, as the leading profession in the health care world, doctors enjoyed—and to some extent still enjoy—an unusually high level of **professional dominance**: freedom from control by other occupations or groups *and* the ability to control any other occupations working in the same economic sphere (Freidson, 1994; Timmermans and Oh, 2010). Although doctors often supervised, taught, or set licensing standards for members of other health occupations, those other occupations rarely had any say over doctors' work.

THE THREATS TO MEDICAL DOMINANCE

More recently, however, this high level of professional dominance by doctors—otherwise known as **medical dominance**—has come under threat.

The Rise of Corporatization

Until the 1960s, nonprofit or government agencies owned most hospitals and other health care institutions. With the initiation of **Medicare** and **Medicaid**, however, the potential for profits in health care expanded tremendously, leading many for-profit corporations to enter the field, as we saw in Chapter 8 (Starr, 1982:428–432; Timmermans and Oh, 2010). This growth of corporate medicine is known as **corporatization**.

Corporatization has substantially affected the work lives of American doctors. As Americans increasingly have obtained their insurance through **managed care organizations**, doctors have increasingly found employment within those organizations. Passage of the **Affordable Care Act** has also led doctors to take salaried positions as a way of protecting themselves financially from whatever changes that law may bring (Rosenthal, 2014a; Ruggieri, 2014). Meanwhile, hospitals increasingly are buying up private medical practices to expand their position in the market, increase their bargaining power with insurers, and thus generate more profits. Buying medical practices also increases hospitals' profits because primary care doctors who work for a hospital are expected to refer their patients to surgeons who work for the same hospital and who typically conduct surgery only in that hospital (Ruggieri, 2014). Currently, the majority of primary care doctors and about two-thirds of surgeons work as paid employees of hospitals or some other corporate institution, and most of the rest obtain

ETHICAL DEBATE

A Duty to Provide Care?

In 2014, Ebola virus disease began raging across Guinea, Liberia, and Sierra Leone. Soon, reports appeared of high fatality rates, not only among patients but also among doctors, nurses, and aides.

In some ways these deaths were not surprising, given the almost total lack of gloves, masks, and other basic infection-control equipment in some West African clinics and hospitals. When two nurses were infected by an Ebola patient in a Dallas hospital, however, despite having taken what seemed like reasonable precautions, fears understandably spread among the health care community, some of whom began questioning whether they had a duty to care for such patients.

Do health care workers have a duty to provide care, even when their own lives or health might be at risk? This question surfaces whenever epidemics rage, from bubonic plague in the Middle Ages, to HIV/AIDS 40 years ago, to Ebola virus disease these days. It also surfaces during wars and disasters: Some doctors and nurses, for example, rushed into New Orleans hospitals during Hurricane Katrina or into Haitian hospitals after its great 2010 earthquake, while others fled.

According to the AMA, doctors have an obligation to provide care whenever they are competent to do so (Twardowski, 2014). This argument is based on the idea that anyone trained and licensed in a helping profession—especially when that training was heavily subsidized through government funding—has accepted the obligation to pay the nation back by helping others. This argument further suggests that the duty to care is highest among those trained in specialties such as infectious disease and lowest for those who receive far less subsidy or training, such as nursing aides.

But is that duty to care absolute? Most importantly, do nations and health care systems have a duty to protect health care workers whenever possible? And if

their patients largely through contracts with managed care organizations (Rosenthal, 2014a; Ruggieri, 2014).

In all of these circumstances, doctors' autonomy has diminished. Administrators have taken over decisions formerly made by individual doctors, such as setting doctors' fees and work schedules, requiring doctors to obtain authorization before scheduling surgeries or prescribing certain medications, and expecting doctors to follow **practice protocols** that establish treatment guidelines aimed at providing the best—but also most cost-effective—treatment for different conditions (McKinlay and Marceau, 2002; Millenson, 1997; Vanderminden and Potter, 2010).

Meanwhile, concern about costs has led corporations to replace doctors with radiation technologists, pharmacists, nurse practitioners, and other allied health personnel. This shift has reduced both doctors' bargaining power with administrators and their power over other health occupations.

The Rise of Government Control

Concern about costs has also led the government to restrict doctors' professional autonomy (Timmermans and Oh, 2010). Because the government pays the bills generated by Medicaid and Medicare, it has a large vested interest in controlling nations and health systems fail to meet this obligation, does that release health care workers from the duty to care?

Similarly, does the duty to care mean that health care workers must use all available techniques, regardless of risk? For example, we might argue that there is an obligation to provide basic supportive care to someone with a dangerous infectious disease, but not to provide treatments that are more likely to infect health care workers than to save patients' lives.

From a practical perspective, the most effective approach may be to put in place the guidelines, training, and technologies that will best protect health care workers. Doing so would not only protect workers and patients from harm but also increase the odds that workers would volunteer for potentially hazardous work. In turn, increasing the numbers of willing and trained workers would help prevent diseases from spreading, thus providing the greatest good for the greatest number of people.

Sociological Questions

1. What views about health care and health care professionals are reflected in the AMA's position on the duty to treat?
2. Does the nation have an obligation to protect health care workers? How is this different from its obligation—or lack of obligation—to protect other workers?
3. Which social groups might argue against a duty to care? Against the funding needed to make that duty reasonably safe?
4. Which of these groups has more power to enforce its view? What kinds of power do they have?
5. What are the intended consequences of mandating a duty to care? What might be the unintended social, economic, political, or health consequences of this policy?

doctors' fees and treatment decisions. To do so, it has established programs such as the **diagnosis-related groups (DRG)** system and the **resource-based relative value scale (RBRVS)**. The DRG system (described in Chapter 8) established preset financial limits for each diagnosis for hospital care under Medicare (and in some states, Medicaid). Because hospitals are not reimbursed for any costs above those limits, they have a vested interest in making sure that doctors stay below the limits. Consequently, hospitals may cut the wages or terminate the contracts of doctors who consistently exceed DRG limits, thus pressuring all doctors in their employ to stay within those limits. Doctors sometimes conclude that they have only two choices: to misreport a patient's diagnosis on the DRG form so they can justify more expensive treatments they believe are necessary or to ignore their own clinical judgment and change their treatment plans to stay within DRG limits.

Whereas DRGs were designed to control Medicare spending on hospital care, RBRVS was designed to control spending on doctors' bills. RBRVS is a complex formula for determining appropriate compensation for medical care under Medicare, based on estimates of the costs and effort required to provide specific services in specific geographic areas. Under this system, incomes of most specialists have declined, whereas those of generalists (other than pediatricians, who receive no Medicare funds) have increased. Although RBRVS legally applies only to Medicare, most other public and private insurance plans also have adopted it, so it now serves, in the words of one observer, as a "de facto national fee schedule" (Sigsbee, 1997).

The Decline in Public Support

Beginning in the 1960s, the rise of the civil rights and feminist movements increased popular emphasis on questioning rather than obeying authorities. These changes helped foster the feminist health movement, the disabled rights movement, and the patients' rights movement, all of which encouraged consumers to take charge of their own health, obtain second opinions, demand more egalitarian relationships with their doctors, and consider using nonmedical health practitioners (Timmermans and Oh, 2010). These movements have stimulated major changes in medical practice, ranging from the sharp decrease in use of general anesthesia during childbirth to the routine use of informed consent forms before patients receive experimental drugs.

Other structural changes also have reduced patients' willingness to accept medical dominance (Timmermans and Oh, 2010). The rise of managed care—and the political backlash against it—has fostered a steady stream of news stories about doctors who supposedly withhold needed care from patients to meet the dictates of managed care organizations. Such stories have left patients less willing to trust their doctors' advice or motives.

Meanwhile, because of the shift from fee-for-service medicine toward insurance-paid medicine, far fewer consumers enjoy long-standing relationships with **primary care doctors** (i.e., doctors in family practice, internal medicine, or pediatrics who are typically the first doctors individuals see when they need medical care). Instead, employers (and consumers) routinely change insurance programs, and insurance programs routinely change their lists of contracted doctors, so consumers often must start new relationships with new doctors. As a result, doctors less often enjoy the sort of trust from their clients that can only develop over time. This loss of trust has been amplified by frequent news stories on the dangers of medical errors and the financial incentives for doctors to recommend costly tests or drugs. Taken together, these forces have led the proportion of Americans who place a "great deal of confidence in the people in charge of running medicine" to drop from 73 percent in 1966 to only 34 percent by 2010 (*Harris Poll*, 2010b).

The rise of the Internet has added impetus to this movement by giving consumers easy access to the medical literature, to literature on nonmedical alternatives, and to others who share their concerns (Shilling, 2001; Vanderminden and Potter, 2010). The federal government has supported this trend; its website at www.healthfinder.org was established specifically to give consumers online access to publications, clearinghouses, databases, other websites, self-help groups, government agencies, and nonprofit organizations related to both allopathic and alternative medicine. Such information can lead consumers to diagnose themselves, challenge their doctors' recommendations, or seek nonmedical care, thus challenging medical dominance.

KEY CONCEPTS

Countervailing Powers

According to researchers, if 2,000 women receive mammograms (used to detect breast cancer) from age 40 to 49, one woman's life may be saved. However, hundreds will erroneously be told they have cancer, and many of them will receive unnecessary biopsies, radiation, or even mastectomies (Keen, 2010). After research documenting these data appeared, the US Preventive Services Task Force issued a recommendation advising *against* mammograms for women in their 40s unless they were at unusually high risk of cancer. The uproar that followed—from doctors, consumers, and others—led the Task Force to replace that recommendation with the suggestion that women in their forties discuss the evidence with their doctors and make their own decisions. The forces allied for and against various mammogram policies illustrate the countervailing powers—including other doctors—that doctors sometimes face.

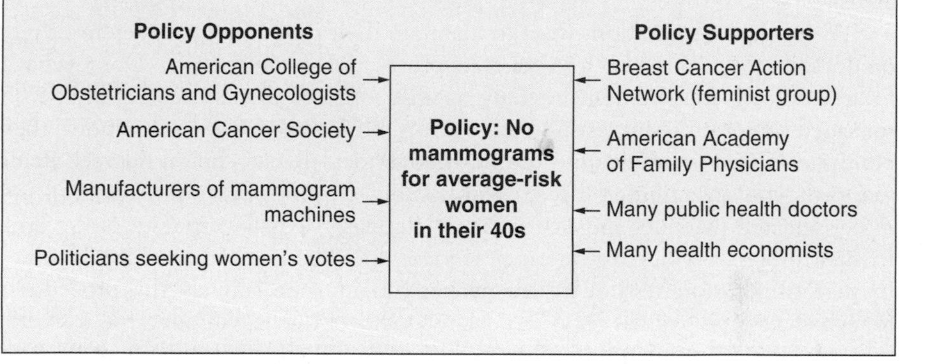

The Decline of the American Medical Association and Countervailing Powers

Medical dominance has also been threatened by the decline of the AMA. Whereas a half-century ago most doctors belonged to the AMA, now at most one-quarter do. Instead, some doctors join more liberal organizations that often oppose the AMA, such as Physicians for Social Responsibility, and many join specialty organizations such as the American College of Obstetricians and Gynecologists. As a result, no one group can speak with the full force of the medical profession behind it.

Meanwhile (and as Chapter 8 discussed), the insurance and pharmaceutical industries have become far more powerful over time. For example, the AMA still controls a larger political lobbying fund than does any other health profession except for dentists and contributed $1.7 million to candidates during the 2008 presidential elections (Center for Responsive Politics, 2011). Those contributions, however, are dwarfed by the combined contributions of pharmaceutical companies, health insurance companies, hospitals, and other health professions. As a result, these other groups sometimes can join together to achieve political goals that the AMA opposes. Taken together, these **countervailing**

powers—the various powerful groups and institutions fighting for control over a given arena such as health care—are actively challenging medical dominance (Light, 2010; Riska, 2010; Timmermans and Oh, 2010). *Key Concepts: Countervailing Powers* illustrates this concept, using the example of recent policy debates regarding mammograms.

THE CONTINUED STRENGTH OF MEDICAL DOMINANCE

Despite declines in autonomy and threats from countervailing powers, medical dominance remains a strong force (Freidson, 1994; Timmermans and Oh, 2010).

As much as corporations want to maintain their profit margins, they must rely on doctors' cooperation both to generate profits and to control costs. Doctors have fought fiercely and often successfully against some of the restrictions built into managed care. As a result, even when doctors work directly for corporations, they continue to enjoy considerable autonomy over day-to-day clinical matters. Practice protocols are rarely enforced, and doctors' treatment recommendations are rarely rejected (Mendel and Scott, 2010). Similarly, when corporate or governmental insurers cut doctors' fees per service, doctors can maintain their incomes by performing more tests or treatments per patient, especially elective procedures for which patients pay out of pocket. Many doctors now heavily advertise cosmetic surgery, laser eye surgery, infertility treatment, and weight loss treatment because these procedures are both high paying and largely free of oversight by insurance, government, or hospital bureaucrats (Sullivan, 2001).

The dominance of doctors relative to other health care occupations also remains largely intact (Freidson, 1994). First, doctors have successfully kept other occupations from any role in regulating medical licensure or practice standards while maintaining a voice in regulating allied health fields. Second, doctors sometimes can use the rise of allied health fields for their own benefit. For example, nurse practitioners now perform many tasks once done by primary care doctors. Yet doctors still often hire, fire, and set the salaries of these nurse practitioners while retaining the option of meeting with any patients they find interesting (and remunerative). Third, even though most managed care organizations, hospitals, and other large health care enterprises are now run by professional administrators, most also have a medical director with considerable power to oppose any administrative dictates that threaten doctors' work conditions or clinical decision making. Finally (and as Chapter 6 discussed), doctors have responded to the challenges posed by alternative health care occupations by adopting practices such as acupuncture, offering services such as vitamin supplements and massage in their offices, and fighting regulatory battles against other alternative occupations (Timmermans and Oh, 2010).

Similarly, although trust in medicine as an institution has declined precipitously in recent years, Americans still consider medicine a highly prestigious field and still strongly trust their own doctors (*Harris Poll*, 2010a, 2010b). Similarly, most patients use the Internet as a *supplement* to medical care rather than as a replacement. Indeed, patients often seek their doctors' assistance in interpreting materials they have found on the Internet, thus reinforcing rather than threatening trust in doctors and medical dominance (Vanderminden and Potter, 2010).

In sum, as Timmermans and Oh (2010: S101) write, "The medical profession has a long track record of deflecting internal and external challenges, appropriating reform attempts, safeguarding its interests, and maximizing profit." Although medical dominance is constantly under challenge, it remains a powerful force in the health care arena.

MEDICAL EDUCATION AND MEDICAL VALUES

Despite the assaults on medical dominance, becoming a doctor remains an attractive option, offering public prestige, the emotional rewards of service, and high incomes. Although applications to medical school declined steadily from 1996 to 2003, they have climbed substantially since then, perhaps because the economic downturn has made a career in medicine seem a safer choice (Association of American Medical Colleges, 2014a). In this section, we look at how doctors-in-training learn both medical knowledge and medical values, and at the consequences of this training for both doctors and patients.

The Structure of Medical Education

Becoming a doctor is not easy. Prospective doctors first must earn a bachelor's degree and then complete four years of training at a medical school. Before they can enter practice, however, and depending on their chosen specialty, they must spend another three to eight years as **residents**. Residents are doctors who are continuing their training while working in hospitals. (The term *intern*, referring to the first year of a residency, is no longer commonly used.) As a result, most don't enter practice until age 30.

For about 84 percent of students, going to medical school means going into debt. The median debt is $180,000, not including undergraduate debts (Association of American Medical Colleges, 2014b). Debt levels have increased substantially over the last 25 years in response to substantial increases in tuition fees.

Becoming a doctor also carries tremendous time costs. Regulations, first adopted after the 1989 death of a patient treated by exhausted residents, now limit first-year residents to working "only" 16 hours per day; second-year residents can still work 24-hour shifts every three days. Even after graduation, doctors typically work long hours. These time pressures, coupled with the financial pressures of training, encourage novice doctors to defer marriage, children, and other personal pursuits and to choose specialties requiring less training over those they otherwise might prefer.

Ethnicity, Sex, Class, and Medical Education

Most medical students—like most college graduates—are from the middle and upper classes. Parents of current medical students (both fathers and mothers) are about three times more likely to have earned a graduate degree than are other Americans (Association of American Medical Colleges, 2010). Moreover, this difference has increased substantially and steadily as tuition has risen (Sacks, 2007).

Compared to their percentage of the population as a whole, Asian Americans are significantly *overrepresented* among medical students, whereas Hispanic Americans and non-Hispanic African Americans are *underrepresented* (Association of American Medical Colleges, 2014a). These ethnic differences reflect social class differences (Asians are more likely to come from affluent homes), cultural differences in the value placed on education, and teachers' stereotypes (at all grades) regarding which students are capable of succeeding in science (Bettie, 2003; Kozol, 2005; Sacks, 2007).

On the other hand, medicine increasingly has opened to women, who now comprise almost half of all first-year medical students (Association of American Medical Colleges, 2014a). However, women still face substantial difficulties in the field, both during their education and afterward. In one survey, for example, 82 percent of female residents reported hearing colleagues or supervisors tell hostile or sexist jokes about women, 62 percent reported receiving overtly sexual comments from coworkers or supervisors, and 22 percent reported receiving sexual advances from coworkers or supervisors (Hinze, 2004). Moreover, women remain concentrated in certain specialties, most of which are relatively low paid (Anspach, 2010). As Figure 11.1 shows, as the percentage of women in a field increases, salaries typically decrease.

Women's concentration in certain specialties reflects a combination of voluntary choices and structural constraints (Hoff, 2010; Riska, 2010). In the past,

FIGURE 11.1 Median Salaries by Percentage Women in Specialty

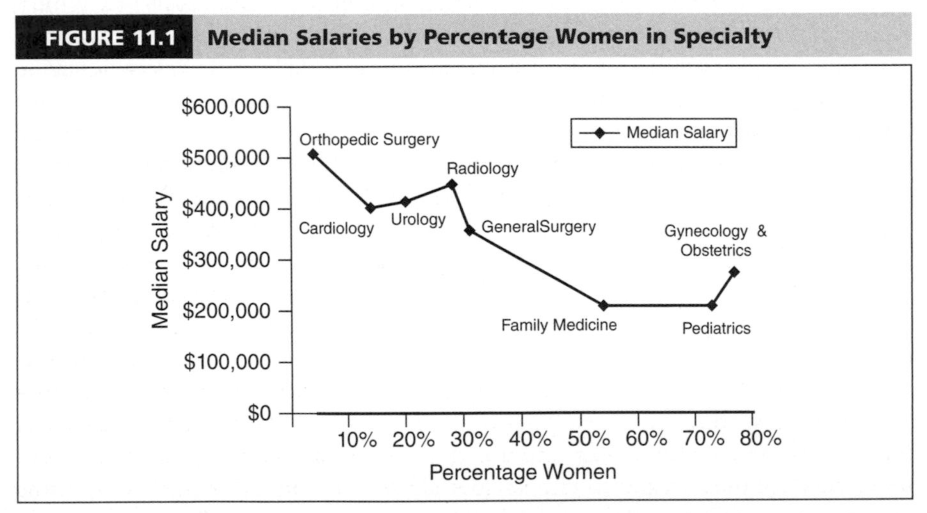

SOURCES: Women Physicians' Congress (2010); Cejka Search (2010).

many believed that the distribution of women within medicine reflected an innate tendency toward fields that emphasized hands-on caring, such as pediatrics. This explanation, however, fails to explain why women are also disproportionately found in radiology (a highly technological field) and pathology (conducting autopsies on dead bodies). Similarly, it can't explain why 30 years ago almost no women worked in obstetrics, and now the field is overwhelmingly female.

In contrast, gender norms do help explain the distribution of women across specialties (Hoff, 2010; Riska, 2010). Those norms sometimes push women *toward* specialties and practice types that offer regular hours and thus make family life more manageable (dermatology versus surgery, salaried versus independent practice). At the same time, women are often pushed *away* from various fields by the lack of doctors willing to mentor them and by senior doctors' belief that women can't succeed in these fields.

Learning Medical Values

Professional socialization refers to the process of learning the skills, knowledge, and values of an occupation. During their long years of training, doctors learn not only a vast quantity of technical information but also a set of **medical norms**—expectations about how doctors should act, think, and feel. As this section describes, the most important of these norms are that doctors should value emotional detachment, trust clinical experience more than scientific evidence, master uncertainty, adopt a mechanistic model of the body, trust intervention more than normal bodily processes, and prefer treating rare or acute illnesses rather than common or chronic illnesses.

Emotional Detachment Undoubtedly, most doctors enter medicine because they want to help others. Yet perhaps the most central medical norm is to maintain emotional detachment from patients (Coulehan and Williams, 2001; Hafferty, 1991). As illustrated by the story that opened this chapter, medical culture values and rewards "strength" and equates emotional involvement or expression with weakness (Hafferty, 1991).

Medical training regularly reinforces emotional detachment, as faculty and students implicitly or explicitly ridicule those who display emotions and question their ability to serve as doctors (Hafferty, 1991). During daily rounds of the wards, faculty members grill residents on highly technical details of patients' diagnoses and treatments. Except in family practice residencies, however, faculty members rarely ask about even the most obviously consequential psychosocial factors. Rounds and other case presentations also teach residents to describe patients in depersonalized language. Residents learn to describe individuals as "the patient," "the ulcer," or "the appendectomy" rather than by name, thus separating the body from the person. The use of medical slang, meanwhile, which peaks during the highly stressful residency years, allows students and residents to turn their anxieties and unacceptable emotions into humor by using terms such as *gomers* for elderly demented patients and *not citizens* for unruly drug addicts. Such terms help doctors vent frustrations regarding the difficulties they face and maintain needed emotional distance but also reinforce disparaging attitudes toward patients (Ofri, 2013). So, too, does language like "The patient *denies* nausea" instead of "Mrs. Clark reports that she does not have nausea."

The structure of the residency years largely prevents residents from emotionally investing in patients (Ofri, 2013). Long hours without sleep often make it impossible for residents to provide much beyond the minimum physical care necessary (Christakis and Feudtner, 1997). When combined with the norm of emotional detachment, such long hours can even encourage doctors to view their patients as foes. As T. M. Luhrmann wrote in his memoir of medical residency:

> I came in one morning to rounds and heard one of my classmates discussing his previous night on call. "Oh," he said, "a woman came in, and we did such and such and such and such but luckily she died by morning." What appalled me was that I understood how he felt: If she had lived, he would have had someone else to take care of (2007:84).

Clinical Experience In addition to teaching doctors certain attitudes toward patients, medical culture also teaches, at a more abstract level, a set of attitudes toward medical care, illness, the body, and what makes humans truly human.

Ironically, given that doctors' prestige rests partly on their scientific training, medical culture values clinical experience more than scientific research and knowledge (Bosk, 2003; Timmermans and Oh, 2010). Students are routinely instructed to value their intuition and their professors' clinical experience over the results of scientific research. At any rate, various reviews of modern medical treatments have found that fewer than half have good scientific support behind them (Naylor, 1995; *British Medical Journal*, 2014; Prasad et al., 2013). High-quality research supporting surgical interventions is even rarer. This partially explains why standard clinical procedure varies enormously across the nation, producing high rates of medical error as well as rates of lumpectomies, prostatectomies, and back-pain surgery that are more than 30 times higher in some states than in others (Center for the Evaluative Clinical Sciences, 1996; Leape, 1994; Wennberg, 2010). Figure 11.2 illustrates these differences with regard to spinal fusions for back pain, a surgery with considerable risk but uncertain benefits: Thirteen percent of patients who receive spinal fusions require another surgery within 30 days (Martin et al., 2014).

Recognition of these wide geographical variations in medical care, combined with concern about rising costs of health care, have led policy makers (including some doctors) to push for the development of **evidence-based medicine**: medical care based on a thorough evaluation of the best available scientific research, especially research using **random samples** and appropriate statistical **controls**. Almost all medical schools now explicitly incorporate evidence-based medicine into their curricula.

The response to evidence-based medicine, however, suggests doctors' continuing commitment to clinical experience as a basis for decision making (Timmermans and Berg, 2003a; Timmermans and Oh, 2010). When doctors are

FIGURE 11.2 Geographic Variations in Use of Spinal Fusion for Back Pain

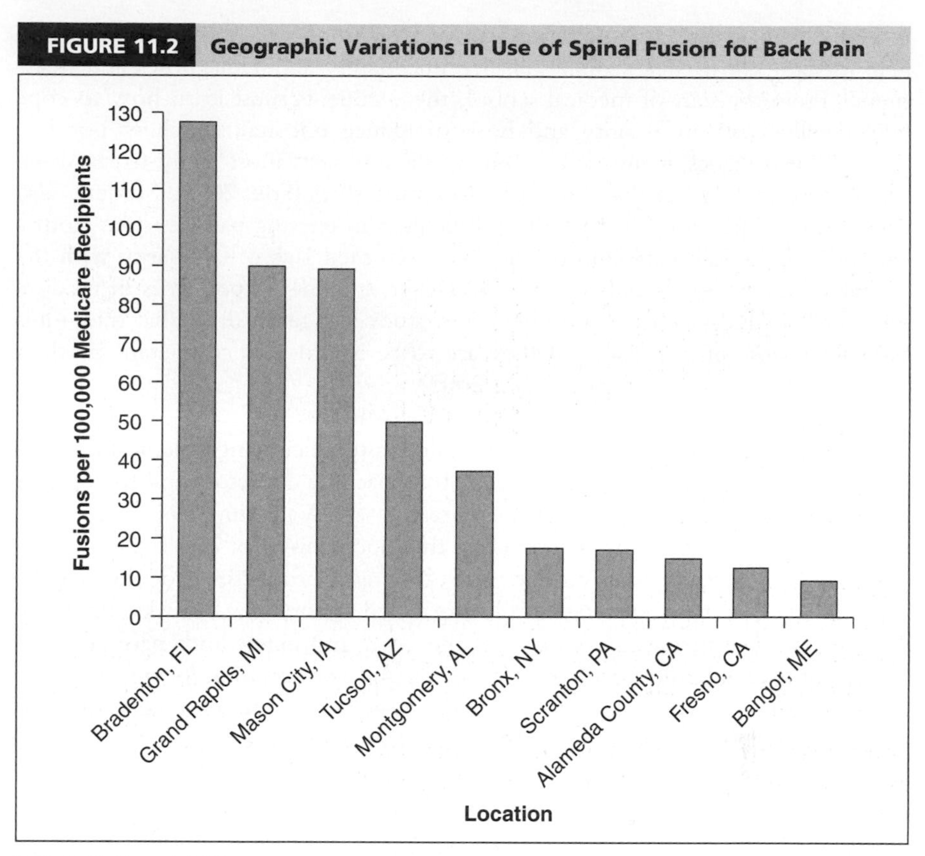

SOURCE: Martin et al., 2014.

working on a case, they rarely have time to obtain the latest research findings on the topic, let alone to evaluate that research fully. Instead, they often must settle for reading a single review or research article. In addition, because practice protocols can't cover all the circumstances in any case and rarely provide an absolute answer, doctors still must rely on their clinical judgment. Finally, medical training and practice remain hierarchical environments in which doctors and medical students are expected to defer to senior doctors and thus are unlikely to challenge orders from senior doctors even if those orders go against practice protocols. On the other hand, because junior doctors increasingly turn to the research literature for answers, more senior doctors must do so as well to retain their reputations and status. In sum, although evidence-based medicine has affected medical care, clinical experience remains the more common basis for decision making.

Mastering Uncertainty One reason medical culture values clinical experience over scientific knowledge is that there is simply too much knowledge for students to learn it all. As a result, students can never be certain that they have treated a patient correctly. Moreover, because the answers to so many medical questions remain unknown, even a student who somehow learned all the available medical knowledge would still face uncertainty about diagnoses and treatments. From the start of medical school, then, students must learn how to cope emotionally with uncertainty and how to reduce uncertainty where possible, through such tactics as primarily studying discrete facts most likely to show up on examinations, rather than broader conceptual issues (Fox, 2000). Students also must learn to question whether their difficulties in treating patients stem from a lack of available knowledge in the field or from their lack of familiarity with the available knowledge. Simultaneously, however, students' experiences in medical school classes and on the wards where they study also teach them that they must hide their sense of uncertainty if they are to be regarded as competent by their professors and patients (Lingard et al., 2003).

Mechanistic Model Along with learning to master uncertainty, medical students also learn to consider the body analogous to a machine or factory and to consider illness analogous similar to a mechanical breakdown (Waitzkin, 1993). For example, medical textbooks routinely describe the biochemistry of cells as a "production line" for converting energy into different products, and the female reproductive system as a hierarchically organized factory that "breaks down" at menopause. The mechanistic model of the body and illness leads naturally to a distrust of natural bodily processes. Doctors learn to always look for signs that the body is breaking down and to view changes in the body as causes or consequences of such breakdowns. As a result, doctors typically view pregnancy as a disease, fight against the effects of aging, use drugs to control minor fevers (the body's natural process for fighting infection), and so on. For example, although the World Health Organization recommends that cesarean sections—a major abdominal surgery—should be performed only when necessary and in no more than 15 percent of births (AbouZahr and Wardlaw, 2001), many US obstetricians now argue that cesareans are safer than vaginal childbirth and encourage women to choose cesarean section even when not medically necessary. In some hospitals, up to *70 percent* of women now deliver via cesareans (Kozhimannil et al., 2013).

Intervention As this example suggests, learning to distrust natural processes is intimately interwoven with learning to value medical intervention. During the first two years of medical school, most students receive only minimal instruction in using tools such as nutrition, exercise, or biofeedback to prevent or treat illness; during the rest of their training, such tools are rarely—if ever—mentioned. Conversely, professors routinely devote much class time to extolling the virtues of technological interventions and encourage students to aggressively seek experience in surgery and the like. Those who do so receive the most approval from their professors.

Emphasis on Acute and Rare Illnesses As a natural corollary of valuing intervention (and a natural result of locating medical training within research-oriented universities), medical culture teaches doctors to consider **acute illness** more interesting than **chronic illness**. This is not surprising given that doctors often can perform spectacular cures for acute illnesses (such as appendicitis) but can do little for chronic illnesses (such as lupus). Similarly, medical culture teaches doctors to consider rare diseases more interesting than common ones because the former require well-honed diagnostic skills and complex tools even if they can't be treated.

The Consequences of Medical Values

In sum, medical training teaches doctors to value emotional detachment, trust their clinical experience, adopt a mechanistic model of illness, rely on interventions, master uncertainty, and prefer working with rare or acute illnesses. Although each of these values serves a purpose, each also can work against the provision of high-quality health care. Emotional detachment can lead doctors to treat patients insensitively and to overlook the emotional and social sources and consequences of illness. In addition, it can cause doctors to feel disdain for patients they consider too emotional. How much emotion a person shows, however, and how that person does so, depends partly on his or her cultural socialization. In contemporary America, women and members of certain ethnic minority groups (such as Jews and Italians) are more likely than are men and nonminorities to display emotion openly. Consequently, these groups are more likely to bear the brunt of doctors' disdain.

Meanwhile, the emphasis on clinical experience, although sometimes useful, can lead doctors to adopt treatments that have not been tested through controlled clinical trials and that lack scientific validity, such as treating ulcers (which are now known to be caused by bacteria) with a bland diet. In addition, the desire for clinical experience sometimes encourages medical students and residents to perform procedures, from drawing blood to doing surgeries, even if they lack sufficient training or supervision or the procedures cause unnecessary pain. Medical students and doctors are most likely to do so if they can define a patient as "training material" rather than as an equal human being. This is most likely to happen when patients are female, minority, poor, elderly, or otherwise significantly different both from the doctors and from the patients on whom those doctors assume they will someday practice.

Mastering uncertainty is necessary if physicians are to retain enough confidence in their clinical decisions to survive emotionally. And presenting an image of authoritative knowledge undoubtedly increases patient confidence and stimulates a placebo effect, if nothing else. At the same time, the desire for certainty—or at least an aura of certainty—also probably contributes to authoritarian relationships with patients. This is particularly problematic when proper treatment really is uncertain. For example, doctors are particularly uncomfortable with patients whose diagnoses are unclear or whose treatment is unsuccessful. Similarly, even though for years evidence indicated that hormone replacement therapy for menopause is dangerous, many doctors—unwilling, perhaps, to give up their aura of certainty—continue to dismiss concerns about these practices and to recommend them to their patients (Prasad et al., 2013).

CONTEMPORARY ISSUES

The Decline in Primary Care

In 1960, primary care doctors accounted for about 50 percent of all US doctors. Currently, they comprise only about one-third of US doctors and only 20 percent of recent medical school graduates (Hoff, 2010). Moreover, even these graduates increasingly choose to specialize *within* primary care, such as in primary care cardiology. The shortage of primary care doctors is already acute in poor and rural areas and is increasing across the nation.

Primary care doctors play a crucial role in any health care system. Patients who receive regular care from a primary care doctor are more likely to receive early diagnosis and treatment for their health problems, thus avoiding potential health catastrophes (and skyrocketing expenses). Similarly, primary care doctors typically coordinate the care of chronically ill patients who receive treatments from multiple doctors, thus reducing the chances that those treatments might interact in dangerous ways.

So why has primary care declined in the United States? First and perhaps most importantly, primary care doctors earn much less than do other doctors, principally because insurers pay less for primary care than for specialty care (Cane and Peckham, 2014). Second, and as this chapter describes, medical culture's emphasis on intervention and rare diseases pushes medical students away from primary care. Similarly, primary care doctors (known in medical slang as "fleas") almost never serve on medical school faculties, whereas surgeons (known as "blades") often do so and often discourage students from entering primary care (Hoff, 2010; Mullan, 2002). Third, many medical students choose to specialize, rather than work in primary care, as a means of reducing the amount of knowledge they must learn and thus managing their own uncertainty about their expertise. Finally, primary care doctors have less control over their schedules and greater pressures to speed through patients than do those in specialty care, making primary care a less appealing line of work.

Given all these forces pushing medical students away from primary care, the field is unlikely to grow unless primary care becomes more central to medical training and unless the financial incentives for primary care increase. The Affordable Care Act does include some provisions aimed at increasing those incentives, but it is too soon to know whether they will have an impact.

The emphasis on working with rare illnesses also creates problems. Most important, this emphasis (along with the emphasis on intervention) has contributed to the oversupply of specialists and undersupply of primary care doctors in the United States (Hoff, 2010; Mullan, 2002). About two-thirds of US doctors are specialists, although most patients require only primary care (US Bureau of the Census, 2013); this topic is discussed further in *Contemporary Issues: The Decline in Primary Care.* Similarly, emphasizing acute illness leads doctors to consider patients with chronic illnesses uninteresting and makes, for example, orthopedic surgery a more appealing field to new doctors than rheumatology (the nonsurgical care of arthritis and related disorders).

Other problems stem from medicine's mechanistic model of the body. This model leads doctors to rely on **reductionistic treatment**. This term refers to treatment in which doctors consider each body part separately from the whole—reducing it to one part—in much the way auto mechanics might replace an inefficient air filter without checking whether the faulty air filter was caused by problems in the car's fuel system. In contrast, sociologists (as well as a minority of doctors) argue for a more holistic image of how the body works and of how illness should be treated (Waitzkin, 1993). **Holistic treatment** refers to treatment that assumes all aspects of an individual's life and body are interconnected. For example, rather than performing wrist surgery on typists who have carpal tunnel syndrome, it might be better to recommend using a wrist rest while typing or changing the height of the typist's desk. And rather than simply excising a tumor when someone has cancer, perhaps doctors and other health care workers should also explore how their patients' social and environmental circumstances contributed to cancer growth and how psychological and financial support might improve their odds of recovery.

Finally, emphasizing intervention can foster overtreatment by encouraging medical students to enter the most interventionist specialties and encouraging doctors to use the most interventionistic tools even when more conservative approaches might better serve patients' interests. For example, research suggests that back pain can best be treated by improving the ergonomics of individuals' work environments rather than by injecting potentially dangerous steroids into patients' spinal columns, yet doctors more commonly suggest the latter rather than the former (Hadler, 2008).

PATIENT–DOCTOR RELATIONSHIPS

From the beginnings of Western medicine, medical culture has stressed a paternalistic value system in which only doctors, and not patients or their families, are presumed capable of making decisions about what is best for a patient. Often, this paternalism is reinforced by patients who prefer to let their doctors make all decisions; indeed, at least part of doctors' efficacy comes simply from patients' faith in doctors' ability to heal. Paternalism is also reinforced by the fact that doctors spend only about 21 minutes per patient visit, which leaves doctors little time either to educate or listen to their patients (Chen, Farwell, and Jha, 2009).

Power and Paternalism Doctors' power over patients is greatest in two situations: (1) when patients are completely incapacitated by coma, stroke, or the like; and (2) when doctors' cultural authority is much greater than that of their patients. For example, faced with a pregnant woman who refuses a cesarean section or a person diagnosed with schizophrenia who opposes hospitalization, courts and hospitals often supports doctors' decisions over their patients' wishes. Finally, doctors' power is higher when interacting with patients who don't share the doctors' language, culture, and social status. In sum, doctors' power depends on their cultural authority, economic independence, cultural differences from patients, and assumed social superiority to patients. As this suggests, and given the demographic composition of contemporary medicine, doctors are most likely to adopt egalitarian interaction patterns with those they consider their equals: white, middle-aged, male, and middle- or upper-class patients (Street, 1991).

Ethnicity, Class, Gender, and Paternalism

Doctors' inclination to make decisions for patients is sometimes bolstered by doctors' racist, sexist, or classist ideas. Doctors are not immune to common stereotypical ideas that label women as flighty, lower-class persons as lazy, and certain ethnic minorities as unintelligent or uncooperative. For example, Danielle Ofri (2013) explains that once medical students begin working on wards, they quickly learn from their medical supervisors to use the term "status Hispanicus" for any Hispanic female patient who is considered too emotional or demanding. The attitudes underlying these stereotypes may encourage doctors to make decisions for such patients instead of involving the patients in decision making.

The attitudes may help to explain ethnic and gender differences in the treatments that doctors offer to patients. (Far less is known about class differences.) As Chapter 3 discussed, African Americans are considerably less likely than whites to receive transplants following kidney failure. Similarly, doctors more often offer life-preserving treatments (including angioplasty, bypass surgery, chemotherapy, and the most effective HIV/AIDS drugs) to whites than to African Americans or other minorities, even when both groups have the same symptoms and insurance coverage (Gross et al., 2008; Nelson, Smedley, and Stith, 2002). In addition, doctors often offer minorities only cheaper, less desirable treatments (e.g., amputating a dangerously infected leg rather than treating it with intensive antibiotic therapy). Similarly, most studies suggest that women are less likely than men to receive high-intensity treatments such as organ transplants and coronary bypasses (Anspach, 2010).

Even when doctors want to provide equal treatment and to enable patients to make their own decisions, cultural barriers can make this difficult. When doctors and patients come from different cultures, doctors may well find it difficult to gain patients' cooperation or to understand patients' beliefs or wishes. Cultural differences are probably greatest when Western-born doctors treat immigrants from non-Western societies. In these circumstances, even the smallest gestures unintentionally can create misunderstanding and ill will. For example, in her observations of US doctors and their Hmong patients who had emigrated from Laos, Anne Fadiman found that

> when doctors conferred with a Hmong family, it was tempting to address the reassuringly Americanized teenaged girl who wore lipstick and spoke English rather than the old man who squatted silently in the corner. Yet failing to work within the traditional Hmong hierarchy, in which males ranked higher than females and old people higher than young ones, not only insulted the entire family but also yielded confused results, since the crucial questions had not been directed toward those who had the power to make the decisions. Doctors could also

> appear disrespectful if they tried to maintain friendly eye contact (which was considered invasive), touched the head of an adult without permission (grossly insulting), or beckoned with a crooked finger (appropriate only for animals) (1997:65).

Taken together, cultural conflicts such as these may lead doctors to conclude that they can't collaborate with such patients and so must make decisions for them.

More broadly, this example illustrates the difficulties that can arise when patients lack cultural health capital. Usually the term *capital* refers to *financial* resources. In contrast, **cultural health capital** refers to *cultural* resources that can facilitate better health care by facilitating better relationships between patients and providers (Shim, 2010). Those resources can include knowledge of basic medical terms, belief in core medical ideas such as the germ theory of illness, and the ability to speak the same language as one's doctors. In addition, cultural health capital increases when patients share their doctors' cultural beliefs, such as the belief that patients should be willing to change their lifestyles for the sake of their health and that doctor–patient interactions should be efficient and unemotional.

Changes in the health care world have made cultural health capital increasingly important (Shim, 2010). Doctors now rarely have long-term relationships with patients but also rarely have enough time to learn how to communicate effectively with any new patients who have little cultural health capital. As a result, working with such patients can feel more frustrating and less rewarding than working with those who have high cultural health capital, and so doctors may consciously or unconsciously choose to spend more time with the latter.

Paternalism as Process

To explore *how* doctors maintain dominance during their meetings with patients, researchers have conducted detailed analyses of conversation patterns between doctors and patients (Karnieli-Miller and Eisikovits, 2009; Waitzkin, 1991). These studies show that doctors typically dominate discussions. Doctors typically indicate when to begin discussing a topic and when the topic is closed, and ask closed-ended rather than open-ended questions, thus making it difficult for patients to raise new topics. For example, a patient might come to a doctor complaining of various problems. The doctor will ask for further details about only some of those problems, typically ignoring how patients' work or living situation might affect their health. The doctor also can ask questions about problems the patient had not mentioned but that the doctor expects to find, thereby defining certain problems as more relevant than others.

Doctors also can reinforce their dominance by the simple tactic of referring to the patient by first name but expecting patients to refer to them by their title ("Dr. Smith"). Finally, when patients must choose between two different treatments or between treatment and "watchful waiting," doctors typically present those options in ways that strongly bias the patient toward accepting whichever course of action the doctor considers best (Karnieli-Miller and Eisikovits, 2009).

Shifting Patient Roles and the Decline of Paternalism

In sum, a variety of factors continue to reinforce paternalistic patient–doctor relationships. Other factors, however, are changing the nature of these relationships. Most importantly, as Boyer and Lutfey (2010:S83) note, "Today's patient role is more often chronic rather than acute; is based on risk of disease rather than existing illness; and requires more active engagement by the patient in monitoring, self-educating, and self-treating over time rather than just seeking treatment from a provider on a one-time basis."

When a patient with a high fever arrives at a hospital and is diagnosed with encephalitis, the doctor in charge will likely begin treatment immediately. In such cases, the patient may be incoherent and unable to give consent to treatment. In contrast, when patients who have lived with diabetes for a decade arrive at a doctor's office, the patients usually have a well-honed sense of their own bodies and of which treatments work best with the fewest side effects. They also may regularly check news reports, websites, and Internet discussion boards to find out what treatments others are using. In these circumstances, communication between doctors and patients often becomes less paternalistic and more a process of negotiation (Boyer and Lutfey, 2010; Heritage and Maynard, 2006).

REFORMING MEDICAL TRAINING

The problems embedded in doctor–patient relationships has led to pressure for reform from both inside and outside medicine. Many doctors now believe that rising rates of malpractice suits largely reflect patients' disenchantment with doctor–patient relationships rather than with the quality of care. In fact, research suggests that even in cases of gross malpractice (such as removing the wrong rib), patients are far less likely to sue if doctors openly admit their error rather than hiding behind their mask of authority. The AMA and the American Hospital Association both now encourage doctors to admit their errors to patients and to their supervisors; hospitals that have aggressively pursued this policy have seen malpractice suits plummet (Sack, 2008a).

Similarly, throughout the United States, medical students and professors are now working to implement innovative programs for integrating more patient-centered perspectives into the medical curriculum. Whereas in the past, students typically spent their first two years studying biology and anatomy and rarely interacting with patients, numerous medical schools are working to introduce students to patients much earlier in their training. At Dartmouth Medical School, for example, students must shadow a community physician once or twice weekly throughout their first year so they see from the start what it means to work with patients. At New York University, students spend part of their first week in medical school listening to lectures by persons living with various diseases. And at Yale Medical School, students are introduced regularly to "patients" (played by actors) whose job is to show them what illness looks like from patients' perspective and to challenge the idea that illness is purely a biological phenomenon.

Cultural competence has also emerged as a commonly cited goal of medical education and practice (Teal and Street, 2009). Cultural competence refers to the ability of health care providers to understand at least basic elements of others' cultures, to recognize the impact of their own cultural identity and biases on their interactions with clients, and thus to provide medical care that better meets their clients' emotional as well as physical needs. Research suggests that culturally competent health care increases the odds that individuals from minority communities will seek health care, feel satisfied with that care, and as a result follow their doctors' recommendations.

The Association of American Medical Colleges, as well as numerous organizations representing various medical specialties, officially endorse including cultural competence in medical training. Increasing numbers of medical students now attend lectures on cultural competence and participate in overseas programs designed to increase their understanding of other cultures (Champaneria and Axtell, 2004). Similarly, hospitals and community clinics increasingly are trying to meet the cultural needs of their patients through such means as inviting Hmong shamans to perform healing rituals in hospital rooms and serving traditional Korean soups to Korean patients (Brown, 2009). In the long run, such efforts may restructure medical culture and doctor–patient relationships.

IMPLICATIONS

Between 1850 and 1950, allopathic medicine attained and then enjoyed unprecedented autonomy and dominance, becoming the premiere example of a profession. In its battles for status with its many nineteenth-century rivals, allopathic medicine benefited from the public's growing respect for scientific knowledge and from the increase over time in the field's scientific foundations. It also benefited from the public's assumption that because allopathic doctors were disproportionately upper-class white men, they must be more competent than the minorities, women, and poorer persons who dominated competing health care fields.

Since the 1950s, however, doctors' social status has declined, and their control over working conditions, relationships with patients, and finances has diminished. Yet doctors continue to have far more autonomy and dominance than do professionals in most other occupations, especially within the health care field. This continued professional dominance—and the continued internecine warfare between medicine and other health care occupations—affects all of us as consumers of health care because it sets the stage on which attempts to improve the health care system must occur. Among other things, as Chapter 12 will show, medical dominance has made it difficult for alternative health care modalities to receive adequate research testing or for tasks to be shifted, when medically justified, from doctors to less expensive occupational groups.

Doctors' professional socialization, too, affects all of us as consumers. In its current form, this process is lengthy, arduous, and expensive, making it difficult if not impossible for many otherwise qualified persons to become doctors and encouraging those who do become doctors to become emotionally hardened or financially driven. (In contrast, in most European nations, the government pays most of the costs of medical training.) To these unintended negative consequences of medical training must be added the problems caused by a medical culture that emphasizes emotional detachment, clinical experience, intervention, mastering uncertainty, and acute and rare illnesses rather than common and chronic illnesses.

As consumers of health care, we all benefit from the extensive training doctors receive. Those benefits, however, must be weighed against the costs we pay when our doctors also learn ways of interacting with patients and thinking about illness that can encourage overly aggressive, scientifically unjustified, or simply discourteous treatment. Only by directly confronting the nature of medical culture can we hope to change medical training and make future doctors better able to meet their patients' needs.

SUMMARY

1. Before the twentieth century, most Americans received their health care from relatives, neighbors, or any of a variety of poorly trained and poorly respected practitioners. Allopathic doctors—the forerunners of modern medical doctors—knew little of science and received minimal training. Their use of "heroic medicine" and ignorance of antisepsis, anesthesia, and drugs left doctors at least as likely to harm as to heal.
2. Doctors achieved professional dominance in the health care world by the 1920s because of the public's growing faith in science; doctors' generally high social status; and, later, improvements in medical education. *Professional dominance* refers to freedom from control by other occupations or groups and ability to control any other occupations working in the same economic sphere.
3. Medicine is considered by sociologists to be a profession because (1) it has the autonomy to set its own educational and licensing standards and to police its members for incompetence or malfeasance; (2) it has technical, specialized knowledge, unique to the occupation and learned through extended, systematic training; and (3) the public believes that doctors follow a code of ethics and are motivated more by a desire to serve than a desire to earn a profit.
4. Medical dominance is now threatened by the rise of corporatization, managed care, and government oversight, as well as the decline in public support and the AMA. It is also threatened by the rise of countervailing powers. Yet although doctors' professional dominance has declined since the 1950s, it remains strong.
5. Through their medical training, students learn a set of cultural norms—to value emotional detachment, trust clinical experience more than scientific

evidence, master uncertainty, adopt a mechanistic model of the body, trust intervention more than normal bodily processes, and prefer working with patients who have rare or acute illnesses rather than those who have common or chronic illnesses.

6. Medical culture teaches a paternalistic value system in which only doctors are presumed capable of making decisions about what is best for a patient, especially if the patient is female, poor, elderly, or minority. Doctors maintain their power when talking with patients by such tactics as asking only close-ended questions, cutting off questions from patients if the doctors consider them irrelevant, and calling patients by their first names while expecting patients to refer to them by their title.
7. Many medical schools have adopted programs to integrate more humanistic perspectives into the curriculum. Schools have focused especially on developing medical students' cultural competence: the ability to understand and work with at least basic elements of others' cultures.

REVIEW QUESTIONS

1. What is the difference between *allopathic* and *homeopathic* doctors?
2. What was medical training like in 1850?
3. What could a doctor offer his patients in 1850? In 1900?
4. What does it mean to say that an occupation is a *profession*?
5. How did doctors achieve professional dominance? What factors have reduced doctors' professional dominance?
6. What are the major medical norms, how do doctors learn them, and how do they affect doctor–patient relationships?
7. What is cultural competence, and why is it important?

CRITICAL THINKING QUESTIONS

1. What factors have helped doctors gain power in American society? What factors are causing them to lose power? On balance, is the power of doctors growing or shrinking?
2. Identify two concepts that you have learned in this course so far and explain why medical students need to be taught these concepts.
3. What is cultural competence? Why is it important, and how might medical schools teach it?

CHAPTER

12

Other Mainstream and Alternative Health Care Providers

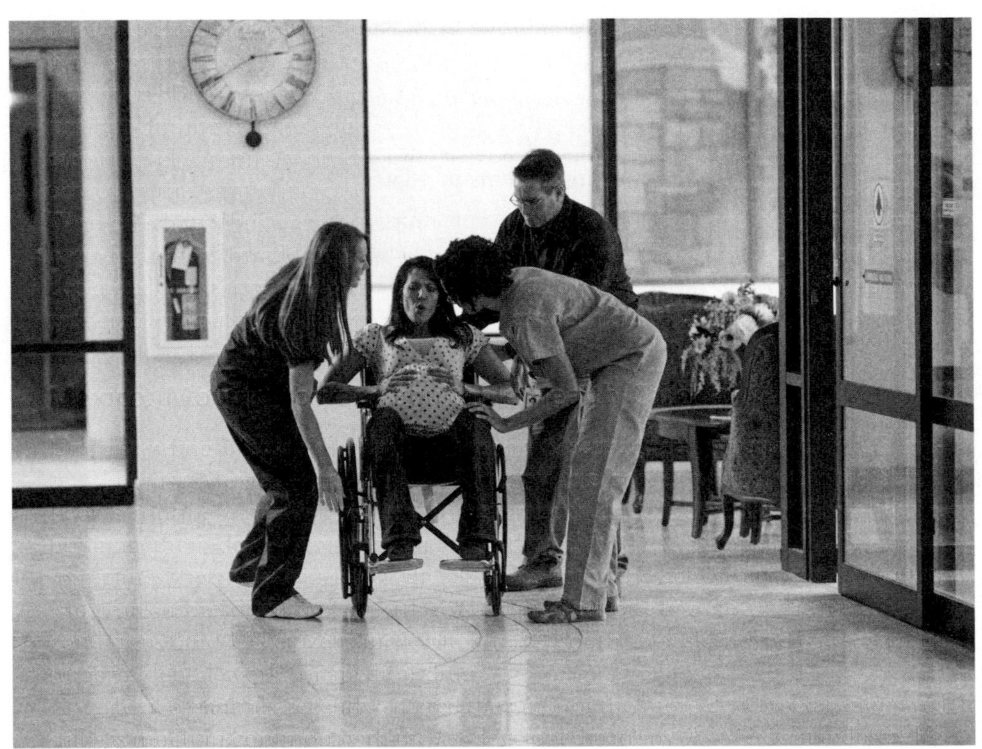

RubberBall Productions/Getty Images

LEARNING OBJECTIVES

After reading this chapter, students should be able to:

- Identify the factors that helped osteopathy and dentistry to gain acceptance as professions.
- Assess the factors that have limited nursing's ability to obtain fully professional status, both historically and currently.
- Describe the consequences of medical dominance for alternative health care providers.

For more than a decade, Juliana van Olphen-Fehr ran an independent practice as a nurse-midwife delivering babies in women's homes. In the following story, she gives us a sense of what it is like to participate in a home birth:

> *Late in the evening, Mona's contractions started getting quite intense. She paced around the room while we watched. She'd sit on the toilet frequently and Dave [her husband] rubbed her back when she was on the bed.... We tried to encourage Dave to go take a nap but he didn't want to leave Mona for a moment. He finally fell asleep in the bed while it was our turn to rub Mona's back. The night moved into early morning. The clock ticked away. We walked and talked.*
>
> *It's amazing how long it takes a baby to be born. As time passes slowly, labor gives one the opportunity to reflect on the process of birth. Each contraction comes and goes, [as] the uterus gets smaller and smaller [and] the baby is massaged down further and further into the pelvis.... Finally, the uterus, getting more powerful as it decreases in size, pushes the baby out of its first cradle, the pelvis, through the vagina, the passageway to life, into the outside world. The mother, feeling more and more pressure, joins the uterus in its expulsive efforts. She bears down gently and involuntarily at first but then more forcefully and purposefully as the baby approaches birth....*
>
> *Mona's labor built up to the point where she started to feel the urge to bear down. Her cervix was completely dilated and I felt the baby's head low in the vagina. She squatted while she pushed during the contractions and walked during the break between them. She found it most comfortable to lean on the banister in her hallway while she pushed.... Dave was still behind her, supporting her hips. I encouraged her to push while I got under her to monitor the baby's heartbeat.*
>
> *Finally, the head appeared. Dave was behind Mona, sitting on the floor, I was beneath her in the front. Together we had our hands around the baby's head, supporting it as we coaxed her to push the baby out slowly. A beautiful little boy was born into Dave's and my hands. I held the baby as Dave eased Mona back*

onto his lap. His arms were around her as they both welcomed the baby into their arms. My birth assistant covered all three of them with blankets to keep the baby warm with their body heat. We turned the light low so the baby would open his eyes. In happy exhaustion, we sat back and through tears watched this family fall in love with each other (Van Olphen-Fehr, 1998:111–113).

Van Olphen-Fehr's story helps us see both why women choose to become homebirth midwives and why health care consumers might choose a nontraditional option like home birth. Since this story took place, however, unaffordable insurance premiums have forced virtually all nurse-midwives to abandon independent practice and to work instead only under direct physician supervision. This situation illustrates the problems faced by nonmedical health care workers in trying to achieve professional status in a system characterized by **medical dominance**.

In this chapter, we first look at the history of three occupations—nursing, osteopathy, and dentistry—and show how each has sought a niche for itself within mainstream health care. We then consider the history of four occupations that, to a greater or lesser extent, remain outside of mainstream health care—chiropractic, direct-entry midwifery, *curanderismo*, and traditional acupuncture. Chiropractic illustrates how, despite medical dominance, an alternative health care occupation can secure a role for itself primarily by limiting its services to a narrow field, and the other three examples show how occupations can remain marginal to the health care system, unable to successfully combat medical dominance.

MAINSTREAM HEALTH CARE PROVIDERS

Nursing: The Struggle for Professional Status

In everyday conversations, Americans often seem to equate health care workers with doctors. Similarly, although many sociologists have researched doctors, few have researched nurses. Yet nurses form the true backbone of the health care system, and hospital patients quickly learn that it is nurses who make the experience miserable or bearable and whose presence or absence often matters most. The history of nursing demonstrates how the drive toward professional status, or **professionalization**, can be especially difficult for a "female" occupation.

The Rise of Nursing Before the twentieth century, most people believed that caring came naturally to women and therefore that families could always call on any female relative to care for any sick family member (Reverby, 1987). Hospitals, meanwhile, relied for custodial nursing care on the involuntary labor of lower-class women who were either recovering hospital patients or inmates of public **almshouses**. These beginnings in home and hospital created the central dilemma of nursing: Nursing was considered a natural extension of women's character and duty rather than an occupation meriting either respect or rights (Reverby, 1987). Nevertheless, increasingly during the nineteenth century, unmarried and widowed women sought paid work as nurses in both homes and hospitals. Few of them, however, had any training.

The need to formalize nursing training and practice did not become obvious until the Crimean War of the 1850s, when the Englishwoman Florence Nightingale demonstrated that trained nurses could alleviate the horrors of war (Reverby, 1987). The acclaim Nightingale garnered for her war work enabled her subsequently to open new training programs and establish nursing as a respectable occupation.

Like most of her generation, Nightingale believed that men and women had inherently different characters and thus should occupy "separate spheres," playing different roles in society. To Nightingale, women's character, as well as their duty, both enabled and required them to care for others. She thus conceived of caring as nursing's central role. In addition, because her war work had convinced her of the benefits of strict discipline, she created a hierarchical structure in which nurses and nursing students would follow orders from their nursing supervisors. This structure, she hoped, would provide nurses with a power base within women's separate sphere parallel to that of doctors within their sphere. These principles became the foundation of British nursing. A few years later, when the US Civil War made the benefits of professional nurses obvious to Americans, these principles were also adopted by American nursing.

Harper's Weekly, January 21, 1871/National Library of Medicine

Nurses first won the respect of the American public during the Civil War.

By the early twentieth century, hospital administrators discovered that running a nursing school gave a hospital ready access to cheap labor, and so hospitals across the nation began opening such schools. Within these hospital-based schools, education was secondary to patient care, and many schools had neither paid instructors nor libraries. Students worked on the wards for 10 to 12 hours daily, with work assignments based on hospital needs rather than on educational goals. Formal lectures or training, if any, occurred only after other work was done.

This exploitative training system stemmed directly, if unintentionally, from the Nightingale model and its emphasis on caring and duty. As historian Susan Reverby notes (1987:75), "Since nursing theory emphasized training in discipline, order, and practical skills, the ideological justification explained the abuse of student labor. And because the nursing work force was made up almost entirely of women, altruism, sacrifice, and submission were expected and encouraged."

Those women who, by the beginning of the twentieth century, sought to make nursing a profession by raising educational standards, establishing standards for licensure or registration, and improving the field's status found their hands tied by the nature of the field (Malka, 2007; Reverby, 1987). According to Reverby, to raise its status, nursing reformers

> had to exalt the womanly character and service ethic of nursing while insisting on the right of nurses to act in their own self-interest, yet not be "unladylike." They had to demand higher wages commensurate with their skills and position, but not appear "commercial." Denouncing the exploitation of nursing students as workers, they had to forge political alliances with hospital physicians and administrators who perpetrated this system of training. While lauding character and sacrifice, they had to measure it with educational criteria in order to formulate registration laws and set admission standards. In doing so, they attacked the background, training, and ideology of the majority of working nurses. Such a series of contradictions were impossible to reconcile (1987: 122).

Political weaknesses also hamstrung nurses' attempts to increase their status. Like other women, few white nurses could vote until 1920, and most nonwhite nurses could not do so until considerably later. Moreover, nurses faced formidable opposition from doctors and hospitals that feared losing control over this cheap workforce. Nevertheless, by the 1920s, most states had adopted licensing laws for nursing schools and nurses. But most laws were weak and poorly enforced, so nurses' status remained somewhat marginal until World War II (Judd, Sitzman, and Davis, 2009; Malka, 2007).

Rising Education and Professional Status Since World War II, nursing leaders have focused on increasing educational requirements for entering nursing as a means of achieving professional status (Judd et al., 2009; Malka, 2007). Beginning in the 1960s, the American Nurses Association promoted the development of two- and four-year college-based nursing programs and lobbied to make college education a requirement for nursing. The new college-based programs quickly proved popular because changing social norms encouraged women to seek a college education in the hopes of improved employment opportunities.

This emphasis on higher education has reinforced nursing's hierarchical structure. At the bottom of the hierarchy are nursing assistants (described in Chapter 10) and **licensed practical nurses (LPNs)**, who receive at most one year of classroom and clinical training and who provide mostly custodial care to patients. Neither nursing assistants nor licensed practical nurses have the autonomy, status, or independent knowledge base that sociologists consider crucial for meeting the definition of a profession, nor is it likely that they will ever do so.

The situation is more complex for **registered nurses (RNs)**, individuals who have received at least two years of nursing training and met national licensure requirements. In past decades, most RNs received their training through hospital-based programs unaffiliated with colleges or universities. These days, most RNs hold associate's degrees in nursing from community colleges, and an increasing number hold bachelor's degrees in nursing from four-year colleges or universities.

Without a doubt, these nurses enjoy more autonomy and status than in the past (Malka, 2007). Many large hospitals now have a Director of Nursing who makes largely independent decisions regarding the nursing staff and whose status is parallel to that of the Director of Medicine. Meanwhile, in many wards, nurses and doctors work together in relatively egalitarian teams; this is especially true in emergency departments, intensive care units, and operating rooms, where quick decisions and good rapport between doctors and nurses are crucial (Carmel, 2006).

Continuing Daily Struggles Despite the increasing autonomy that rising levels of education have brought, asserting professional status remains a daily struggle for many nurses (Gordon, 2005). Doctors continue to determine much of nurses' everyday working conditions: who works when, where, and for how much money. And doctors sometimes include abusive treatment as part of these conditions: Almost all nurses who responded to a 2002 national survey reported either experiencing or witnessing incidents in which doctors screamed at nurses, hit or threw things at nurses, abusively criticized them, or otherwise made it difficult for them to function (Rosenstein, 2002).

Even when doctors don't abuse nurses, they often underscore the status difference between them (Gordon, 2005). Most doctors expect to be referred to by their title—"Dr. Smith"—while referring to nurses by their first names or simply as "my nurse." They rarely read nurses' notes on patients' charts, eat with nurses in hospital cafeterias, include nurses in discussions on hospital rounds, or invite nurses to important meetings about patients. Meanwhile, in what is referred to as the **doctor–nurse game**, experienced nurses are still expected to subtly instruct inexperienced doctors in how to treat patients without revealing the doctors' ignorance to onlookers (Gordon, 2005). For example, an experienced surgical nurse might subtly suggest what the doctor should do by placing certain instruments on the table or by telling the patient, step by step, what the doctor is about to do. Similarly, nurses often do the work of doctors—prescribing drugs, tests, or physical therapy—when doctors are unavailable, but the doctors often reinforce their own status by telling others that the nurses are simply following the doctors' known preferences. Even when patients' lives are saved by nurses' quick actions, doctors typically receive the credit from patients, administrators, and other doctors (Gordon, 2005).

Changing Gender Roles and Professionalization One important factor that may affect nursing's ability to gain professional status is the changes in gender roles in the broader society. Over the past three decades, as women gained entry to other fields, intelligent and motivated women increasingly chose to enter medicine, pharmacy, or biological research instead of nursing (*New York Times*, 1999). For the same reason, nursing attracted fewer white students and middle- or upper-class students.

That said, the economic downturn that began in 2008, coupled with changes in gender roles, has increased nursing's appeal for men as well as women. Currently, men comprise about 11 percent of registered students in bachelors and master's degree programs in nursing (American Association of Colleges of Nursing, 2012). Because nursing is so strongly identified with femininity, working as a nurse presents men with a serious conflict between their gender identity and their work identity. Men typically respond to this conflict by stressing the differences between what they do and traditional nursing—deemphasizing nurturing while emphasizing their technical skills, quick thinking, or use of physical strength. As a result, men are disproportionately represented in areas considered "masculine" such as operating rooms and emergency departments and underrepresented in areas such as pediatrics (Snyder and Green, 2008).

Structural Changes and Professionalization Structural changes in health care have also affected nurses' lives and professional status. Since the 1970s, **corporatization** and the resulting emphasis on cost control have resulted in worse working conditions and decreased job satisfaction for most hospital-based nurses. To save costs, hospitals try to release patients before their insurance coverage ends, leaving only the sickest patients in the hospital. Yet to keep their staffing costs as low as possible, hospitals now hire considerably fewer RNs per patient than they used to (Gordon, 2005). Thus, the typical hospital ward now has fewer nurses but sicker patients than in the past. As a result, nurses' satisfaction has declined while deaths and injuries among patients have risen (American Nurses Association, 2014).

Other changes have also worsened nurses' position. First, because RNs can perform more tasks more efficiently than LPNs, hospitals now save money by assigning RNs many of the labor-intensive, menial tasks formerly performed by LPNs. Because RNs remain responsible for many administrative and skilled technical tasks, this shift has both deprofessionalized their daily work and dramatically increased their workload (Aiken, Sochalski, and Anderson, 1996; Brannon, 1996; Gordon, 2005). Second, hospitals increasingly save money by hiring nurses temporarily and without benefits or by moving full-time nurse employees from ward to ward as needed. As a result, nurses have considerably less control than in the past over their schedules and the nature of their work, and are less able to choose the people they will work with. Third, hospitals have saved costs by shifting services from inpatient wards to less expensive outpatient clinics, where fewer RNs are needed, RN salaries are lower, and their work is less prestigious (Norrish and Rundall, 2001). Finally, nurses are increasingly pressured to work back-to-back shifts and longer hours (often unpaid). Taken together, these factors have resulted in high dropout rates from nursing careers (Gordon, 2005).

Advanced Practice Nursing and Professional Status Although most nurses continue to struggle for professional status, few would doubt that **advanced practice nurses** have achieved it. These nurses typically hold masters degrees that license them to work as nurse-anesthesiologists, as nurse-practitioners, or in other specialized fields. As of 2015, however, all new advanced practice nurses will be required to earn a Doctorate in Nursing Practice. In addition, nurses who want to work as nursing professors or researchers must earn a different sort of doctorate that focuses on research training.

The number of doctoral-trained nurses has soared over the last decade. These nurses now run their own classrooms, laboratories, research journals, research grants, and licensing boards; generate knowledge parallel to that generated by medical doctors; earn substantial salaries; enjoy autonomy, status, and public respect similar to that of other professionals; and have the legal right to call themselves *doctors*. Moreover, about half of the states allow them to prescribe at least some medications and to work essentially as **primary care** providers. This autonomy and status is justified by numerous research studies, published in major medical journals, that have found care provided by advanced practice nurses to be as good as or better than that provided by doctors (Mundinger et al., 2000; Sakr et al., 1999).

Not surprisingly, however, the American Medical Association (AMA) is fighting against this trend and fighting to maintain its dominance (Harris, 2011). The case of nurse-midwifery illustrates these pressures. **Nurse-midwives** are registered nurses who additionally earn nationally accredited graduate degrees in midwifery. The earliest nurse-midwives, beginning in the 1920s, practiced primarily in poor or rural areas with few doctors and enjoyed considerable autonomy. Beginning in the 1960s, a growing number of nurse-midwives worked largely independent of doctors in homes and clinics, providing a true alternative to medicalized childbirth. These days, however, changes in insurance costs and regulations (supported by doctors) have made it virtually impossible for nurse-midwives to work independently. Almost all nurse-midwives now attend deliveries only in hospitals and only when doctors expect the delivery to be routine, uninteresting, and poorly paid. Yet research indicates that care by nurse-midwives (at home or in hospitals) is at least as safe as care by doctors in hospitals (MacDorman and Singh, 1999; Sandall et al., 2013).

On the other hand, nurse-midwives have legal authority to practice and to write prescriptions in all 50 states. The majority of states require private health insurers to reimburse nurse-midwives for their services, and all states reimburse midwives for serving **Medicaid** clients. In 2012, nurse-midwives attended 8 percent of all US births (American College of Nurse-Midwives, 2014). In sum, nurse-midwives have gained considerable autonomy and public recognition as well as an established place for themselves in the health care system. Their ability to gain greater professional status and independence from medical control, however, remains restricted.

Osteopathy: A Parallel Profession

Osteopathy exemplifies a health care occupation that has achieved professional status almost equal to that of medicine. Osteopaths function as **parallel practitioners**, performing basically the same roles as **allopathic doctors** while retaining professional autonomy and at least remnants of a fundamentally different ideology about illness causation (Gevitz, 1988). The history of osteopathy demonstrates the benefits and costs of gaining professional status in the face of medical dominance.

Nineteenth-Century Roots Osteopathy was founded by Andrew Taylor Still, a self-taught allopathic doctor (Miller, 1998). In 1864, three of his children died from meningitis. These deaths, coupled with his belief that all drug use was immoral, provoked Still to investigate alternatives to allopathic medicine. The system Still eventually developed drew on the popular contemporary concept of "magnetic healing" (Miller, 1998). **Magnetic healers** theorized that an invisible magnetic fluid flowed through the body and that illness occurred when that flow was obstructed, unbalanced, inadequate, or excessive. They believed that by moving their hands along patients' spinal cords, they could correct problems in the magnetic fluid and thus cure illness. Still adopted this theory essentially intact, although he attributed health and illness to problems in the flow of blood rather than the flow of magnetic fluid.

During the next few years, Still also studied the work of local bonesetters, whose work consisted primarily of setting broken and dislocated bones and joints and secondarily of treating joint problems through extending and manipulating limbs. Still's experiences convinced him that such manipulations could cure a wide variety of illnesses.

Combining magnetic healing and bonesetting, Still concluded that disease occurs when misplaced bones, especially in the spinal column, interfere with the circulation of blood. He named his new system of spinal manipulation *osteopathy*, from the Greek words for "bone" and "sickness." After the germ theory of disease became widely accepted, Still incorporated it into his theory by arguing that spinal problems predispose individuals to infections and that correcting spinal problems can help the body fight infection. To date, no research has demonstrated clearly whether osteopathic treatment has any effect, whether positive or negative. (The same, of course, can be said for most drugs and procedures used by allopathic doctors, as we saw in Chapter 11.)

Professionalizing Osteopathy In 1892, Still established the American School of Osteopathy and began accepting students for a four-month course of instruction. Five years later, in 1897, he helped found the American Osteopathic Association (AOA). The AOA fought hard to obtain professional recognition and autonomy for the field through increasing educational standards and gaining state approval for independent osteopathic registration boards. In addition, like the AMA and organizations for other emerging professions, the AOA adopted a code of ethics to help convince the public and the state that they were reputable practitioners. Such codes of ethics still play a role in maintaining the reputations of professions and in policing the behavior of professionals, as *Ethical Debate: Pharmacists and Conscience Clauses* discusses.

The AOA's fight for professional recognition proved highly successful. By 1901, and despite strong opposition from doctors and medical societies, 15 states legally recognized osteopathy. By 1923, osteopathic colleges required as many years of education as medical colleges, and all but two states licensed osteopaths. Nevertheless, although threats from allopathic medicine have failed to eliminate osteopathy, changes from within raise questions about its future as an independent field. By the 1920s, most osteopaths had concluded that to compete with allopathic doctors, they would have to offer a similar range of patient services. As a result, osteopaths increasingly treated patients with acute illnesses as well as those with chronic illnesses. In addition, osteopathic colleges continued to teach spinal manipulation but added courses in surgery and obstetrics, often taught out of medical textbooks. By the end of the decade, in a major break with its founder, the AOA mandated that osteopathic colleges provide a course in "supplementary therapeutics," including drugs. Thus, osteopathy began moving toward a merger with allopathic medicine (Miller, 1998).

Despite these changes, many allopathic doctors still disdained osteopaths. Although osteopathic education had improved, it had not kept up with the changes in allopathic education, leading many states to grant only restricted privileges to osteopaths. To combat this problem, the AOA adopted a series of reforms between 1935 and 1960, including requiring three years of college for admission to osteopathic colleges; improving the curriculum, facilities, and faculty at those colleges; and strengthening internship programs at osteopathic hospitals. Because of these changes, by 1960, osteopaths had received unrestricted privileges to practice in 38 states.

The Waning of Osteopathic Identity Despite these reforms, osteopaths still lacked the professional autonomy and status of allopathic doctors, who outnumbered them by at least 20 to one throughout the 1900s. This situation led osteopaths in California, the state where osteopathy was most entrenched, to strike a bargain in 1962 with their allopathic counterparts. Ninety percent of California osteopaths agreed to dissolve their ties with the AOA, stop using their osteopathic degrees, and accept new medical degrees. The California osteopathic hospitals and colleges agreed to become allopathic institutions, and the state osteopathic organization agreed that the state would stop issuing osteopathic licenses.

Although at the time many osteopaths worried that this move would weaken osteopathy, the reverse proved true (Gevitz, 1998). Many allopathic and osteopathic doctors alike opposed the merger, making any further mergers unlikely. In addition, the continuing professional problems of the former

ETHICAL DEBATE

Pharmacists and Conscience Clauses

Sarah Johnson works as a pharmacist in a chain drug store in rural Washington state. She loves having the opportunity to help people deal with their health problems and enjoys her status as a competent, valued professional. Recently, though, her manager told her that she must stock and dispense Plan B, a drug that can be used after unprotected sex to prevent pregnancy by either preventing sperm from fertilizing an egg or (much less often) preventing a fertilized egg from implanting in a woman's uterus. Because federal regulations now allow Plan B to be sold without a prescription to anyone over age 17 but require pharmacists to keep it behind the counter, as the store's pharmacist she would have to physically hand the drug to any customers who request it and would have full responsibility (unmediated by a doctor) to instruct them in using it.

To many Americans, Plan B is a life saver that protects individuals from unwanted pregnancies (including those caused by rape). To others, it's a life killer. Although many who oppose abortion believe that life doesn't start until after a fertilized egg is implanted in the uterus, others believe that life begins as soon as an egg is fertilized. To these individuals, Plan B is just another form of abortion, and abortion is just another form of murder. Moreover, Plan B seems particularly reprehensible because it allows individuals to end pregnancies quickly, cheaply, and safely, thus making abortion far more palatable and feasible than it might otherwise be.

Since abortion was legalized by the Supreme Court in 1973, most states have passed "conscience clauses" that permit health care students and professionals to opt out of learning or performing abortions or other tasks (such as sterilization or physician-assisted suicide) that they consider unethical (Berlinger, 2008). More recently, pharmacists and others who oppose abortion have successfully pressed some states to pass conscience clauses that allow pharmacists to opt out of personally providing any drugs they consider unethical. Conversely, other states have passed laws that, at a minimum, require pharmacists to inform customers of other pharmacies that do stock and sell these drugs.

California osteopaths convinced osteopaths elsewhere that merging wouldn't end their problems. Thus, interest in pursuing a broader merger did not develop. Meanwhile, both federal and state legislators and regulators interpreted AMA support for the merger to mean that osteopathic and allopathic doctors were essentially equivalent. Partly as a result, by the 1970s, osteopaths had received unrestricted privileges in all 50 states and now have essentially the same relationship with insurance providers as do allopathic doctors. The number of osteopaths practicing in the United States has more than doubled over the past 30 years.

Osteopathy, then, no longer faces serious threats from the outside. Its existence remains threatened, however, by its success (Miller, 1998). Like allopathic doctors, to become specialists osteopaths must complete a residency. As of 2015, graduates from osteopathic schools must meet the same standards as graduates of allopathic schools to obtain a residency—a change which may well move the two fields closer (Berger, 2014). Meanwhile, although osteopaths occasionally use spinal manipulation, they generally use the same treatment modalities as allopaths. As a result, ties among osteopaths have faded while those to allopathic

Pharmacists who believe on religious grounds that Plan B causes abortions and is therefore immoral argue that forcing them to provide the drug goes against their right to religious freedom (Flynn, 2008:105). Even if not phrased as a religious belief, they argue, professionals must have the right to refuse work that they consider unethical: If, for example, we expect military doctors to refuse orders to aid torturers, we should also not only allow but expect pharmacists to refuse orders that they believe aid murderers.

Others, however, argue that such refusals by pharmacists go against pharmacists' professional code of ethics (Flynn, 2008:105). According to the American Pharmacists Association's code, pharmacists are expected to "respect the autonomy and dignity of each patient." When a pharmacist refuses to dispense a drug that a customer requests, that pharmacist is implicitly deciding what is best for the customer and imposing his or her own moral and religious view on that customer. Moreover, in rural areas, and especially for poor customers who lack ready transportation, refusing to provide a drug may make it difficult or even impossible for a customer to obtain the drug in a timely manner (Berlinger, 2008).

In these situations, the integrity, autonomy, and religious freedom of customers and pharmacists are necessarily at odds.

Sociological Questions

1. What social views and values about medicine, society, and the body are reflected in the debate over pharmacists and conscience clauses? Whose views are these?
2. Which social groups are in conflict over this issue? Whose interests are served by requiring pharmacists to dispense drugs? By allowing them to refuse to do so?
3. Which of these groups has more power to enforce its view? What kinds of power do they have?

doctors have grown. At the same time, the virtual elimination of differences between allopathic and osteopathic treatment and theory has reduced osteopaths' sense of a strong separate identity.

On the other hand, the growth of the consumer health movement and the rise of interest in alternative medicine have given a new burst of life to osteopathy. Since 1980, the number of osteopathic schools has doubled, and several are now based in state universities (Berger, 2014). Modern consumers are increasingly sympathetic to osteopaths' orientation toward patient care, which in general is more holistic and humanistic than that found among allopathic doctors. In addition, consumers increasingly have sought less interventionistic treatments, such as osteopathic manipulation, either instead of or in addition to allopathic treatment. Osteopaths also pride themselves on their commitment to serving poorer populations, which is stressed far more in osteopathic schools than in allopathic schools (Berger, 2014).

In sum, the history of osteopathy demonstrates the benefits of achieving full professional status as well as the difficulties a parallel health care profession can face in maintaining an independent identity when it no longer faces discrimination from the medical world and when the ideological justification for its separate existence wanes.

Dentistry: Maintaining Independence

Like osteopathy, dentistry has maintained a professional status nearly equal to that of medicine. Unlike osteopathy, though, dentistry's independent identity and status have remained intact over the decades (Adams, 1999). Dentists face little if any challenge from doctors over their sphere of influence, while maintaining considerable control over the dental hygienists who work under them.

Building a Profession A main reason for dentistry's independence is that, unlike osteopathy, it is *not* a parallel occupation (Adams, 1999). Rather, from the start its focus has been on oral health, an area that medical doctors lost interest in by the early twentieth century. As a result, it posed little threat to the medical profession's territorial ambitions. Moreover, dentistry did not challenge core medical beliefs about scientific research or about the causes of health and illness (such as belief in the germ theory of illness). Similarly, although dentistry moved into universities more slowly than medicine did, by the 1930s all dental education was provided at fully accredited universities (Schulein, 2004). As a result, doctors were less inclined to attack dentistry than they were to attack faith healers and others who rejected their basic beliefs and knowledge.

It also helped that by the early twentieth century dentistry was dominated by middle-class white men (Adams, 1999). As a result, dentists and doctors brought equal social status to any legal battles during the early period of professionalization. In addition, the American Dental Association was founded just a few years after the AMA, so dentists had a group to fight on their behalf by the time the AMA began actively challenging the legal status of other competing health care occupations.

A Profession Apart By the mid-twentieth century, dentistry, like medicine, was an accepted and respected profession. The fact that, by this point, dentists needed four years of college and four years of post-graduate education before they could practice certainly helped their status. (Dentists would additionally need two or three years of internship training to enter the subspecialties that emerged later in the century.) In addition, dentists' status was augmented by their growing expertise. In earlier eras, dentists could only extract broken, infected, or eroded teeth. By the mid-twentieth century, however, dentists could fill cavities, make crowns for teeth, prevent cavities through cleaning and fluoride treatments, craft bridges, surgically implant replacement teeth, and straighten teeth. All these developments not only increased the scope of dental practice but also added to dentists' prestige (and incomes).

In one important way, though, the fates of dentists and doctors diverged. In 1965, Medicare and Medicaid were founded to provide affordable health care to the poor, the disabled, and the elderly. Dental care was not included in either program, as it was considered less necessary than medical care and costs were low enough that many individuals could afford to purchase it on their own. In addition, the American Dental Association opposed adding dental coverage to Medicare, although they supported adding it to Medicaid to provide poor children with preventive care (Waldman, Truhlar and Perlman, 2005). As a result, far fewer Americans have dental insurance than have medical insurance. Thus dentists have so far largely avoided the regulations, oversight, and time-consuming paperwork required by insurance companies. The Affordable Care Act is expected to significantly increase dental coverage among children but to have only a modest impact on adults; the impact on dentists remains unknown (American Dental Association, 2013).

Dentists are considerably more likely than doctors to work in private practices. Their median income of about $150,000 is far below that of doctors, but they work fewer hours and enjoy considerably more autonomy (Bureau of Labor Statistics, 2014). Similar to doctors, many are now increasing their incomes and preserving their autonomy by offering cosmetic procedures, which are not covered by insurance companies (Exley, 2009). As a result, interest in dentistry as a profession has grown, and the number of dental schools has increased over the last 15 years (Hoover, 2014).

ALTERNATIVE HEALTH CARE PROVIDERS

The occupations described to this point all basically share allopathic medicine's understanding of how the body works, and all enjoy significant roles within the mainstream health care system. The occupations described in the remainder of this chapter are sufficiently divorced from mainstream American medicine to be considered **alternative** or **complementary therapies**—neither taught in medical schools nor widely used by doctors, even if they sometimes are covered by health insurance.

With a few exceptions (such as chiropractic, direct-entry midwifery, and acupuncture), little is known about the effectiveness of alternative healing techniques, which include meditation, reflexology, faith healing, herbal therapies, and colonics. Because allopathic medicine has dominated the American health care system for so long, researching alternative therapies has been all but impossible. Scientific testing requires large investments of time and money, generally available only from the government, universities, or pharmaceutical companies. Until recently, researchers who wanted to study alternative techniques found it nearly impossible to obtain funding, especially from pharmaceutical companies, which have no reason to fund research on herbs or techniques that they can't patent. In addition, researchers who studied these techniques faced great difficulties in getting their results published in the prestigious medical publications that set the standards for health care practice.

In 1992, however, and in a major break with past policy, the US Congress voted to establish within the National Institutes of Health a unit now known as the National Center for Complementary and Alternative Medicine. The major impetus for this legislation came from former California Congressman Berkley Bedell, who had experimented with alternative therapies after his doctors diagnosed him with terminal cancer. His apparently successful experiences convinced him that such treatments warranted wider study and use. Bedell's success in getting this legislation passed reflects legislators' recognition of both the soaring costs of mainstream medical care and the growing public interest in alternative health care.

Interest in alternative healing is growing not only among American consumers but also among allopathic doctors. Most medical schools now require some coursework in alternative medicine—often called "integrative medicine" in the medical world (Loviglio, 2005)—and many students choose to take electives in the area as well. In addition, growing numbers of doctors attend conferences and workshops on alternative therapies or even run alternative therapy centers at major hospitals and medical schools (Baer, 2010).

In the rest of this chapter, we examine four alternative health care occupations. The first two, chiropractic and lay midwifery, at least sometimes use the language of science to justify their work. The other two occupations, *curanderismo* and traditional acupuncture, draw on traditional beliefs unrelated to the Western scientific worldview.

Chiropractors: From Marginal to Limited Practitioners

Unlike osteopaths, **chiropractors** have fully retained their unique identity. The history of chiropractic illustrates how **marginal practitioners**, who treat a wide range of physical ailments and illnesses but have low social status, can become, like podiatrists, optometrists, and dentists, **limited practitioners**—nonmedical health care workers who gain greater social acceptance by confining their work to a limited range of treatments and bodily parts (Wardwell, 1979). *Key Concepts: Limited and Marginal Health Care Occupations* illustrates this distinction.

Early History The roots of chiropractic closely mirror those of osteopathy. Chiropractic was founded in 1895 by Daniel David Palmer, who coined the term from the Greek words for "hand" and "practice." Like Still, the founder of osteopathy, Palmer studied magnetic healing and spinal manipulation and

KEY CONCEPTS

Limited and Marginal Health Care Occupations

		Limited Range of Care	
		Yes	**No**
Marginal social position	**Yes**	Lay midwives	Traditional healers
	No	Chiropractors	Allopathic doctors

concluded that spinal manipulation could both prevent and cure illness. However, Still argued that spinal problems foster disease by restricting blood flow, whereas Palmer argued that spinal problems foster disease by restricting nerves.

In 1896, Palmer founded the first chiropractic school to teach his techniques of spinal manipulation. By 1916, about 7,000 chiropractors had opened practices; by 1930, that number had more than doubled as schools opened around the country (Wardwell, 1988:159, 174).

The Fight against Medical Dominance The American medical establishment greeted the emergence of chiropractic with the same hostility it had demonstrated toward osteopathy. To eliminate these competitors, the AMA and its regional organizations during the 1930s and 1940s filed lawsuits—many of them successful—against more than 15,000 chiropractors for practicing medicine without a license.

To further restrict chiropractic, the AMA pressed for legislation requiring prospective chiropractors to pass statewide basic science examinations written by allopathic-controlled boards. Ironically, this requirement strengthened rather than weakened chiropractic by forcing the field to raise its previously low educational standards. (As with early allopathic and osteopathic schools, early chiropractic schools accepted essentially all who could pay tuition and offered only a few months of training.) Standards improved most dramatically during the 1940s, when the National Chiropractic Association established accrediting standards for schools and when tuition money from veterans studying chiropractic under the federal GI Bill provided the funds schools needed to meet those standards. Since 1968, all chiropractic schools have required two years of college for admission, and most states require four years of chiropractic schooling for licensure.

Similarly, chiropractic in the end benefited from allopathic medicine's legal war against it. When **Medicare** first began in 1965, Congress bowed to pressure from the AMA and voted that Medicare would not cover services by chiropractors (or by clinical psychologists, social workers, physical therapists, and others in competition with doctors). Outraged chiropractic patients responded with a massive public letter-writing campaign, which led Congress in 1972 to pass legislation extending Medicare coverage to chiropractic services despite the lack of scientific research available at the time on its effects. This set the stage for state legislatures to require other insurance plans to reimburse for chiropractic care, at least in certain situations (Wardwell, 1988:179).

In 1974, the last of the 50 states passed legislation licensing chiropractors. Yet organized medicine continued to limit the ability of chiropractors to practice freely. In addition to fighting legislation designed to allow chiropractors to receive private insurance reimbursement, the AMA banned contact between chiropractors and allopaths, making it impossible for chiropractors and allopaths to refer patients to each other. In response, chiropractors and their supporters filed antitrust suits in the late 1970s against the AMA, various state medical associations, the American Hospital Association, and several other representatives of organized medicine (as well as the AOA), alleging that these organizations had restrained trade illegally. Chiropractors and their defenders eventually won or favorably settled out of court all the suits. As a result, overt opposition to chiropractic ended.

Current Status These changes have allowed chiropractors to solidify their social position. One recent national random survey found that 8.5 percent of US residents had used chiropractic (or osteopathic) manipulation in the previous 12 months (Peregoy et al., 2014). (This survey did not differentiate between chiropractors and osteopaths, but a parallel survey conducted in 2002 found that 7.5 percent reported using chiropractors alone.) Public interest in chiropractic continues to increase, and the job outlook is above average for the field. About 44,000 chiropractors work in the United States, most in solo or group practice (Bureau of Labor Statistics, 2014). The mean annual income for chiropractors is $66,000—considerably below physicians' incomes but for a much shorter workweek and with much less, and less expensive, education required to enter the field (Bureau of Labor Statistics, 2014). These figures alone suggest chiropractic's success.

That success, however, is bounded by chiropractors' status as limited practitioners. Insurers now often pay for chiropractic services—about half of the people who use chiropractic services have full or partial coverage—but usually will do so only for treating specific conditions in specific ways (Tindle et al., 2005). State licensure laws sometimes set similar limits, as does patient demand: Despite many chiropractors' desire to treat a broader range of problems, most patients visit them solely for treatment of acute back, head, or neck pain. In addition, only 5 percent of chiropractic patients are referred by a medical doctor, clearly indicating that doctors don't regard chiropractors as colleagues (Mootz et al., 2005).

Nevertheless, chiropractors continue to push for a wider role in health care. Many chiropractors believe spinal problems underlie all illness and that spinal manipulation can cure most health problems, from asthma to cancer. As a result, they believe they can serve effectively as **primary care** providers and now advertise heavily that they offer care for the whole family throughout the life course. This has stimulated new conflicts with mainstream medicine, especially because a significant minority of chiropractors oppose medical treatments, drugs, and vaccination (Campbell, Busse, and Injeyan, 2000). *Contemporary Issues: Vaccine Refusal* discusses the risks incurred when children are not vaccinated.

Current research suggests that chiropractic care may provide slight help to those with acute lower back pain but is unlikely to help others (Cherkin et al., 1998; Hadler, 2008). Nor does it seem likely that future research will identify more benefits given that the basic principles of chiropractic simply don't mesh with scientific understandings of human biology.

Direct-Entry Midwives: Limited But Still Marginal

The history of direct-entry midwifery shows the difficulties members of an occupation face in gaining acceptance as limited practitioners when the occupation draws only from socially marginal groups—in this case, women, often from minority groups. Although until the twentieth century, **direct-entry midwives**

CONTEMPORARY ISSUES

Vaccine Refusal

As memories of measles and other infectious disease epidemics have faded, growing numbers of parents have decided against vaccinating their children (Omer et al., 2009; Steinhauer, 2008). These decisions may stem from Internet rumors, religious or philosophical beliefs, or skepticism regarding modern medicine (sometimes fueled by chiropractors). Although scientific support for vaccination is overwhelming (Roush et al., 2007; Stratton, Wilson, and McCormick, 2002), parents and practitioners may nonetheless question the safety of vaccinating young children with dead or weak strains of disease-causing viruses. Others believe in vaccination but question the wisdom of simultaneously injecting children with multiple vaccinations. Still others believe—based on an article published in the prestigious *British Medical Journal* that was later proved fraudulent—that vaccination can somehow cause autism (Steinhauer, 2008; Godlee, Smith, and Marcovitch, 2011).

The rise in vaccine refusal has led to outbreaks of measles and other diseases in the past few years. Although a disease such as measles, for example, may seem to be merely a nuisance, some who become infected develop ear infections, pneumonia, and encephalitis, and some of the infected become permanently deafened or even die as a result. Moreover, whenever an unvaccinated child becomes infected with a disease, he or she can spread the disease to adults whose vaccinations have worn off with time and to children who can't be vaccinated because their immune systems are too weak (either because of illness or chemotherapy or because they are younger than one year old). Currently, all states allow parents to opt out of vaccination for religious or medical reasons (such as preexisting illness), and a growing number of states allow parents to opt out based on any personal beliefs.

(i.e., midwives who lack nursing degrees) delivered the majority of American babies, by 2005, direct-entry midwives delivered less than 1 percent of all babies (Martin et al., 2007). In this section, we consider how these changes came about and how direct-entry midwives have attempted to regain their lost position.

The Struggle to Control Childbirth Until well into the nineteenth century, Americans considered childbirth solely a woman's affair (Wertz and Wertz, 1989). Almost all women gave birth at home, attended by a direct-entry midwife or by female friends or relatives. Although a few local governments during the colonial era licensed midwives, licensure laws did not survive past US independence, so anyone who wanted to call herself a midwife could practice essentially without legal restrictions. Unlike nurse-midwives, who did not exist until the twentieth century, these **direct-entry midwives** had no formal training, but rather learned their skills through experience and, sometimes, through informal apprenticeships. Typically, they served only women from their own geographic or ethnic community. Doctors (all of whom were men) played almost no role in childbirth because Americans suspected the motives of any men who worked intimately with female bodies (Wertz and Wertz, 1989:97–98). Moreover, doctors had little to offer childbearing women beyond the ability to destroy and remove the fetus when prolonged labor threatened to kill the mother. Midwives, meanwhile, could offer only patience, skilled hands, and a few herbal remedies.

During the late nineteenth century, Americans' willingness to have doctors attend childbirths gradually increased, as did doctors' interest in doing so. As described in Chapter 11, nineteenth-century allopathic doctors faced substantial competition not only from each other but also from many other kinds of practitioners. As a result, doctors attempted to expand into various fields, from pulling teeth to embalming the dead to assisting in childbirth (Starr, 1982:85). Doctors considered assisting in childbirth especially crucial because they believed that families who came to a doctor for childbirth would stay with him for other services (Wertz and Wertz, 1989:55).

As Americans' belief in science and medicine grew during the late nineteenth century, medical assistance in childbirth became more socially acceptable among the upper classes (Starr, 1982:59). Many women supported this change because it allowed them to obtain painkillers from doctors without feeling guilty for circumventing the biblical command to bring forth children in pain (Wertz and Wertz, 1989:110–113). In addition, because midwifery was not a respectable occupation for Victorian women, by the late nineteenth century, middle- and upper-class women seeking a childbirth attendant had only two options: lower-class lay midwives or doctors of their own social class. Having a doctor attend one's childbirth thus could both reflect and increase one's social standing (Leavitt, 1986: 39; Wertz and Wertz, 1989). Ironically, however, doctors probably threatened women's health more than did midwives; although inexperienced or impatient midwives certainly could endanger women, doctors more often used surgical and manual interventions that could cause permanent injuries or deadly infections (Leavitt, 1983:281–292, 1986:43–58; Rooks, 1997).

Doctors' desire to obtain a monopoly on childbirth care led them, beginning in the mid-nineteenth century, to voice opposition to midwives. These attacks escalated substantially in the early twentieth century (Sullivan and Weitz, 1988:9–14). Recent waves of immigrants had swelled the ranks of midwives and made them more visible and threatening to doctors, whose status, especially in obstetrics, remained low. Moreover, doctors now needed the business of poor women as well as wealthier women because the rise in scientific medical education had created a need for poor women patients who could serve as both research subjects and training material.

To expand their clientele, doctors attempted through speeches and publications to convince women that childbirth was inherently and unpredictably dangerous and therefore required medical assistance. In addition, doctors played on contemporary prejudices against immigrants, African Americans, and women to argue that midwives were ignorant, uneducable, and a threat to American values and that therefore midwifery should be outlawed. For example, writing in the *Southern Medical Journal*, Dr. Felix J. Underwood, the director of the Mississippi Bureau of Child Hygiene, described African American midwives as "filthy and ignorant and not far removed from the jungles of Africa, with its atmosphere of weird superstition and voodooism" (1926:683).

Although these campaigns cost midwives many clients, they had little effect on the law. Many members of the public, and even many doctors (particularly those in public health), believed that trained midwives could provide satisfactory care, at least for poor and nonwhite women who couldn't afford doctors' services. Consequently, laws passed during this era tended to have quite lenient provisions. In the end, however, imposing lenient laws, rather than laws requiring upgraded midwifery training and skills, resulted in the deterioration of midwifery and its virtual elimination. The only exceptions were in immigrant and nonwhite communities in the rural South and Southwest, where traditional midwives continued to conduct home births until at least the 1950s (Sullivan and Weitz, 1988:13–14).

The Resurgence of Direct-Entry Midwifery By the second half of the twentieth century, childbirth had moved almost solely into hospital wards under medical care. Although childbearing women were grateful for the pain relief and safety that doctors promised, all too often women nonetheless found the experience painful, humiliating, and alienating. Despite the absence of scientific support for such practices, doctors routinely shaved women's pubic areas before delivery, strapped them on their backs to labor and delivery tables (the most painful and difficult position for delivering a baby), isolated them from their husbands during delivery and from their infants afterward, and gave them drugs to speed up their labors or make them unconscious—all practices that scientific research would eventually find unnecessary or dangerous (Sullivan and Weitz, 1988).

Objections to such procedures sparked the growth of the natural childbirth movement during the 1960s and 1970s and forced numerous changes in obstetric practices. Most hospitals, for example, now offer natural childbirth classes. Critics, however, argue that the real purpose of these classes is to make women patients more compliant and convince them that they have had a natural childbirth as long as they remain conscious even if their doctors use drugs, surgery, or forceps (Sullivan and Weitz, 1988:39).

By the late 1960s, many women had concluded that hospitals would never offer truly natural childbirth (Sullivan and Weitz, 1988:38–39). As a result, a tiny but growing number of women chose to give birth at home. For assistance, they turned to sympathetic doctors and to female friends and relatives, some of whom were nurses. Over time, women who gained experience in this fashion might find themselves identified within their communities as midwives. This new generation of direct-entry midwives who attend almost solely home births reflects the broader revolt against medicalized birth (Sullivan and Weitz, 1988:23–59).

Working as a direct-entry midwife means long and uncertain hours with little pay. Most midwives, however, are motivated by humanitarian and philosophical concerns rather than by financial gain (Simonds, Rothman, and Norman, 2007). Although midwives recognize the need for obstetricians to manage the complications that occur in about 10 percent of births, they fear the physical and emotional dangers that arise when obstetricians use interventionist practices, developed for the rare pathological case, during all births. Like nurse-midwives, direct-entry midwives strongly believe in the general normalcy of pregnancy and childbirth and in the benefits of individualized, holistic maternity care in which midwife and client work as partners.

No national laws set the status of direct-entry midwives. Direct-entry midwifery is now legal in 28 states. It is either illegal or has an unclear legal status in the other states (Midwives Association of North America, 2013). In states where midwifery is illegal, midwives run the risk of prosecution for practicing medicine without a license and for child abuse, manslaughter, or homicide if a mother or baby suffers injury or death. Yet research strongly suggests that home births conducted by experienced direct-entry midwives working with low-risk populations are as safe as or safer than doctor-attended hospital births, even taking into account the small number of midwifery clients who develop problems needing medical attention (Johnson and Daviss, 2005; Lewis, 1993).

In states where direct-entry midwifery is legal, midwives typically must abide by regulations restricting them to "low-risk" clients (such as women younger than age 35) and restricting the techniques they can use (such as forbidding them from suturing tears after deliveries). These midwives typically must have a backup doctor and must transfer their clients to medical care if the doctor so orders. Thus, legalization has given midwives some degree of freedom to practice in exchange for limited subordination to medicine (Sullivan and Weitz, 1988:97–111).

Despite evidence such as this, medical opposition to direct-entry midwifery remains strong and public support weak, although insurance companies do cover midwifery services in some states. Thus direct-entry midwives, even where legal, can't claim to have achieved social acceptance even as limited practitioners.

Curanderos

Curanderos are folk healers who function within Mexican and Mexican American communities (DeBellonia et al., 2008; Perrone, Stockel, and Krueger, 1989; Roeder, 1988). In the United States, curanderos are used primarily by immigrants, as well as by some US-born Mexican Americans, especially those who live in close-knit communities in the Southwest. Those who use curanderos rarely reject modern medicine. Instead, they seek curanderos when medical care fails to cure illness, when distance or poverty limits access to medical care, or when fear of deportation keeps them from seeking medical care, as well as because such curanderos share their language and their cultural beliefs (Favazza, 2013; Sack, 2008b). Use of curanderos appears to have risen because of the downturn in the US economy and crackdowns on undocumented immigrants (Sack, 2008b). The former has made it difficult for some to afford mainstream health care, whereas the latter has led some to avoid mainstream health care for fear that they will be reported to immigration authorities.

Theories and Treatments Curanderos recognize both Western categories of disease, such as colds, and categories of illness unique to Hispanic culture, such as *susto* (DeBellonia et al., 2008; Roeder, 1988). A common diagnosis, *susto* refers to an illness that occurs when fright "jars the soul from the body, in which case treatment consists of calling the soul back" (Roeder, 1988:324). Curanderos also sometimes trace illness to supernatural forces such as *mal de ojo*, or the evil eye.

Curanderos treat illness in a variety of ways, including herbal remedies, massage, prayer, and rituals designed to combat supernatural forces. They believe illness reflects all aspects of an individual's life—biology, environment, social setting, religion, and supernatural forces—and thus must be treated holistically. As a result, curanderos often spend considerable time listening to their clients. The successes curanderos sometimes achieve in treating their clients' illnesses thus derive not only from their knowledge of herbs and the healing powers of their clients' faith but also from the simple healing power of a sympathetic listener. At the same time, curanderos' lack of scientific knowledge can threaten health when, for example, they use folk remedies that contain mercury, lead, or other toxins (DeBellonia et al., 2008).

Becoming a Curandero Individuals become curanderos through apprenticeships, typically with family members. Successful curanderos find that their practices evolve gradually from part-time work, paid primarily in goods and services, to more or less full-time, cash businesses.

The story of Gregorita Rodriguez, a *curandera* (female curandero) living in Santa Fe, New Mexico, who specializes in massage treatments, illustrates this process:

> Gregorita traces her own career as a *curandera* back to her grandmother, Juliana Montoya, who taught Gregorita's aunt, Valentina Romero, the art of *curanderismo*. When any of Gregorita's seventeen children became ill, she took them to her Aunt Valentina for treatment. *La curandera* taught Gregorita, encouraging her by asking, "Why don't you learn? Look, touch here." Using her children's bellies as a classroom, Gregorita felt the different abdominal disorders and learned how to manipulate the intestines to relieve the ailments.
>
> Another of her patients during this learning period was her husband. Responding to his complaints, Gregorita said, "Maybe I can do something for you." Mr. Rodriguez replied, "No, no, no! You are not going to boss me!" So, off he went to see Aunt Valentina, who was elsewhere delivering a baby. Finally, Gregorita got her chance. Her husband was desperate and allowed her to learn, all the time howling about how much she was hurting him. "Cranky," she described him, "especially when I felt a big ball in his stomach and had to work very hard. Slow, slow, I fixed him and he got better. When he went to my aunt, she said he was okay now. After that I treated my husband and one of my sisters and then her family. That's the way it started" (Perrone et al., 1989:108–109). After this incident, others began coming to Gregorita for treatment, and she soon found herself accepted as a curandero.

The Impact of Medical Dominance Because they lack any recognized training in health care, curanderos can't legally charge fees or bill insurance companies for payment. Some work for free, others for fees ranging from $10 to $100. Most keep a low profile, obtaining clients only through word of mouth and only within the Hispanic community. As this suggests, even a folk healer who appears to function completely outside the bounds and control of the Western scientific world can't avoid its authority altogether.

That said, some hospitals and clinics in heavily Hispanic areas now invite curanderos to perform ritual healings (or "cleanings") for their patients. Doing so, they hope, will reduce patients' stress levels, increase their satisfaction with care, and thus increase the likelihood that they will recommend the hospital to others (Brown, 2009). As this suggests, curanderos can't completely escape medical dominance, but as the Hispanic population grows, neither will medical institutions be able to totally reject curanderos.

Acupuncturists

Acupuncture is one of the oldest forms of healing known. Its recorded history goes back 2,000 years, with strong prehistorical evidence going back to the Bronze Age. Only recently, however, has it gained traction in Western societies.

Theories and Treatments Acupuncturists' ideas regarding health and illness bear even less relationship to Western medicine than do curanderos' ideas. Like all traditional Chinese medicine, acupuncture is based on the concept of *chi*. This concept, which has no Western equivalent, refers to the vital life force, or energy. Health occurs when *chi* flows freely through the body, balanced between *yin* and *yang*, the opposing forces in nature. Because any combination of problems in the mind, body, spirit, social environment, or physical environment can restrict *chi*, treatment must be holistic.

Following this theory, traditional Chinese healers consider both symptoms and diagnosis unimportant and focus instead on unblocking *chi*. Acupuncture is based on the theory that *chi* runs through the body to the different organs in channels known as meridians, which have no Western equivalents. To cure a problem in the colon, for example, acupuncturists apply needles to the index finger, which they believe connects to the colon via a meridian. In this way, they believe, they can stimulate an individual's *chi* and direct it wherever it is needed. Acupuncturists decide on treatment through taking a complete history, palpating the patient's abdomen, measuring his or her blood pressure, and reading the 12 pulses recognized by Chinese medicine.

Acupuncture is still used extensively in China, both alone and in conjunction with Western medicine, and is used increasingly in the West. The World Health Organization (2003) considers acupuncture effective and safe for relieving anxiety, panic disorders, insomnia, postoperative and dental pain, and nausea caused by pregnancy or chemotherapy.

The Impact of Medical Dominance Widespread American interest in acupuncture began during the 1970s, when the People's Republic of China first opened to US travelers. Early travelers brought back near-miraculous tales of acupuncture anesthesia and treatment. Because American doctors had no scientific model that could account for acupuncture's effects, these tales threatened their position and worldview (Wolpe, 1985). As a result, various well-known doctors publicly denounced acupuncture, claiming it worked only as a placebo or only because Chinese stoicism or revolutionary zeal allowed them to ignore pain even though acupuncture also had worked on animals and on Western travelers to China.

To remove this threat to their cultural authority, doctors endeavored to control the definition, study, and use of acupuncture (Wolpe, 1985). This proved relatively easy because, unlike chiropractic or osteopathy, acupuncture at the time had few American supporters. Consequently, in their writings and public pronouncements, doctors could strip acupuncture of its grounding in traditional Chinese medical philosophy and define it simply as the use of needles to produce anesthesia. Pressure from medical organizations led the National Institutes of Health to adopt a similar definition in funding research on acupuncture. At the same time, pressure from doctors led most states to adopt licensure laws allowing any doctors, regardless of training, to practice acupuncture but forbidding all others, no matter how well trained, from doing so except under medical supervision. Thus, for many years, most traditional acupuncturists in the United States worked illegally within Asian communities.

During the past decade, however, as acceptance of alternative healing traditions has increased, the position of acupuncturists has improved. Some insurance companies will reimburse nondoctors for acupuncture treatments, and most states now allow nondoctors to perform acupuncture, although some of these states require medical supervision or require acupuncturists to be licensed by medically dominated boards. Use of acupuncture remains rare: only 1.4 percent of US residents report using it in a one-year period (Barnes et al., 2008). Surprisingly, about half of acupuncture users report receiving some insurance coverage for it (Tindle et al., 2005). Despite this small indication of increasing acceptance, these figures suggest that acupuncture remains a marginal therapy and occupation, posing little threat to medical dominance.

IMPLICATIONS

As the discussions in this chapter have suggested, the health care arena is much broader than we usually recognize. Many alternatives to medical treatment exist far beyond those discussed herein. Most of these alternatives function not so much in opposition to mainstream health care as in parallel, with those seeking care jumping back and forth across the tracks. For example, a woman might deliver her first child with a doctor, her second with a nurse-midwife, and her third with a direct-entry midwife, whereas a man who experiences chronic back pain might see a chiropractor or acupuncturist either before, after, or in addition to seeing a medical doctor.

This chapter has highlighted the factors that help health care occupations gain professional autonomy in the face of medical dominance. Timing certainly seems to play a role: Those occupations that emerged before medical dominance became cemented, such as osteopathy and chiropractic, have proved most successful. Social factors, too, consistently seem important: Health care occupations with roots in and support from higher-status social groups have a better chance of winning professional autonomy than do those with lower-status roots and supporters.

Other occupations seem to retain some autonomy—if a marginal position in the health care arena—because they pose little threat to medical dominance. Curanderos, for example, attract a small clientele of poor Mexicans and Mexican Americans who might not be able to pay for medical care or to communicate effectively with medical doctors anyway. Doctors thus have little incentive to eliminate curanderos' practices. Acupuncturists, on the other hand, have attracted not only Asians and Asian Americans but also well-educated whites—including individuals with the skills and resources to publicize the virtues of acupuncture. Consequently, doctors have had a far greater vested interest in restricting acupuncturists' practices and in co-opting acupuncture for their own purposes.

Not surprisingly, developing professional autonomy seems most difficult for those, such as nurses, who work directly under medical control. In contrast, groups such as chiropractors have considerably more leeway to develop their practices without interference from medical doctors.

Finally and ironically, strict licensing laws, even when devised by doctors opposed to an occupation's growth, in the end can help occupations gain professional autonomy by forcing them to increase standards and thereby enabling them to gain additional status and freedom to practice.

To date, medical doctors have succeeded in retaining their professional autonomy and dominance partly because of their greater ability to provide scientific data supporting their theories and practices—or at least to convince the public that they have such data. It remains to be seen whether increased federal support for research on alternatives will increase scientific credibility and public support for these practices.

SUMMARY

1. Nursing as a field has tried to improve its status primarily by increasing educational requirements. It has been held back by its status as a "female" occupation and by public expectations that women are naturally caring and thus don't need professional salaries, professional status, or good work conditions in exchange for their caregiving.
2. The emphasis on higher education has reinforced nursing's hierarchical structure and alienated nurses who lack higher degrees. Meanwhile, corporatization and the emphasis on cost cutting have worsened nurses' working conditions.
3. Advanced practice nurses (most of whom are also registered nurses) hold postgraduate degrees in specialized nursing fields. They now enjoy autonomy, status, and public respect similar to that of many other professionals, and serve in some cases essentially as primary care providers.

4. Osteopathy has achieved professional status almost equal to medicine by adopting beliefs and practices almost identical to that of medical doctors. As a result, osteopathy's independent identity has been threatened. In contrast, chiropractors have retained their unique identity and gained occupational status by increasing educational standards and remaining limited practitioners, although some chiropractors continue to seek a wider role in health care.
5. Alternative or complementary therapies are therapies that are neither widely used by doctors nor typically taught in medical schools, even if they sometimes are covered by health insurance. Little is known about the effectiveness of these therapies. Interest in alternative healing is growing not only among American consumers but also among allopathic doctors.
6. Before the nineteenth century, direct-entry midwives delivered almost all American babies. Direct-entry midwives lost this status because of growing public belief in science, competition from doctors, their low status as women, and the lack of strict licensing and educational requirements. Direct-entry midwifery reappeared in response to dissatisfaction with medicalized childbirth, but it remains a marginalized occupation with very few clients despite evidence of its safety.
7. Curanderos are folk healers with no legal standing and no formal training who function within Mexican and Mexican American communities. They recognize both Western and non-Western categories of disease and treat illness holistically.
8. Acupuncturists believe that illness is caused when the body's *chi*, best translated as "vital life force," becomes unbalanced or blocked. Western doctors have tried to limit the use of acupuncture by controlling its definition, study, and use. Use of acupuncture and insurance reimbursement for it are growing, as is evidence that it is effective for certain conditions.

REVIEW QUESTIONS

1. How did the early history of nursing make it difficult for nurses to increase their status or improve their working conditions?
2. How have nurses attempted to professionalize? What factors have limited their success?
3. How have changes in the health care system affected nurses' occupational status and position?
4. How did osteopaths attempt to professionalize? What factors enabled them to succeed? What price has osteopathy paid for its success?
5. How did dentistry succeed in retaining professional status and independence from medical dominance?

6. To what extent and in what ways have chiropractors succeeded in improving their occupational status?
7. How and why did doctors gain control over childbirth?
8. What factors led to the growth of direct-entry midwifery?
9. How do individuals become curanderos? How does medical dominance affect their work and their lives?
10. How have doctors attempted to control acupuncture? What factors have helped or hindered them in this attempt?

CRITICAL THINKING QUESTIONS

1. The term *doula* refers to a woman (there are no men) who provides physical, emotional, and educational support to pregnant women during labor and delivery. There are no national standards for certification for this new health care occupation. Compare doulas with nurses and explain why you think doulas will or won't find it difficult to achieve professional status.
2. Assume that ten years from now all registered nurses will have Bachelor's degrees. Explain one reason why you think this will change the status of nursing as an occupation and one reason why it *won't* change the occupation's status.
3. How are alternative healers affected by medical dominance?

CHAPTER

13

Issues in Bioethics

David McNew/Getty Images

LEARNING OBJECTIVES

After reading this chapter, students should be able to:

- Describe the key cases in the history of bioethics.
- Evaluate the issues involved in several contemporary bioethics debates.
- Understand the ways in which bioethics has become institutionalized.
- Assess the impact and limitations of bioethics on research, medical education, and clinical practice.

In January 1996 my brother-in-law, Brian, was injured in a catastrophic industrial accident that left him with second- and third-degree burns over 95 percent of his body, as well as strong indications that his throat and lungs had been seriously burned.

Brian's accident occurred literally in sight of a major hospital with a regional burn unit, and he was brought to the hospital within minutes. Following the accident, Brian remained in a strange limbo between life and death—unconscious although not comatose, and kept alive by aggressive medical treatment and an ever-increasing assortment of drugs and machines. Burned everywhere except his genitals and the soles of his feet, bandaged from head to toe with only his face showing, and swollen grotesquely, Brian's appearance was literally nightmarish; no one who saw him slept well afterward. Each day brought minor crises, and each week brought a major crisis that made death seem imminent—as indeed it was, for Brian died three and a half weeks after the accident.

The severity of Brian's injuries immediately made me wonder whether it might be best to treat only his pain and let him die a natural death. Brian had never written a living will (a document specifying the circumstances in which he would no longer want medical treatment), but he had told his wife, Lisa, that he would not want to live if his quality of life was ever compromised substantially. Questions about whether treatment made sense became increasingly salient to the family as the days passed; his lungs, stomach, and kidneys failed; and bacterial, viral, and fungal infections assaulted his body.

Because Brian remained unconscious throughout his hospital stay, legally Lisa was authorized to make treatment decisions for him. The doctors acknowledged that the final decisions were up to Lisa and that they could not ethically or legally proceed without her informed consent. In practice, however, they kept decision-making authority to themselves by defining certain decisions as purely technical matters not requiring Lisa's consent, shaping her treatment decisions by providing information selectively, ignoring her decisions when they disagreed with them, cutting off her questions when they found them uncomfortable, and telling her that withholding treatment was unethical and hence out of the question. Although some nurses indicated quietly to Lisa that her concerns were valid, the hospital's pastoral counselors and social workers urged Lisa to trust the doctors' judgment.

In the end, Brian's condition began deteriorating so rapidly and completely that the doctors had no further treatments to try. Around the same time, a new resident who took Lisa's concerns seriously joined the staff. A long conversation with him helped Lisa both to understand the doctors' perspective and to express her own view. When this resident recommended that she consent to withdrawing the drug that kept Brian's heart beating, Lisa agreed. Brian died that night (Weitz, 1999).

Medicine, nursing, and other health care professionals have long recognized that health care should be based on ethical principles. The Hippocratic Oath, for example, written in about 400 BC, instructed doctors to take only actions that would benefit their patients and to forswear euthanasia, seducing patients, and divulging patients' secrets. As Brian and Lisa's story suggests, however, in practice, health care still sometimes falls short of meeting ethical principles. In this chapter, we explore the history of **bioethics**, the study of all ethical issues involved in the biological sciences and health care, and analyze how bioethics has—and has not—affected American health care and medical research.

To some students and faculty, it might seem odd to include a chapter on bioethics in a sociology textbook. Yet the issues raised by bioethics are sociological ones, for many revolve around the impact of power differences among social groups (most importantly, between physicians and patients). Even when exploring the same issues, however, bioethicists and sociologists use different lenses. Robert Zussman, a sociologist who has studied bioethics extensively, succinctly summarizes the difference:

> Medical ethics may be thought of as the normative study of high principles for the purpose of guiding clinical decisions. In contrast, the sociology of medical ethics may be thought of as the empirical study of clinical decisions for the purpose of understanding the social structure of medicine. Clearly then, medical ethicists and sociologists of medical ethics travel much of the same terrain, but they do so traveling in different directions (1997:174).

HISTORY OF BIOETHICS

Since its beginning in 1848, the **American Medical Association (AMA)** has required its members to subscribe to its code of ethics. The code, however, speaks more of medical etiquette—proper relations among doctors—than of medical ethics or, more broadly, bioethics. Indeed, throughout the nineteenth century and well into the twentieth century, doctors' ideas regarding bioethics remained vague, and their commitment to bioethics remained weak. Although doctors undoubtedly would have identified relieving human suffering as their primary goal, both in their research and in clinical practice, they sometimes behaved in ways that would horrify modern doctors and bioethicists. For example, Dr. J. Marion Sims, considered the father of modern obstetrics, achieved fame during the 1840s for developing a surgical procedure to correct vesicovaginal fistulae, tears in the wall between a woman's vagina and bladder usually caused by overaggressive medical intervention during childbirth (Barker-Benfield, 1976). Women who suffered these fistulae could not control the leakage of urine and often had to withdraw from social life altogether because of odor and the resulting social shame. To develop a surgical cure, Sims bought black women slaves who had fistulae and then operated on them as many as 30 times each, in an era before antibiotics and antisepsis and with only addictive drugs for anesthetics. When Sims announced his new surgical technique, the medical world and the public greeted him with acclaim. No one questioned his ethics.

A century later, Nazi doctors working in German concentration camps used prisoners whom they considered less than human for equally barbaric—and even less justifiable—experiments. The world's response to these experiments would mark the beginnings of modern bioethics.

The Nazi Doctors and the Nuremberg Code

In 1933, the German people voted the Nazi party and Adolf Hitler into power. At that time, Germany's medical schools and researchers were respected worldwide, and its health care system was considered one of the world's best (Redlich, 1978).

Shortly after coming to power, the Nazi government passed the Law for the Prevention of Congenitally Ill Progeny (Lifton, 1986). This law required the sterilization of anyone considered likely to give birth to children with diseases that doctors considered genetic, including mental retardation, schizophrenia, manic depression, epilepsy, blindness, deafness, or alcoholism. Under this law, government-employed doctors sterilized between 200,000 and 300,000 persons. Two years later, in 1935, the government passed the Law to Protect Genetic Health, prohibiting the marriage of persons with certain diseases.

Both these laws reflected a belief in **eugenics**, the theory that the population should be "improved" through selective breeding and birth control. The eugenics movement has had many followers throughout the Western world. By 1920, 25 US states had passed laws allowing the sterilization of those believed (usually incorrectly) to carry genes for mental retardation or criminality. Several states also passed laws forbidding the marriage of persons with illnesses considered genetic (Lifton, 1986).

As the power of the Nazis grew in Germany and as public response to their actions both within and outside Germany proved mild, the Nazis adopted ever bolder eugenic actions (Lifton, 1986; Redlich, 1978). Beginning in 1939, the

Nazis began systematically killing patients in state mental hospitals. Doctors played a central role in this program, selecting patients for death and supervising their poisoning with lethal drugs or carbon monoxide gas. Doctors and nurses also watched silently while many more patients starved to death. In total, between 80,000 and 100,000 adults and 5,000 children died (Lifton, 1986). Doctors played similar roles in Nazi concentration camps where millions of Jews, Gypsies, and others died (Lifton, 1986; Redlich, 1978). In addition, doctors working in these concentration camps (including university professors and highly respected senior medical researchers) performed hundreds of unethical experiments on prisoners—such as studying how quickly individuals would die when exposed to freezing cold and whether injecting dye into prisoners' eyes would change their eye color. Doctors also used prisoners to gain surgical experience by, for example, removing healthy ovaries or kidneys or creating wounds on which to practice surgical treatments.

Box 13.1 Principles of the Nuremberg Code

1. The voluntary consent of the human subject is absolutely essential....
2. The experiment should be such as to yield fruitful results for the good of society, unprocurable by other methods or means of study, and not random and unnecessary in nature.
3. The experiment should be so designed and based on the results of animal experimentation and a knowledge of the natural history of the disease or other problem under study that the anticipated results will justify the performance of the experiment.
4. The experiment should be so conducted as to avoid all unnecessary physical and mental suffering and injury.
5. No experiment should be conducted where there is an *a priori* reason to believe that death or disabling injury will occur....
6. The degree of risk to be taken should never exceed that determined by the humanitarian importance of the problem to be solved by the experiment.
7. Proper preparations should be made and adequate facilities provided to protect the experimental subject against even remote possibilities of injury, disability, or death.
8. The experiment should be conducted only by scientifically qualified persons. The highest degree of skill and care should be required through all stages of the experiment of those who conduct or engage in the experiment.
9. During the course of the experiment, the human subject should be at liberty to bring the experiment to an end....
10. The scientist in charge must be prepared to terminate the experiment at any stage if he has probable cause to believe.... that a continuation of the experiment is likely to result in injury, disability, or death to the experimental subject.

SOURCE: *Trials of War Criminals before the Nuremberg Military Tribunals under Control Council Law No. 10* (1949).

After the Nazi defeat, the Allied victors prosecuted 23 of these doctors for committing "medical crimes against humanity," eventually sentencing 7 to death and 9 to prison (Lifton, 1986). These decisions constituted the basis for what is now known as the **Nuremberg Code**, a set of internationally recognized principles regarding the ethics of human experimentation (see Box 13.1). The code requires researchers to have a medically justifiable purpose; do all they can to protect their subjects from harm; and ensure that their subjects give **informed consent**, that is, voluntarily agree to participate in the research with a full understanding of the potential risks and benefits.

The Rise of Bioethics

Because the trials received relatively little publicity in the United States and because Americans typically viewed Nazi doctors as *Nazis* rather than as doctors, few drew connections between Nazi practices and American medical practices (Rothman, 1991). As a result, discussion of bioethics remained largely dormant in the years following the Nuremberg Trials. During the 1960s and early 1970s, however, ethical questions regarding medical care and research in the United States became topics of popular discussion.

Bioethics and the New Technologies One reason for rising concern was the rise of new technologies. Of particular importance was the development of kidney dialysis, a technology that could keep alive persons whose kidneys had failed (Fox and Swazey, 1974). Demand for dialysis far outstripped supply, forcing selection committees made up of doctors and in some cases lay people to decide who would receive this life-saving treatment and who would die. Forced to choose from among the many who, on medical grounds, were equally likely to benefit from the treatment, these committees frequently based their choices on social criteria such as gender, age, apparent emotional stability, social class, and marital status. When news of these committees' work reached the public, the resulting outcry led to new federal regulations designed to allocate kidney dialysis more fairly.

Whereas the dialysis debate focused on the right to gain access to life-saving technologies, the case of Karen Quinlan focused attention on the **right to die**, which essentially translates to the right to refuse medical technologies, from feeding tubes to heart/lung machines. At the age of 21, after ingesting a combination of drugs at a party, Quinlan fell into a coma. Initially, her parents encouraged her doctors to do everything possible to keep her alive and restore her health. Once her parents learned that she had suffered extensive brain damage and would never regain any mental or physical functioning, however, they asked that she be removed from life support and allowed to die. When the doctors refused, the parents took their fight to the courts. After almost a year of legal battles, Quinlan's parents won the right to remove her from the mechanical respirator that had kept her alive.

The Quinlan case gained enormous public attention and sympathy for the right to die and highlighted the problems involved in having too much, rather than too little, access to medical care and technology. In addition, the Quinlan case signaled both the entry of lawyers and the legal system into health care decision making and the problems with using the courts to decide such intensely personal issues (Bosk, 2010).

Bioethics and Medical Research During the same years, concern grew not only about the ethics of medical *practice*, but also about the ethics of medical *research*. These concerns were initially brought to public attention in 1966, when Henry Beecher (1966) published an article in the prestigious *New England Journal of Medicine* describing 22 recent research studies, published in top medical journals, that had relied on ethically questionable methods. In one study, for example, soldiers sick with streptococcal infections received experimental treatments instead of penicillin, causing 25 soldiers to develop rheumatic fever. In another, doctors inserted catheters into the bladders of healthy newborns and X-rayed them, without parental consent, to study how bladders worked.

To determine the frequency of such studies, Beecher looked at 100 consecutive research studies published in a prestigious medical journal. In 12 of the 100 studies, researchers had not told subjects of the risks involved in the experiments or had not even told the subjects that they were in an experiment. Yet no journal reviewer, editor, or reader had questioned the ethics of these studies.

Beecher's article sent ripples of concern through both the medical world and the general public as news of the article spread through the mass media. This public concern translated into pressure on Congress and on the US Public Health Service (PHS), at the time the nation's major funder of medical research. To demonstrate to Congress that it could handle the problem itself and to keep public concern from turning into budget cuts, the PHS in 1966 published guidelines for protecting human subjects in medical research (Rothman, 1991).

The Willowbrook Hepatitis Study Concern about medical research was further heightened when the Willowbrook hepatitis story and the Tuskegee syphilis study burst into the news in the early 1970s. Willowbrook State School, run by the state of New York, was an institution for mentally retarded children. Conditions in Willowbrook were horrendous, with children routinely left naked, hungry, and lying in urine and excrement. As a result, hepatitis, a highly contagious, debilitating, and sometimes deadly disease, ran rampant among the children and to a lesser extent the hospital staff.

In 1956, to document the natural history of hepatitis and to test vaccinations and treatments, two pediatrics professors from New York University School of Medicine began purposely infecting children with the disease. In addition, to test the effectiveness of different dosages of gamma globulin, which the researchers knew offered some protection against hepatitis, they injected some children with gamma globulin but left others unvaccinated for comparison. The children's parents had consented to this research but had received only vague descriptions of its nature and potential risks.

The researchers offered several justifications for their work. First, they argued, the benefits of the research outweighed any potential risks. Second, they had infected the children only with a relatively mild strain of the virus and therefore had decreased the odds that the children would become infected with the far less common but considerably more dangerous strain that also existed in the school. Third, the children who participated in the experiments lived in better conditions than did the others in the institution and therefore were protected against the many other infections common there. Fourth, the researchers argued that the children would probably become infected with hepatitis anyway, given the abysmal conditions in the institution. Finally, the researchers believed they should not be held accountable because the parents had given permission. Using these arguments, the researchers had obtained approval for their experiments from the state of New York, the Willowbrook State School, and New York University. Over a 15-year period, they published a series of articles based on their research, without any reviewers, editors, or readers raising ethical objections.

In 1970, however, the ethical flaws of these experiments were exposed in the popular media and in medical journals. These experiments, many argued, violated the basic principle of informed consent. The parents had not given truly *voluntary* consent because they could get their children admitted to Willowbrook only by allowing them to participate in the hepatitis experiments. In addition, parents had not given truly *informed* consent because researchers had not told them that gamma globulin could provide long-term immunity to hepatitis. Opponents of the study also questioned why the researchers experimented on children, who could not give informed consent, rather than on hospital staff. Finally, opponents questioned why the researchers—who, after all, were pediatricians—had chosen to take advantage of this "opportunity" to study hepatitis rather than trying to wipe out the epidemic. This debate exploded in the New York media and, in the ensuing public outcry, the research ground to a halt.

The Tuskegee Syphilis Study Similar questions arose in 1972, when the Tuskegee Syphilis Study made headlines (Jones, 1993). Begun by the federal PHS in 1932, the study, which was still underway, was intended to document the natural progression of untreated syphilis in African American men. At the time the study began, medical scientists understood the devastating consequences of untreated syphilis in whites (including, in its later stages, neurological damage and heart disease). But reflecting the racist logic of the times, the scientists suspected its progression took a different and milder form in African Americans.

For this study, researchers identified 399 desperately poor and mostly illiterate African American men, all with untreated late-stage syphilis, who lived in the Tuskegee, Alabama, area. The men were neither told they had syphilis nor offered treatment. Instead, researchers informed them that they had "bad blood" (a local term for a wide variety of ailments) and offered them free health care, transportation to medical clinics, free meals on examination days, and payment of burial expenses—enormous inducements given the men's extreme poverty—if the men would participate in the study.

At the time the study began, treating syphilis was difficult, lengthy, and costly. The development of penicillin in the early 1940s, however, gave doctors a simple and effective treatment. Yet throughout the course of the study, researchers not only did not offer penicillin to their subjects but also kept them from receiving it elsewhere. During World War II, researchers worked with local draft boards to prevent their subjects from getting drafted into the military, where the subjects might have received treatment. When federally funded venereal disease treatment clinics opened locally, researchers enlisted the support of clinic doctors to keep research subjects from receiving treatment. Similarly, they enlisted the cooperation of Tuskegee's all-white medical society to ensure that no local doctor gave penicillin to their subjects for any other reason.

The Tuskegee Syphilis Study, which treated African American men as less-than-human guinea pigs, was not the work of a few isolated crackpots. Rather, it was run by a respected federal agency with the cooperation of state and county medical associations and even of doctors and nurses from the local Tuskegee Institute, a world-renowned college for African Americans. Over the years, more than a dozen articles based on the study appeared in top medical journals without anyone ever questioning the study's ethics. Yet the study patently flouted the Nuremberg Code and, after 1966, the PHS's own research ethics guidelines. Nevertheless, the study did not end until a 1972 newspaper exposé caused public outrage. By that time, at least 28 and possibly as many as 100 research subjects had died of syphilis, and an unknown number had succumbed to syphilis-related heart problems (Jones, 1993).

CONTEMPORARY ISSUES IN BIOETHICS

As health care has evolved, so have questions about bioethics. Indeed, there are far too many bioethics issues for one chapter—or even one book—to cover. This section provides a sampling of currently simmering debates in bioethics.

Reproductive Technology

One area that has sparked considerable debate since the late 1970s is **reproductive technology**, or medical developments that allow doctors to control the process of human conception and fetal development. Reproductive technology first came to the public's attention in 1978 with the birth of Louise Brown, the world's first "test-tube baby." Louise's mother was unable to conceive a baby because her fallopian tubes, through which eggs must descend to reach sperm and be fertilized, were blocked. Using a technique known as *in vitro fertilization*, her doctors removed an egg from her body, fertilized it with her husband's sperm in a test tube, and then implanted it in her uterus to develop. Nine months later, Louise Brown was born.

Louise Brown's birth raised questions about how far doctors should go in interfering in the normal human processes of reproduction. Subsequent cases raised even trickier questions. For example, courts have had to decide whether fetuses should be placed for adoption when the biological parents have died and whether custody of fetuses after divorce should go to the parent who wants the fetuses implanted or the one who wants them destroyed. More recently, doctors and others have debated whether couples should be allowed to hire women to carry their fetuses to term for them, whether postmenopausal women should be allowed to have a baby using another woman's egg, and whether doctors should be allowed to combine genes from a man and *two* women into one embryo to avoid transmitting genetic defects carried by one of the women.

More broadly, these cases have raised basic questions regarding the morality of intervening so directly in the process of human reproduction, including whether individuals are harmed or helped by having access to such technologies. Those who favor the new reproductive technologies argue that the technologies give couples greater control over their destinies. Those who oppose the new technologies, on the other hand, argue that these technologies seduce couples into spending enormous amounts of time and money in a usually futile effort to have children who share their genes, rather than finding other ways (such as adoption) to create meaningful lives for themselves. Opponents also question whether these technologies encourage the idea that children are purchasable commodities and the idea that, for the right price, prospective parents can guarantee they will get "perfect" children (Rothman, 1986).

More recently, increasing use of *in vitro* fertilization and related technologies has contributed to a rise in women carrying multiple fetuses and consequently to an increase in premature births. About 12 percent of US babies are now born prematurely (Martin, Osterman, and Sutton, 2010; Saul, 2009).

When a woman learns she is carrying multiple fetuses, she (and her partner, if she has one) must decide either to abort some of the fetuses or to risk having twins, triplets,.... or octuplets. Regardless of their previous feelings on abortion, deciding to abort is very difficult for anyone who has struggled to have a child.

If the woman continues with multiple fetuses, the entire family, at some level, is at risk. The woman may be confined to bed for months to avoid miscarrying, placing her under significant physical, psychological, and financial stress if she can no longer work. Childbirth, too, is especially dangerous for both mothers and babies during multiple births. Moreover, because more than 50 percent of twins and more than 85 percent of triplets are born prematurely, many need ferociously expensive intensive care, die within the first year, or suffer permanent mental or physical defects, often resulting in long-lasting emotional strains, financial strains, or time burdens on the family (Saul, 2009).

Enhancing Human Traits

The past 30 years also have witnessed growing concern about the ethics of medical interventions designed to enhance human traits. No clear definition of such **enhancements** exist, but the term is used to refer to techniques used to improve human traits beyond a level generally considered normal rather than to treat conditions considered deviant or defective. This is a necessarily subjective definition because individual judgments regarding what is normal vary greatly. Nevertheless, we would probably all acknowledge a qualitative difference between providing cosmetic surgery to a person with a severely burned face versus providing it to a professional model who desires more prominent cheekbones.

Similarly, there is a qualitative difference between using psychotropic drugs to avoid schizophrenic hallucinations versus using them to improve one's exam grades—a process psychiatrist Peter Kramer (1993) refers to as "cosmetic psychopharmacology."

Ethical questions regarding enhancements have increased as their use has increased (Whitehouse et al., 1997). Is it ethically justifiable for individuals to improve their offspring through genetic preselection or fetal surgery, and if so, will those who don't use these technologies become a "genetic underclass?" Should health insurance cover drugs such as Viagra, which helps men achieve erections and can improve quality of life perhaps beyond the norm for a given age? Should health insurance cover cosmetic (as opposed to reconstructive) surgery, and should doctors promote surgeries (such as liposuction) whose benefits are purely cosmetic and whose potential risks include death? Should psychotropic drugs be prescribed to individuals who don't have diagnosable mental illnesses but who want to be more sociable, alert, or assertive? And is it ethical to provide potentially harmful medical care to enhance some individuals while others still lack basic medical care? Finally, some have questioned whether enhancements provide unethical advantages. If Olympic athletes are forbidden from taking drugs to improve their performance, why are waitresses allowed to get breast implants to generate more tips and businesspeople allowed to take Ritalin to improve their concentration? Conversely, is it ethical to restrain the options of those who would provide or purchase such services?

Resource Allocation and the Right to Refuse to Treat

For many years, policy analysts, researchers, and ethicists have raised questions about whether the health care system distributes resources such as drugs, medical care, and surgical care in a just and ethical way. This debate takes two forms: deciding which *individuals* should get care and deciding which *procedures* should be funded regardless of individual patient.

In the United States, decisions about *who* should get care mostly occur through implicit rationing: Anyone who can afford health care gets it, anyone who can't afford it doesn't (Callahan, 1998). In contrast, and as we saw in Chapter 9, in other developed nations access to health care is considered a right. The Affordable Care Act (ACA) is an important step in that direction.

But even when individuals can afford health care, decisions still sometimes need to be made about whether they *should* have access to that care (Wicclair and White, 2014). This situation occurs whenever doctors or others consider a treatment to be futile. For example, imagine an 80-year-old woman with a life-threatening infection and heart condition that have left her too ill to speak for herself. Now imagine that her husband demands heart surgery even though her doctors believe her odds of surviving the surgery and regaining a decent quality of life are very slim. In cases like this one, the individual's right to autonomy (in this case, with the husband legally speaking for his wife) must be balanced with doctors' ethical responsibilities both to this patient (whom they believe will be harmed by continuing her suffering) and to other patients (given that any resources devoted to futile care for one patient will not be available to patients who might benefit more).

As initially proposed, the ACA aimed to reduce these conflicts by reimbursing doctors for time spent discussing end-of-life care with patients and families. This plan was dropped from the ACA after opponents labeled such discussions "death panels." However, in 2014 the AMA proposed adding end-of-life counseling to the list of services for which Medicare would reimburse doctors.

Debate over which *procedures* should be made available first came to the fore in 1989 with passage of legislation establishing the Oregon Health Plan (OHP), which offered free care to all Oregonians who were too poor to purchase insurance, but not poor enough to get **Medicaid** (Saha, Coffman, and Smits, 2010). To do so, OHP would first list all possible health care services in order of priority based on effectiveness and costs as well as public priorities and values. It would then prospectively contract with **managed care organizations (MCOs)** to purchase services for OHP members, beginning at the top of its priority list and working its way down until it reached its budget limit. Thus, based on the budget available in any given year, expensive, lower priority services (such as heart transplants) might be eliminated, but low-cost, highly effective services (such as vaccinations) would be funded, and no individuals would be dropped from the program. (Unfortunately, the OHP did have to drop members after its budget was cut sharply in 2003, but the priority list remains in place.)

The OHP legislation marked the first time that a US governmental body explicitly rationed health care—deciding in advance that some procedures simply cost too much to provide. The explicit use of rationing resulted in an outcry across the country, both from those who considered it discrimination against persons with disabilities and those who believed it was unethical to ration care only for the poor. More recently, Arizona officials were met with national outrage when they announced in 2010 that the state's health plan for the poor would no longer pay for certain organ transplants that typically have little impact on life expectancy. Tellingly, the ACA did not adopt this approach.

Stem Cell Research

The rise of new technologies has also led to considerable concern over the use of stem cells and the associated technique of cloning (Dunn, 2002). Stem cells are naturally occurring human cells that have the ability to grow into numerous types of cells. Although no successful treatments have yet been developed from stem cells, researchers hope someday to use them to replace defective cells in individuals with diseases such as diabetes and Parkinson's disease.

There are two ways to grow stem cells. First, scientists can grow stem cells in the laboratory after harvesting them from adults or from fetal blood left in a woman's blood system after giving birth. No ethical issues have been raised about this use of stem cells, which now accounts for about half of all research in this area (Kolata, 2004). Second, scientists can grow stem cells from embryos. To do so, researchers fertilize human eggs with sperm in a laboratory to turn them into embryos. They then leave the embryos for a week or so until each has grown into a few hundred cells and then extract their stem cells (thus destroying the embryos). Alternatively, researchers can replace the nucleus from an unfertilized human egg with a cell nucleus taken from a donor's skin or muscle, artificially stimulate this egg (instead of fertilizing it) so it develops into an embryo, and then extract its stem cells. This second process is a form of cloning because the embryo will be genetically identical to the donor.

To many opponents of stem cell research, the destruction of human embryos to harvest stem cells is the same as killing humans. Other critics argue that producing human cells to treat other humans is too close to selling human beings and human body parts. This is particularly worrisome because heavy political opposition to stem cell research has shifted much of this research to the for-profit sector, where it escapes most regulation. Others object specifically to the use of cloning to produce stem cells on the grounds that it is only a matter of time before some doctors begin using cloned embryos to create cloned babies. They wonder whether in the future babies will be "farmed" and "harvested" to match parents' images of the perfect baby.

Supporters of human stem cell research argue that its potential benefits outweigh its potential problems (Dunn, 2002). Most of the support for this research has come from persons who hope stem cells will provide a cure to the diseases that affect them or their loved ones. Supporters also argue that destroying an artificially created embryo that has no potential to grow into a human being unless it is somehow implanted in a woman's uterus is not morally equivalent to destroying a human being. Finally, with regard to cloning, supporters argue that many women who want babies already have donor eggs implanted in their uteruses and that few would choose to use cloned eggs because the chances of success are so low. (So far, no researcher has been able to keep a cloned egg alive for more than a few days, much less for a nine-month pregnancy.) For all of these reasons, supporters of stem cell research argue that instead of trying to eliminate this research, we should adopt regulations to ensure that it is conducted ethically.

Athletes and Concussions

In recent years, numerous studies have demonstrated the potentially devastating consequences of concussions for athletes, in sports ranging from soccer to cheerleading (Kirschen et al., 2014). Concussions can cause severe headaches, memory loss, dementia, and depression, and can be a major factor in suicides. The dangers multiply when players are very young (so their brains are still developing), when they are hit with great force (as in football or bull-riding), when they suffer multiple concussions over time, and when they are quickly returned to play after a concussion or hard hit to the head. For these reasons, the American Academy of Neurology has officially declared that doctors who care for athletes are ethically required "to safeguard the current and future physical and mental health of [those] patients" (Kirschen et al., 2014:352). This means that doctors must put the interests of players above those of teams, schools, and any other organization or individual for whom they work.

One can easily argue that those same ethical obligations extend to coaches, trainers, nurses, parents and anyone else who participates in deciding whether a player should return to the field—although of course the ethical obligation falls heaviest on those with the most medical knowledge. In addition, some argue that schools and athletic organizations have an ethical obligation to provide appropriate testing and treatment to injured athletes and to ensure that players don't lose their scholarships if they must stop playing to protect their health. Finally, some argue that sports fans who understand the risks have an ethical obligation to make their views known—whether through boos, letters, or decisions against buying tickets—when players are returned to the field too soon (Kaminer, 2012).

INSTITUTIONALIZING BIOETHICS

Concern about bioethics has led to the development of formal mechanisms to ensure that health care and health research will be conducted ethically. In this section, we look at four of those mechanisms: hospital ethics committees, institutional review boards, professional ethics committees, and community advisory boards.

Hospital Ethics Committees

The origins of hospital ethics committees can be traced to the 1950s. As noted earlier, many hospitals used committees to select patients for kidney dialysis. Similarly, hospitals routinely used committees to decide which women

Luc Novovitch/Alamy

Schools, leagues, and coaches as well as doctors may have an ethical obligation to protect athletes from brain damage stemming from sports-related concussions.

merited abortions on medical grounds. At the time, the legal status of abortion was unclear, and the moral status of abortion was just starting to become a public issue (Luker, 1984).

Other hospital ethics committees arose in the aftermath of the 1982 "Baby Doe" case, in which parents of a newborn who was mentally retarded and had a defective digestive system decided that they did not want the defect corrected by surgery. The doctors complied with their decision, and the baby died six days later. When news of the case broke, a public furor arose. These days, most large hospitals have ethics committees or consultants available to review any cases considered ethically problematic.

Institutional Review Boards

Although universities and hospitals began establishing committees to review research ethics in the 1960s, such committees did not become common until the 1970s. In the aftermath of the Tuskegee scandal, Congress in 1974 created the National Commission for the Protection of Human Subjects of Biomedical and Behavioral Research. The Commission's reports laid the groundwork for current guidelines regarding research ethics. That same year, the National Research Act mandated the development of **institutional review boards (IRBs)** charged with reviewing all federally funded research projects involving human subjects. Such boards now exist at all universities and other research institutions, and monitor social science as well as medical research.

In recent years, though, pharmaceutical research has increasingly shifted from federally funded projects in hospitals and universities to for-profit projects funded by pharmaceutical companies and conducted by research organizations, individual doctors, or the pharmaceutical companies themselves. To oversee this research, for-profit, *commercial* IRBs have emerged, either run by or under contract with pharmaceutical companies or other research organizations (Lemmens and Freedman, 2000).

The conflict of interest involved in such IRBs is obvious. When employees of a pharmaceutical company review their company's research, they know that their company's success depends on getting that research approved. Similarly, those who work for commercial IRBs know that they are unlikely to get future contracts from pharmaceutical companies unless they approve those companies' research proposals.

Professional Ethics Committees

Many professional organizations now also have ethics committees that establish guidelines for professional practice. The American Fertility Society, for example, has published a statement of principles regarding the moral status of human embryos created in laboratories, and the ethics committee of the American College of Obstetrics and Gynecology has published guidelines regarding the ethics of selectively aborting fetuses when a woman who has used fertility drugs becomes pregnant with multiple fetuses.

Community Advisory Boards

The most recent development in this area is the emergence of community advisory boards (CABs). The purpose of CABs is to bring individuals from the community together with health care providers to make difficult bioethical decisions regarding both research and treatment (Quinn, 2004). For example, CABs may be asked to represent patients in treatment decisions when the patients are unconscious or incompetent and family members are unavailable.

The use of CABs to evaluate research designs is linked to the rise of genetic research. Typically, we think of genetic testing as an individual decision: Should someone whose mother died of breast cancer or whose sister has Down syndrome get a genetic test to ascertain her own risk of having or passing on these diseases? But genetic testing also has implications for communities. Genetic tests can stigmatize an entire community (when, for example, African Americans were first identified as having higher risks of sickle cell anemia), can challenge ideas about who *belongs* to a community (when genetic differences are found within a community), and can challenge a community's ideas about its origins (when, for example, Native American legends locating tribal origins in the Americas clash with genetic findings suggesting Asian origins). For these reasons, researchers have begun involving communities in discussions of research priorities, research design, and the dissemination of research findings. This leaves open, however, the very large question of who constitutes a community and who should decide for it.

THE IMPACT OF BIOETHICS

The growth of the bioethics movement and the institutionalizing of bioethics in US hospitals and universities have made ethical issues more visible than ever before. Articles on bioethics, virtually nonexistent before the 1960s, now appear routinely in medical journals while in both the clinical and research worlds, ethics committees have proliferated.

These developments have led some observers to conclude that the bioethics movement has fundamentally altered the nature of medical work. According to historian David Rothman:

> By the mid-1970s, both the style and the substance of medical decision-making had changed. The authority that an individual physician had once exercised covertly was now subject to debate and review by colleagues and laypeople. Let the physician design a research protocol to deliver an experimental treatment, and in the room, by federal mandate, was an institutional review board composed of other physicians, lawyers, and community representatives to make certain that the potential benefits to the subject patient outweighed the risks. Let the physician attempt to allocate a scarce resource, like a donor heart, and in the room were federal and state legislators and administrators to help set standards

> of equity and justice. Let the physician decide to withdraw or terminate life sustaining treatment from an incompetent patient, and in the room were state judges to rule, in advance, on the legality of these actions. (1991:2)

Other observers, however, contend that the impact of the bioethics movement has been more muted (Annas, 1991). These critics argue that hospital, research, community, and professional ethics committees, like the earlier hospital abortion committees, exist primarily to offer legal protection and social support to researchers and clinicians, not to protect patients or research subjects. Furthermore, they argue, although clinicians have become more concerned with *documenting* their allegiance to ethics guidelines, they have not become any more concerned with *following* those guidelines. The following sections evaluate the impact of bioethics on health care research, medical education, and clinical practice.

The Impact on Research

According to ethicist George Annas, the bioethics movement, as institutionalized in research ethics boards and committees, has affected medical research only slightly. In his words, the

> primary mission [of research ethics committees] is to protect the institution by providing an alternative forum to litigation or unwanted publicity.... [For this reason] its membership is almost exclusively made up of researchers (not potential subjects) from the particular institution. These committees have changed the face of research in the U.S. by requiring investigators to justify their research on humans to a peer review group prior to recruiting subjects. But this does not mean that they have made research universally more "ethical." In at least a few spectacular instances, these committees have provided ethical and legal cover that enabled experiments to be performed that otherwise would not have been because of their potentially devastating impact on human subjects. (1991:19)

As an example, Annas cites the case of "Baby Fae" (not her real name), who died in 1984 soon after doctors replaced her defective heart with a baboon's heart. Although all available evidence indicated that cross-species transplants could not succeed, the doctors who performed the surgery had received approval from their hospital's IRB. A subsequent review found that Baby Fae's parents had not given truly informed consent because the doctors had not suggested seeking a human transplant, had disparaged available surgical treatments, and had unreasonably encouraged the parents to believe that a baboon transplant could succeed.

Similarly, the requirement that medical research be approved by bioethics committees has in some cases led researchers to confuse ethics with regulation. For example, in one recent study, neuroscientists were asked, "What role do ethical concerns play in how you set up the research?" Nearly half gave answers like the one given by the following scientist:

> I don't think it really affected anything I do. It made me think about things that I wouldn't have otherwise thought about, like insurance and what happens if things go wrong.... But in terms of the actual study, I think I just set it up as I wanted to do it, and then obviously sent it off to Ethics and it was all fine. So there was nothing that Ethics and the ethical procedure stopped me from doing that I wanted to do (Brosnan, et al., 2013:1138).

For scientists such as this one, meeting regulatory requirements can become a substitute for thinking about difficult ethical issues.

Lack of resources and conflicts of interest also limit the effectiveness of IRBs. IRB members are unpaid volunteers who typically must review between 300 and 2,000 proposed experiments yearly and who, in many cases, have vested interests in approving research proposals so their institutions can obtain research funding (Hilts, 1999). Meanwhile, final responsibility for overseeing IRBs falls to the federal Office of Protection from Research Risks, which is far too understaffed to thoroughly review human subjects research.

Nevertheless, and despite the limitations of IRBs and research ethics committees, the rise of bioethics has curbed the most egregious abuses of human subjects. According to David Rothman:

> The experiments that Henry Beecher described could not now occur; even the most ambitious or confident investigator would not today put forward such protocols. Indeed, the transformation in research practices is most dramatic in the area that was once most problematic: research on incompetent and institutionalized subjects. The young, the elderly, the mentally disabled, and the incarcerated are not fair game for the investigator. Researchers no longer get to choose the martyrs for mankind (1991:251).

In fact, the balance has shifted to such an extent that we now sometimes read news stories not of researchers pressuring individuals to become research subjects, but rather of desperately ill individuals pressuring researchers to accept them as research subjects for experimental treatments. At the same time, the shift toward for-profit drug testing (described in *Contemporary Issues*: "*Guinea Pigging*") has raised new concerns not only about the safety of research subjects but also about the credibility of the research enterprise itself.

The Impact on Medical Education

One obvious result of the bioethics movement has been the incorporation of ethics training into medical education, with courses on ethics now common at US medical schools. As critics have noted, however, those courses too often are divorced from real life, aimed at teaching students ethical principles and legal norms through classroom lectures rather than at teaching students how to negotiate the everyday ethical dilemmas they will face. Moreover, such courses often assume that students who are already undergoing socialization to medical culture still can identify ethically problematic aspects of that culture (Hafferty and Franks, 1994). Finally, ethics courses can't compensate for the ways in which ethics are discounted in the "hidden curriculum" of medical practice and culture. For example, a structure that expects students both to provide care for patients *and* to learn techniques on patients without the patients' knowledge inherently teaches students to view patients at least partly as objects rather than as subjects. From this perspective, only through "the integration of ethical principles into the everyday work of both science and medicine" can we expect new doctors to adopt more ethical approaches to care (Hafferty and Franks, 1994:868).

CONTEMPORARY ISSUES

"Guinea Pigging"

The term *guinea pigging* first entered the mainstream American vocabulary in January 2008, when an article by that name was published in *New Yorker Magazine* (Elliott, 2008). *Guinea pigging* refers to healthy individuals (overwhelmingly poor, sometimes students) who participate in clinical drug trials for pay.

In the past, most participants in drug trials were either medical students and personnel who at least intellectually understood the risks they faced or persons struggling with illnesses who might benefit from the drugs they tested. Over the past ten years, however, as drug testing and development have exploded and have largely shifted from nonprofit to for-profit operations, the need to quickly find large numbers of research subjects has led to the widespread use of healthy subjects for pay in early trials of drugs. (If the drugs prove safe with healthy subjects, they are then tested for efficacy on ill subjects.)

These human guinea pigs can earn up to several thousand dollars for participating in research studies that can last weeks or months. The risks they face, though, can be high. The obvious dangers come from the drugs themselves. In March 2006, for example, six volunteers who participated in tests of a potential treatment for immune disorders were almost killed, and apparently all are now permanently disabled (Elliott, 2008). In addition, testing sometimes involves invasive and potentially dangerous procedures such as biopsies or endoscopies. Moreover, most clinical trials don't cover medical costs—let alone compensation for pain or lost wages—when volunteers are injured or become ill as a result of the experiments. Participating in drug trials can also be extremely unpleasant, requiring subjects to wear rectal probes, experience food or sleep deprivation, live for weeks in hospital-like environs, or the like.

In addition to the risks faced by volunteers, the public is also placed at risk when drugs are tested in these circumstances. When subjects participate because they need money, they may feel no qualms about ignoring research protocols, such as sneaking food or alcohol when they are supposed to fast or abstain. Similarly, when researchers are employed by for-profit corporations, they may be inclined to interpret results optimistically or to recruit homeless alcoholics who need money rather than spending the time needed to recruit a more representative sample.

The Impact on Clinical Practice

Relatively few studies have looked at the impact of the bioethics movement on clinical practices. One series of studies looked at the impact of New York's 1987 law establishing formal policies for writing "don't resuscitate" orders (orders forbidding health care workers from intervening if the lungs or heart of a terminally ill patient stops functioning). These studies found that after the law's passage, doctors significantly altered how they *documented* their actions, but not how they *acted* (Zussman, 1992:162). Similarly, studies have found that hospitals sharply limit access of patients, family, and nonmedical staff to ethics consultations. As a result, consultations primarily function to provide additional institutional support to doctors confronted by families or patients they consider disruptive, such as those who challenge doctors' decisions regarding how aggressively to treat a patient (Kelly et al., 1997; Orr and Moon, 1993). These findings have led researchers to conclude that the true purpose of ethics consultations is to reinforce doctors' power.

The most extensive study of the impact of bioethics on clinical practice appears in *Intensive Care: Medical Ethics and the Medical Profession* (1992) by sociologist Robert Zussman. Zussman spent more than two years observing and interviewing in the intensive care units of two hospitals. His research suggests both the impact and the limitations of the bioethics movement.

Although cases such as Karen Quinlan's and Baby Doe's might suggest that doctors often want to use aggressive treatment despite the objections of patients and families, Zussman found that on intensive care wards, the reverse is usually the case. Knowing that most of their patients will die, doctors on these wards often hesitate before beginning aggressive treatment, which might only escalate costs, increase their work as well as their patients' suffering, and prolong the dying process. Patients and their relatives, however, often face a sudden and unexpected medical crisis. Unable to believe the situation hopeless, they demand that health care workers "do everything." In these situations, the doctors Zussman studied expressed allegiance to the principle that families have the right to make decisions regarding treatment. In practice, however, doctors found ways to assert their discretion, if no longer the authority they had in years past.

Doctors asserted their discretion in several ways. First, doctors made decisions without asking the family on the assumption that the family would agree with their decisions. Second, doctors sometimes ignored a family's stated decisions, arguing that it was cruel to force a family to make life-or-death decisions that they might later regret. Third, doctors might respect a family's wishes, but only after first shaping those wishes through providing information selectively. This information included defining the patient as terminally ill or not—a highly significant designation because ethical guidelines permit health care workers to withhold or terminate treatment only for terminally ill patients. Fourth, when doctors failed to shape a family's wishes, the doctors could discount those wishes on the grounds that the family was too emotionally distraught to decide rationally.

Finally, and perhaps most important, doctors continued to assert their discretion by defining the decision to withhold treatment as merely a technical problem and thus defining family members' wishes as irrelevant. For example, doctors might acknowledge families' general wishes regarding how aggressively treatment should proceed but then define each specific intervention as a technical decision best left to doctors.

Summing up his findings, Zussman writes:

> The picture I have drawn corresponds neither to an image of unbridled professional discretion nor to one of patients' rights triumphant. As many observers of contemporary medicine have argued, the discretion of physicians in clinical decisions (like the discretion of professionals in other fields) depends on their ability to make successful claims to the exclusive command of technical knowledge. Yet, while... physicians.... make such claims, they don't always succeed either in convincing themselves that they are legitimate or in converting them to influence over patients and their families, for the claims of physicians are met by the counterclaims of patients and, more important, families.... The institutionalization of patients' rights, in law and in hospital policy.... empower[s] families when they do insist on doing everything. In such a situation, physicians may continue to exercise considerable influence and enjoy considerable discretion. By no means have they been reduced to the role of technicians and nothing more. But at the same time, they must, at the very least, take the wishes of patients and families into account (1992:159–160).

IMPLICATIONS

As we have seen, bioethics and sociology have much in common. At a very basic, if typically unacknowledged, level, bioethics, like sociology, is about power. The abuses of the Nazi doctors, for example, not only illuminate the horrors possible when ethical principles are ignored but also illustrate both how social groups can obtain power over others and how individuals can be harmed or even killed when this happens. Conversely, sociology, in similarly unacknowledged ways, is at a basic level an ethical enterprise. Hidden assumptions about what society *should* be like and how society *should* be changed often underlie abstract, technical sociological discussions. Such assumptions often draw on philosophies regarding justice, autonomy, human worth, and other basic ethical issues. Yet in the same way that bioethicists often ignore the sociological implications of their work, sociologists often ignore the ethical implications of the questions they ask, the research they conduct, and the findings their research generates.

It seems, then, that bioethicists and sociologists can provide each other with broader perspectives that can only enrich our understanding of both fields—encouraging bioethicists to see not only individual cases but also broader social and political issues and encouraging sociologists to see the world and their work in it as an ethical as well as a political and intellectual enterprise. These are issues that all of us should keep in mind as we seek our place in the world.

SUMMARY

1. Bioethics is the study of ethical issues in biological sciences and health care. Many of the issues bioethicists ponder revolve, whether explicitly or not, around the use and impact of power—a central concern of sociologists.
2. Although US doctors have been required to subscribe to a Code of Ethics since 1848, in the past many doctors conducted research and practiced medicine in ways that would horrify modern bioethicists. One example was Dr. J. Marion Sims's use of African American slaves as surgical guinea pigs.
3. After the Nazi defeat, Allied countries prosecuted 23 doctors as part of the broader war crimes trials popularly known as the Nuremberg Trials. These trials resulted in the development of the Nuremberg Code, a set of internationally recognized principles regarding the ethics of human experimentation.
4. Interest in bioethics in the United States grew substantially during the 1960s and 1970s, sparked by popular and medical articles on new technologies (such as kidney dialysis and machines that could keep people "alive" even if brain dead) and on unethical medical research practices (such as the Willowbrook hepatitis experiments and the Tuskegee syphilis research).
5. Contemporary bioethical issues include reproductive technology, the enhancement of human traits (through, for example, cosmetic surgery or memory-enhancing drugs), stem cell research, resource allocation and the right to refuse to treat, and the obligation to protect athletes from the long-term consequences of concussions.
6. In recent years, various institutional mechanisms have developed to ensure that bioethical principles will be followed in health care and health research, including hospital, research, community, and professional ethics committees. An important new development is the use of for-profit research ethics committees.
7. The growth of the bioethics movement and the institutionalizing of bioethics in US hospitals and universities have made ethical issues more visible than ever before. These changes have increased concern among doctors about bioethics, but it is not clear whether they have truly increased doctors' commitment to acting ethically.

REVIEW QUESTIONS

1. What is the Nuremberg Code, and how and why did it come into existence?
2. What factors led to the emergence of the bioethics movement in the late 1960s?

3. Why do researchers now consider the Tuskegee Syphilis Study and the Willowbrook hepatitis experiments to have been unethical?
4. What are the ethical problems involved in the new reproductive technology? In enhancements? In stem cell research? In resource allocation? In protecting athletes from physical harm?
5. What impact has bioethics had on health care and on health research?

CRITICAL THINKING QUESTIONS

1. How can US health care be made more ethical?
2. What role does doctors' professional dominance play in creating ethical problems in medical care? What role does bioethics, as institutionalized in the American health care system, play in limiting doctors' professional dominance?
3. What notable similarities and differences do you see between the behavior of the doctors in the Tuskegee syphilis experiments and in the Nazi genocide?
4. You now hold a position of power in our health care system. (You choose which position.) What three changes could you realistically attempt to make so that health care would be provided more ethically? Justify your decisions.

Glossary

accommodation A technique individuals use to smooth interactions between themselves and those they consider potential sources of trouble, as well as one used between the latter group and other people.

acquired immunodeficiency syndrome (AIDS) A latter phase of HIV infection that is marked by major health problems. See *HIV/AIDS.*

active voluntary euthanasia Ending the life of individuals who, because of illness or disability, have requested that they be killed.

actuarial risk rating A system in which insurers try to maximize their financial gain by identifying and insuring only populations that have low health risks.

acute Anything that has a sudden and recent onset, such as acute illness or acute pain.

acute disease Any disease that strikes suddenly and disappears rapidly (within a month or so). Examples include chicken pox, colds, and influenza.

ADA See *Americans with Disabilities Act.*

advanced practice nurses Individuals who, after becoming registered nurses, additionally receive specialized postgraduate training. Includes nurse-midwives and nurse practitioners. See *registered nurses.*

Affordable Care Act (ACA) Passed by Congress in 2010 and designed to reduce the number of uninsured Americans, while working within the existing health care system.

age-adjusted rates Epidemiological data that have been manipulated using standard statistical techniques to eliminate any effects that arise because some populations include more older or younger persons than do others. Age adjustment allows us to compare populations with different age distributions.

agency The ability of individuals to make their own choices, free of any limitations placed on them by other people, culture, or social forces. Similar to the concept of free will.

AIDS Acquired immunodeficiency syndrome. See *HIV/AIDS.*

allopathic doctors Nineteenth-century forerunners of contemporary medical doctors. Also known as "regular" doctors.

almshouse An institution, also known as a poorhouse, in which all public wards, including orphans, criminals, the disabled, and the insane, received custodial care.

alternative therapies Treatments rarely taught in medical schools and rarely used in hospitals.

AMA See *American Medical Association.*

American Medical Association (AMA) The main professional association for medical doctors.

Americans with Disabilities Act (ADA) Federal law, passed in 1990, that outlaws discrimination against individuals with disabilities in employment, public services (including transit), and public accommodations (such as restaurants, hotels, and stores). It requires that existing public transit systems and public accommodations be made accessible, along with all new public buildings and major renovations of existing buildings.

assistant doctors Chinese health care workers who receive three years of postsecondary training, similar to that of doctors, in both Western and traditional Chinese medicine.

balance bill To bill patients for the difference between the amount their insurance will pay for a given procedure and the amount the doctor would normally charge for that procedure.

big data Huge studies that pull together multiple sets of data from entire populations.

bioethics The study of all ethical issues involved in the biological sciences and health care.

biomedicine The social institution that combines medicine, science, and technology.

blaming the victim Process through which individuals are blamed for causing their problems.

Blue Cross A group of private companies offering insurance that reimburses individuals primarily for the costs of hospital care, not including doctors' bills. Blue Cross insurance is often offered and bought in conjunction with Blue Shield insurance. See *Blue Shield.*

Blue Shield A group of private companies offering insurance that reimburses individuals primarily for the costs of receiving care from doctors, especially care received in hospitals. Blue Shield insurance is often offered and bought in conjunction with Blue Cross insurance. See *Blue Cross.*

body project The intense focus on shaping one's bodies to meet cultural norms, to such an extent that doing so becomes central to one's identity and life.

capitation A system in which doctors are paid a set annual fee for each patient in their practice, regardless of how many times they see their patients or what services the doctors provide for their patients.

chiropractors Health care practitioners who specialize in spinal manipulation, who trace illness and disability to misalignments of the spine, and who believe spinal manipulation can cure a wide range of acute and chronic health problems.

chronic Anything that continues over a long period, such as chronic illness or chronic pain.

chronic disease Disease that develops in an individual gradually or is present from birth and that will probably continue at least for several months and possibly until the person dies. Examples include muscular dystrophy, asthma, and diabetes.

cognitive norms Socially accepted rules regarding proper ways of thinking. For example, someone should not think that he is Napoleon or that his radio is sending him secret messages from outer space.

commercial insurance Insurance offered by companies that function on a for-profit basis.

commodification Process of turning people into products that can be bought or sold.

community rating A system for calculating insurance premiums in which each individual pays a premium based on the average health risk of his or her community as a whole.

complementary therapies Treatments rarely taught in medical schools and rarely used in hospitals.

compliance Individuals' willingness to follow the advice of health care workers.

concurrent sexual partners Two or more sexual partners held by an individual during a given, overlapping time period.

conflict perspective View that society is held together by power and coercion, with dominant groups imposing their will on subordinate groups.

contested illness Any collection of distressing, painful symptoms that occur together and that lay people assume constitute an illness even though many doctors disagree.

control A process through which researchers statistically eliminate the potential influence of extraneous factors. For example, because social class and race often go together, researchers who want to investigate the impact of social class have to be sure that they are not really seeing the impact of race. To study the impact of social class on mental illness, therefore, researchers would have to look separately at the relationship between social class and mental illness among whites and then at the relationship among blacks to control for any effect of race.

copayment Under some forms of health insurance, a fee that individuals must pay each time they see a health care provider. Fees can range from nominal sums to 20 percent of all costs.

corporatization The growing role of investor-owned corporations in the health care field.

cost shifting Raising prices charged some individuals for services received in order to make up for losses incurred when services are provided to other individuals who cannot or will not pay for services.

countervailing powers The various powerful groups and institutions fighting for control over a given arena such as health care.

cultural competence The ability of health care providers to understand at least basic elements of others' culture and thus to provide medical care in ways that better meet clients' emotional as well as physical needs.

cultural health capital Cultural resources that can facilitate better health care by facilitating better relationships between patients and providers, including knowledge of basic medical terms, acceptance of medical concepts such as the germ theory of illness, belief in cultural values held by most doctors, such as the benefits of efficiency, and the ability to speak the same language as one's doctors.

cumulative inequality (CI) theory This theory argues that inequality 1) primarily results from social systems (rather than individual choices) and 2) causes health problems that accumulate over the lifetime.

cumulative stress burden The sum of an individual's acute and chronic stresses over the life span.

curanderos Folk healers who function within Mexican and Mexican American communities.

death brokering The process through which medical authorities make deaths explainable, culturally acceptable, and individually meaningful.

deductible Under some forms of health insurance, the minimum dollar amount of health care expenses that individuals must pay annually out of pocket before the insurance plan will begin covering any of their expenses.

defensive medicine Tests and procedures that doctors perform primarily to protect themselves against lawsuits rather than to protect their patients' health.

deinstitutionalize To remove individuals (such as mentally retarded and mentally ill persons) from large institutions and return them to the community.

demedicalization The process through which a condition or behavior becomes defined as a natural condition or process rather than an illness.

depersonalization The process through which an individual comes to feel less than fully human or comes to be viewed by others as less than fully human.

depoliticize To define a situation in a way that hides or minimizes the political nature of that situation.

deviance Behavior that violates a particular culture's norms or expectations for proper behavior and therefore results in negative social sanctions. See *negative social sanctions*.

diagnosis-related groups (DRGs) System established by the federal government that sets, for all Medicaid and Medicare patients and for each possible diagnosis, an average length of hospital stay and cost of inpatient treatment. Under the DRG system, hospitals are paid the established cost for each patient with a given diagnosis, regardless of the actual cost of treatment.

Diagnostic and Statistical Manual of Mental Disorders (DSM) Manual published by the American Psychiatric Association and used by mental health workers to assign diagnoses to clients. Generally, this manual must be used if mental health workers want to obtain reimbursement for their services from insurance providers.

differential An adjective referring, in sociological writing, to a situation in which one group has more or less of something than another. For example, different social classes in the United States have differential access to health care.

disability Restrictions or lack of ability to perform activities resulting from physical limitations or from the interplay among those limitations, social responses, and the built or social environment.

discrimination Differential and unequal treatment grounded in prejudice. See *prejudice*.

disease A biological problem within an organism.

doctor–nurse game "Game" in which the nurse is expected to make recommendations for medical treatment in such a way that the recommendations appear to have come from the doctor.

DRG See *diagnosis-related groups*.

DSM See *Diagnostic and Statistical Manual of Mental Disorders*.

dysfunctional Refers to anything that threatens to undermine social stability.

employer mandate A legal requirement that each employer offer health insurance to its employees and pay a specified percentage of the costs.

endemic Referring to diseases that appear at a more or less stable rate over time within a given population.

enhancements Techniques deemed to improve human traits beyond a level generally considered normal rather than to treat conditions considered deviant or defective. This distinction is artificial but occasionally useful.

entrepreneurial system A system based on capitalism and free enterprise.

environmental racism The disproportionate burden of environmental pollution experienced by racial and ethnic minorities.

epidemic Either a sudden increase in the rate of a disease or the first appearance of a new disease.

epidemiological transition The shift from a society burdened by infectious and parasitic diseases and in which life expectancy is low to one characterized by chronic and degenerative diseases and high life expectancy.

epidemiology The study of the distribution of disease within a population.

epigenetic effect The combined effect of genes and environments on a trait or disease.

eugenics The theory that the population should be "improved" through selective breeding and birth control.

evidence-based medicine The use of medical therapies whose efficacy has been confirmed by large, randomized, controlled clinical studies. See *control.*

family leave programs Programs that allow individuals to take time off from work without risking their jobs to care for family members. Some programs offer paid leave; others offer only unpaid leave.

fee-for-service The practice of paying doctors for each health care service they provide, rather than paying them a salary.

fee-for-service insurance Insurance that reimburses patients for all or part of the costs of the health care services they have purchased.

feeling norms Socially defined expectations regarding the range, intensity, and duration of appropriate feelings and regarding how individuals should express those feelings in a given situation.

feeling work Efforts made by individuals to avoid being labeled mentally ill by making their emotions match social expectations. Individuals can (1) change or reinterpret the situation that is causing their unacceptable feelings; (2) change their emotions physiologically through drugs, meditation, biofeedback, or other methods; (3) change their behavior, acting as if they feel more appropriate emotions than is actually the case; or (4) reinterpret their feelings, telling themselves, for example, that they are only tired rather than worried.

feminization of aging The fact that women comprise a larger proportion of the elderly than of younger age groups; the steady rise in the proportion of the population who are female at each successive age.

fetal rights The growing body of legal, medical, and public opinion holding that fetuses have rights separate from and sometimes contrary to those of their mothers.

financially progressive Describes any system in which poorer persons pay a smaller proportion of their income for a given good or service than do wealthier persons.

financially regressive Describes any system in which poorer persons pay a larger proportion of their income for a given good or service than do wealthier persons.

Flexner Report A report on the status of American medical education produced in 1910 by Abraham Flexner for the Carnegie Foundation. This report identified serious deficiencies in medical education and helped to produce substantial improvements in that system.

for-profit, private hospitals Hospitals run with the primary goal of producing a profit each year for shareholders.

formulary Official list of drugs that doctors in a managed care organization can prescribe without special authorization. See *managed care.*

functionalism View of society as a harmonious whole held together by socialization, mutual consent, and mutual interests.

fundamental-cause theory A theoretical perspective that argues that, in each time and place, those with greater access to resources will experience better health because they will be better able to use whatever resources are available to protect their health.

gender The social categories of masculine and feminine and the social expectations of masculinity and femininity.

gender convergence The ways in which men and women's lives along with social expectations for how men and women should behave have become more similar over time.

geneticization The shift toward increasingly defining genes as the cause of human disease, behavior, and differences.

global health The idea that health and illness needs to be understood as a global process rather than something contained within individual nations. This includes recognizing the how international, national, and local organizations can affect health and how health and disease may have similar roots in social forces around the world.

globalization The process through which ideas, resources, and persons increasingly operate within a worldwide rather than local framework. For example, the globalization of tourism means that US tourists now consider Africa a plausible destination.

Great Confinement The shift from the 1830s on, in both Europe and the United States, toward confining mentally ill persons in large public institutions instead of in almshouses, small private "madhouses," or family homes.

habitual dispositions Routine, almost instinctual, attitudes regarding the merit of various behaviors that might harm or preserve health.

health belief model Model predicting that individuals will follow medical advice when they (1) believe they are susceptible to a particular health problem, (2) believe the health problem they risk is a serious one, (3) believe compliance will significantly reduce their risk, and (4) do not perceive any significant barriers to compliance.

health care convergence Ways in which international health care systems become increasingly similar over time because of similar scientific, technological, economic, and epidemiological pressures.

health lifestyle theory A theory that attempts to predict why groups adopt patterns of healthy or unhealthy behavior by showing how demographic circumstances and cultural memberships combine with socialization and experiences to produce both life chances and life choices. These life chances and choices in turn lead to habitual dispositions toward healthy or unhealthy behaviors, which then lead to actual behaviors.

health maintenance organizations (HMOs) Organizations that provide health care based on prepaid group insurance. Patients pay a fixed yearly fee in exchange for a full range of health care services, including hospital care as well as doctor's services.

health project The intense focus on actively protecting one's health to such an extent that doing so becomes central to one's identity and life.

health social movements Informal networks of individuals who band together to collectively challenge health policy, politics, beliefs, or practices.

heroic medicine System of treatment used by allopathic doctors before about 1860 that emphasized curing illnesses by purging the body through bloodletting, causing extreme vomiting, or using repeated laxatives and diuretics. See *allopathic doctors*.

Hispanic paradox The relatively high life expectancy and low infant mortality apparently enjoyed on average by Hispanic Americans despite their overall lower social class status.

HIV (human immunodeficiency virus) The virus that causes AIDS.

HIV/AIDS The term that summarizes all stages of disease in humans caused by HIV infection. The disease harms individuals' health by gradually destroying their body's immune system.

HMO See *health maintenance organization*.

holistic treatment Treatment based on the premise that all aspects of an individual's life and body are interconnected—that, for example, to treat an individual with cancer effectively, health care workers must look at all organs of the body, not only the one that currently has a tumor, as well as at the individual's psychological and social functioning.

home health aides Workers, typically untrained, who provide essentially custodial care within individuals' homes.

homeopathic doctors Popular nineteenth-century health care workers who treated illnesses with extremely diluted solutions of drugs that at full strength would produce symptoms similar to those caused by the illnesses.

hospices Institutions designed to meet the needs of dying people.

human immunodeficiency virus See *HIV*.

illness The social experience of having a disease.

illness behavior The process of responding to symptoms and deciding whether to seek diagnosis and treatment.

illness behavior model A model that predicts the circumstances in which individuals are most likely to seek medical care. According to this model, individuals are most likely to do so if their symptoms are frequent or persistent, visible, and severe enough to interfere with daily activities *and* if they lack alternative explanations for the symptoms.

incidence Number of new cases of an illness or health problem occurring within a given population during a given time period (e.g., the number of children born with Down syndrome in the United States during 2009).

income inequality The *gap* in income between a nation's poorest and wealthiest.

individual mandate A legal requirement that each individual obtain health insurance.

individualism A set of cultural beliefs and practices that encourages the autonomy, equality, and dignity of individuals and that downplays the importance of connections to social groups.

informed consent Voluntary agreement to participate in medical research or to receive a medical procedure or treatment, with a full understanding of the potential risks and benefits.

inpatient Hospital patient who is formally admitted and kept overnight.

institution An enduring social structure that meets basic human needs, such as the family, education, religion, or medicine (taken in its entirety).

institutional review boards (IRBs) Federally mandated committees charged with reviewing the ethics of research projects involving human subjects. No research can be conducted using federal funds unless it first receives IRB approval.

insurance premium The yearly fee individuals pay to purchase insurance.

intersex Adjective used to describe individuals who have both male and female biological characteristics, such as a penis and a uterus or a vagina and an unusually large clitoris.

IRBs See *institutional review boards.*

irregular practitioners Nineteenth-century health care practitioners other than allopathic doctors, such as homeopaths, midwives, botanic doctors, bonesetters, and patent medicine makers.

least developed nations Those less developed nations that have the least gross national product per capita and lowest life expectancy.

less developed nations Nations characterized by a relatively low gross national product per capita. These countries typically have relatively high rates of illiteracy, infant mortality, and other related problems, and their economies rely heavily on a few industries or products.

licensed practical nurses (LPNs) Individuals, not registered nurses, who assist nurses primarily with the custodial care of patients. LPNs usually have completed approximately one year of classroom and clinical training.

life events Any changes that force readjustments in individuals' lives, including marriage or divorce, starting or leaving school, and gaining or losing a job.

life expectancy The average number of years that individuals in a given population and born in a given year are expected to live.

limited practitioners Occupational groups, such as chiropractors and optometrists, that confine their work to a limited range of treatments and certain parts of the body.

LPNs See *licensed practical nurses.*

magic bullets Drugs that prevent or cure illness by attacking one specific etiological factor.

magnetic healers Nineteenth-century health workers who believed that an invisible magnetic fluid flowed through the body and that illness occurred when that flow was obstructed, unbalanced, inadequate, or excessive. Their treatments consisted of moving their hands along patients' spinal cords to "free" blocked magnetic fluid.

managed care A system that controls health care spending by monitoring closely how health care providers treat patients and where and when patients receive their health care.

managed care organizations (MCOs) Health insurance providers, such as health maintenance organizations, that operate under the principles of managed care.

manufacturers of illness Groups, such as alcohol and tobacco manufacturers, that promote illness-causing behaviors and social conditions.

marginal practitioners Occupational groups such as faith healers that have low social status.

master status A status viewed by others as so important that it overwhelms all other information about that individual. For example, if we know someone as the local scoutmaster, know he is a Republican and likes to play chess, and then learn he is gay, we might start thinking about him and interacting with him solely on the basis of his sexual orientation, essentially forgetting or ignoring the other information we have about him.

MCOs See *managed care organizations.*

Medicaid Joint federal and state health insurance program that pays the costs of health care for people with incomes below a certain (very low) amount. Most Medicaid recipients are poor mothers and their children. Medicaid can cover the costs of both preventive and therapeutic medical care and both inpatient and outpatient hospital care, but details of coverage vary considerably from state to state, with some providing considerably more services than others.

medical dominance Professional dominance by doctors. See *professional dominance.*

medical model of disability A model of disability that assumes that disability stems solely from forces within the individual mind or body rather than from constraints built into the environment or into social attitudes.

medical model of illness The way in which doctors conceptualize illness. This model consists of five doctrines: that disease is deviation from normal, specific and universal, caused by unique biological forces, analogous to the breakdown of a machine, and defined and treated medically through a neutral scientific process.

medical model of mental illness A model of mental illness assuming that (1) objectively measurable conditions define mental illness; (2) mental illness stems largely or solely from something within individual psychology or biology; (3) mental illness will worsen if left untreated but might improve or disappear if treated promptly by a medical authority; and (4) treating someone who might be healthy is safer than not treating someone who might be ill.

medical norms Expectations doctors hold regarding how they should act, think, and feel.

medicalization Process through which a condition or behavior becomes defined as a medical problem requiring a medical solution or through which the definition of an illness is broadened to cover a wider population.

Medicare Federal insurance, based on the Social Security system, that offers hospital insurance and medical insurance to those older than age 65 and to persons with permanent disabilities.

miasma According to pre–twentieth-century doctors, disease-causing air "corrupted" by foul odors or fumes.

minority group Any group that is considered inferior and subjected to differential and unequal treatment solely because of its physical or cultural characteristics.

moral status A status that identifies in society's eyes whether a person is good or bad, worthy or unworthy.

moral treatment A nineteenth-century practice aimed at curing persons with mental illness by treating them with kindness and giving them opportunities for both work and play.

morbidity Symptoms, illnesses, injuries, or impairments.

more developed nations Nations characterized by a relatively high gross national product per capita. These countries typically have diversified economies and low rates of illiteracy, infant mortality, and other related problems.

mortality Deaths.

mortification A process, occurring in total institutions, through which a person's prior self-image is partially or totally destroyed and replaced by a personality suited for life in the institution. See *total institutions.*

national health insurance A system in which all citizens of a country receive their health coverage from a single governmental insurance plan.

National Health Service (NHS) A system in which the government directly pays all costs of health care for its citizens.

negative social sanctions Punishments, ranging from ridicule to execution, meted out to those considered deviant by society.

neoliberalism a socioeconomic philosophy that encourages free trade and private enterprise, discourages government involvement in social services, and promotes the idea that each individual has the freedom *and the responsibility* to make wise consumer choices.

neonatal infant mortality Deaths of infants during the first 27 days after birth.

NHS See *National Health Service.*

normalize To make something seem like the normal course of events. In the context of medical error, this refers to emphasizing how medical errors can happen to anyone. In the context of mental illness, this refers to explaining to oneself and others how unusual behavior is not really a sign of mental illness.

norms Social expectations for appropriate behavior.

Nuremberg Code A set of internationally recognized principles regarding the ethics of human experimentation that emerged during the post–World War II Nuremberg trials for medical crimes against humanity. The code stipulates that researchers must have a medically justifiable purpose; do all within their power to protect their subjects from harm; and ensure that their subjects give voluntary, informed consent.

nurse-midwives Registered nurses who receive additional formal, nationally accredited training in midwifery.

nursing assistants Individuals, often untrained, who provide basic custodial care for patients, most often in nursing homes and hospitals. See *nursing homes.*

nursing homes Facilities that primarily provide nursing and custodial care to many individuals over a long period of time. Skilled nursing homes also provide some medical care.

outpatient Hospital patient who is neither formally admitted nor kept overnight.

pandemic A worldwide epidemic. See *epidemic.*

parallel practitioners Occupational groups, such as osteopaths, that perform basically the same roles as allopathic doctors while retaining occupational autonomy. See *allopathic doctors.*

passive euthanasia When health care workers allow patients to die through inaction.

performance norms Socially accepted rules for how a person should perform his or her roles. For example, we expect mothers to keep their children clean and paid workers to arrive on time each day.

physician extenders Health care providers who have less education than physicians but who can, at lower costs, take over some of the tasks traditionally done by physicians.

physician-assisted death When doctors help patients to end their lives.

placebo Anything offered as a cure that has no known biological effect.

placebo effect The process through which *belief* in a drug's effectiveness leads patients to *experience* physical benefits from a drug (such as decrease in pain).

positive social sanctions Rewards of any sort, from good grades to public esteem.

postneonatal infant mortality Deaths of infants between day 28 after birth and 11 months after birth.

potentially ill Individuals who have been identified as being at a higher than average risk of illness.

power Refers to the ability to get others to do what one wants, whether willingly or unwillingly.

practice protocols Guidelines that establish norms of care for particular medical conditions under particular circumstances based on careful review of clinical research.

prejudice Unwarranted suspicion or dislike of individuals because they belong to a particular group.

prevalence Total number of cases of an illness or health problem within a given population at a particular point in time (e.g., the number of persons living in the United States who have hepatitis). This includes both those newly diagnosed and those diagnosed earlier who still have the disease.

primary care Health care provided by physicians (such as family care doctors) and others who are trained to offer treatment and prevention services when individuals first seek health care and, ideally, as part of an ongoing provider–patient relationship.

primary care doctors Doctors in family or general practice, internal medicine, and pediatrics who are typically the first doctors individuals see when they need medical care.

primary practice; primary practitioners See *primary care doctors.*

privatization of health care The shift toward encouraging the private purchase of health care; the private, for-profit practice of medicine; and, in general, the operation of market forces in health care.

profession An occupation that (1) has the autonomy to set its own educational and licensing standards and to police its members for incompetence or malfeasance; (2) has its own technical, specialized knowledge that is learned through extended, systematic training; and (3) has the public's confidence that it follows a code of ethics and works more from a sense of service than a desire for profit.

professional dominance A profession's freedom from control by other occupations or groups and ability to control other occupations working in the same sphere. Only priests, for example, can decide whether someone can become a priest, and priests control the training and work responsibilities of lay religious workers in their churches.

professional socialization The process of learning the skills, knowledge, and values of an occupation.

professionalization Process through which an occupation achieves professional status.

pseudodisease Conditions diagnosed as disease based solely on test results but that will never cause health problems for the diagnosed individual.

public hospitals Hospitals established by state and federal governments to provide services to groups that would not otherwise receive care.

random samples Samples selected in such a way that each member of a population has an equal chance of being selected. When a sample is randomly selected, we can be fairly certain that the selected individuals will represent the population as a whole well.

rates Proportions of populations that experience certain circumstances.

RBRVS See *resource-based relative value scale.*

reductionistic treatment Treatment based on the assumption that each part can be treated separately from the whole in the same way that an air filter can be replaced in a car without worrying whether the problem with the air filter has caused or stemmed from problems in the car's electrical system.

registered nurses (RNs) Individuals who have received at least two years of nursing training and passed national licensure requirements. In everyday conversation, the word *nurse* generally means *registered nurse.*

regular doctors Nineteenth-century forerunners of contemporary medical doctors. Also known as allopathic doctors. See *allopathic doctors.*

reliability The likelihood that different people using the same measure will reach the same conclusions.

remedicalization The process through which mental illness is increasingly regarded by doctors and others as rooted in biology and amenable only to biological treatments.

reproductive technology Medical developments that offer control over human conception and fetal development.

residents Individuals who have graduated medical school and received their MD degrees but who are now engaging in further on-the-job training needed before they can enter independent practice.

resource-based relative value scale (RBRVS) A complex formula designed to curb the costs of Medicare by limiting reimbursement to doctors to the estimated actual costs of services in a particular geographic area.

respite care Any system designed to give family caregivers a break from their responsibilities.

right to die The right to make decisions concerning one's own death.

risk society Any society in which the risks from potentially dangerous modern technologies are interwoven with the economy and thus are commonplace and accepted.

RNs See *registered nurses.*

self-diagnosis The process through which individuals try to diagnose themselves.

self-fulfilling prophecy A situation in which individuals become what they are expected to be. For example, when it is assumed that no girls can throw a ball properly, girls might never be taught to do so, might never think it is worth trying on their own, and thus might never learn to do so.

serial sexual partners Sexual relationships that occur one after another (rather than overlapping) in an individual's life.

sex The biological categories of male and female, to which we are assigned based on our chromosomal structure, genitalia, hormones, and so on. Generally, individuals are considered male if they have XY sex chromosomes and female if they have XX sex chromosomes.

sick role The set of four social expectations in Western society regarding how society should view sick people and how sick people should behave. First, the sick person is considered to have a legitimate reason for not fulfilling his or her normal social role. Second, sickness is considered beyond individual control, something for which the individual is not held responsible. Third, the sick person must recognize that sickness is undesirable and work to get well. Fourth, the sick person should seek and follow medical advice.

sickness funds German insurance programs offered by nonprofit groups to serve a given occupation, geographic location, or employer. Otherwise known as social insurance.

single-payer system A health care system in which a single government health insurance organization covers all residents of a nation.

social capital The resources available to an individual through his or her social network. Social capital is typically measured by some combination of the number of people with whom one has close personal relationships combined with what types of resources one can access through those relationships.

social class The combination of an individual's education, income, and occupational status or prestige; some researchers use only one of these indicators to measure social class, but others combine two or more indicators.

social construction Ideas created by a social group, as opposed to something that is objectively or naturally given.

social control Means used by a social group to ensure that individuals conform to social norms and to ensure that the existing balance of power among groups is maintained. Social control can be formal (such as execution or commitment to a mental hospital) or informal (such as ridicule or shunning). See *norms.*

social control agents Individuals or groups of individuals who have the authority to enforce social norms, including parents, teachers, religious leaders, and doctors. See *norms.*

social drift theory A theory holding that lower-class persons have higher rates of illness because middle-class persons who become ill drift over time into the lower class.

social epidemiology The study of the distribution of disease within a population according to social factors (such as social class, use of alcohol, or unemployment) rather than biological factors (such as blood pressure or genetics).

social insurance See *sickness funds.*

social networks The webs of social relationships that link people to each other as friends, relatives, acquaintances, coworkers, and so on.

Social Security Federally funded program that, since 1935, has provided financial assistance to formerly employed adults with mental or physical disabilities as well as to elderly adults, blind individuals, and children with disabilities.

social stress theory A theory holding that lower-class persons have higher rates of mental illness because of the stresses of lower-class life.

sociological model of disability A model that defines disabilities as restrictions or lack of ability to perform activities resulting largely or solely either from social responses to bodies that fail to meet social expectations or from assumptions about the body reflected in the social or physical environment.

sociological model of illness A way of thinking about illness, common among sociologists, that argues that illness is a subjective, moral, and political label. It is *subjective* in that individuals may reasonably differ on whether something should be labeled illness. It is *moral* in that those labeled ill are often regarded as inferior to others. It is *political* in that some groups have more power than others to decide what should be defined as illness.

sociological perspective A perspective regarding human life and society that focuses on identifying social patterns and grappling with social problems rather than on analyzing individual behavior and finding solutions for personal troubles.

sociology in medicine An approach to the sociological study of health, illness, and health care that focuses on research questions of interest to doctors.

sociology of medicine An approach that emphasizes using the area of health, illness, and health care to answer research questions of interest to sociologists in general. This approach often requires researchers to raise questions that could challenge medical views of the world and power relationships within the health care world.

stakeholder mobilization Organized political opposition or support by groups with vested interest in the outcome.

stereotypes Oversimplistic assumptions regarding the nature of group members, such as assuming that African Americans are unintelligent.

stigma Any personal attribute that would be deeply discrediting if it becomes known.

street doctors Chinese health care workers with little formal training who work in urban outpatient clinics under the supervision of a doctor, offering primary and basic emergency care as well as health education, immunization, and assistance with birth control.

stress Situations that make individuals feel anxious and unsure how to respond, the emotions resulting from exposure to such situations, or the bodily changes occurring in response to these situations and feelings.

structure The social forces around us, including cultural pressures, economic standing, gender expectations, presence of absence of necessary resources, and so on. When used as the opposite of *agency*, refers to the concept that individual choices are limited by all of these social forces.

structural violence Social arrangements that are both deeply embedded in the politics, culture, or economy of a society and that harm individuals or keep them from reaching their full potential.

symbolic interactionism A theoretical perspective arguing that identity develops as part of an ongoing process of social interaction. Through this process, individuals learn to see themselves through the eyes of others, adopt the values of their community, and measure their self-worth against those values.

technological imperative Belief that technology is always good, so any existing technological interventions should be used.

technology Any human-made object used to perform a task, or a process using such objects. For example, the term *technology* can refer both to the overall process of kidney dialysis and to the specific pieces of equipment used in that process.

total institutions Institutions in which all aspects of life are controlled by a central authority and in which large numbers of like-situated persons are dealt with *en masse*. Examples include mental hospitals, prisons, and the military.

toxic agents Any substances that can harm or kill people or other organisms.

unintended negative consequences Unplanned, harmful effects of actions that had been expected to produce only benefits.

universal coverage Health care systems that provide access to health care for all legal residents of a nation.

utilization review A system in which insurance companies require doctors to get approval before ordering certain tests, performing surgery, hospitalizing a patient, or keeping a patient hospitalized more than a given number of days.

validity The likelihood that a given measure accurately reflects reality and measures what researchers believe it measures.

village doctors Chinese agricultural workers who receive a few months of training in health care and provide basic health services to members of their agricultural production team.

voluntary hospitals Hospitals that are financially based in voluntarism, or charity, rather than a profit motive. Same as non-profit institutions.

WHO See *World Health Organization.*

World Health Organization (WHO) United Nations organization charged with documenting health problems and improving world health.

References

Abel, Emily K. 2000. *Hearts of Wisdom: American Women Caring for Kin, 1850–1940.* Cambridge, MA: Harvard University Press.

Abelson, Reed, with Patricia Leigh Brown. 2002. "Alternative medicine is finding its niche in nation's hospitals." *New York Times*, April 13: B1.

AbouZahr, Carla, and Tessa Wardlaw. 2001. "Maternal mortality at the end of a decade: Signs of progress?" *Bulletin of the World Health Organization*, 79: 61–573.

Abramson, John. 2004. *Overdosed America: The Broken Promise of American Medicine.* New York: HarperCollins.

Accord Alliance. "Welcome to Accord Alliance." Accordalliance.org, accessed November 2014.

Adams, Tracey. 1999. "Dentistry and medical dominance." *Social Science & Medicine*, 48: 407–420.

Adler, Nancy E., and David H. Rehkopf. 2008. "U.S. disparities in health: Descriptions, causes, and mechanisms." *Annual Review of Public Health*, 29: 235–252.

Aguirre, Jessica C. 2012. "Cost of treatment still a challenge for HIV patients in U.S." http://www.npr.org/blogs/health/2012/07/27/157499134/cost-of-treatment-still-a-challenge-for-hiv-patients-in-u-s, accessed November 2014.

Aiken, Linda H., Julie Sochalski, and Gerard F. Anderson. 1996. "Downsizing the hospital nursing workforce." *Health Affairs*, 15: 88–92.

Aizenman, N. C. 2010. "Doctors shift away from Medicare patients." *Arizona Republic*, November 28: A1+.

Albrecht, Gary L. 1992. *The Disability Business: Rehabilitation in America.* Newbury Park, CA: Sage.

Aldridge, Melissa D., Mark. Schlesinger, Colleen L. Barry, R. Sean Morrison, Ruth McCorkle, Rosemary Hürzeler, and Elizabeth H. Bradley. 2014. "National hospice survey results: For-Profit status, community engagement, and service." *JAMA Internal Medicine*, 174(4): 500–506.

Aleman, Andre, Rene S. Kahn, and Jean-Paul Selten. 2003. "Sex differences in the risk of schizophrenia: Evidence from meta-analysis." *Archives of General Psychiatry*, 60: 565–571.

Allan Guttmacher Institute. 2014. *Induced Abortion in the United States*. New York: Author.

Amaro, Hortensia. 1999. "An expensive policy: The impact of inadequate funding for substance abuse treatment." *American Journal of Public Health*, 89: 657–659.

American Association of Colleges of Nursing. 2012. *New AACN Data Show an Enrollment Surge in Baccalaureate and Graduate Programs Amid Calls for More Highly Educated Nurses*. Washington, DC: Author.

American College of Nurse-Midwives. 2012. *Essential Facts about Midwives*. Silver Spring, MD: Author.

American Dental Association. 2013. Affordable Care Act, dental benefits examined. http://www.ada.org/en/publications/ada-news/2013-archive/august/affordable-care-act-dental-benefits-examined, accessed November 2014.

American Nurses Association. 2014. *Safe Staffing Literature Review*. Silver Spring, MD: Author.

American Pain Society. 2000. *Pain Assessment and Treatment in the Managed Care Environment*. Glenview, IL: Author.

Andersson, Gunnar, Karsten Hank, and Marit Ronsen. 2006. "Gendering family composition: Sex preferences for children and childbearing behavior in the Nordic countries." *Demography*, 43: 255–267.

Aneshensel, Carol S. 2009. "Toward explaining mental health disparities." *Journal of Health and Social Behavior*, 50(4): 377–394.

Angell, Marcia. 2004. *The Truth About the Drug Companies: How They Deceive Us and What to Do About It*. New York: Random House.

Annandale, Ellen C. 2010. "Health status and gender." Pp. 97–112 in *New Blackwell Companion to Medical Sociology*, edited by William C. Cockerham. Chichester, UK: Wiley-Blackwell.

Annas, George J. 1991. "Ethics committees: From ethical comfort to ethical cover." *Hastings Center Report*, 21 (May–June): 18–21.

Ansay, A. Manette. 2001. *Limbo: A Memoir*. New York: HarperCollins.

Anson, Ofra, and Shifang Sun. 2005. *Health Care in Rural China*. Aldershot, UK: Ashgate Publishing.

Anspach, Renee R. 1997. 2010. "Gender and health care." Pp. 229–248 in *Handbook of Medical Sociology*, edited by Chloe E. Bird, Peter Conrad, Allen M. Fremont, and Stefan Timmermans. Nashville, TN: Vanderbilt University Press.

Apovian, Caroline M. 2004. "Sugar-sweetened soft drinks, obesity, and type 2 diabetes." *Journal of the American Medical Association*, 291: 978–979.

Appleby, Julie. 2008. "Survey: Many request drugs advertised on TV." *Arizona Republic*. March 4: A12.

Arias, Elizabeth. 2010. "United States life tables by Hispanic origin." National Center for Health Statistics. *Vital Health Statistics*, 2(152).

Arias, Elizabeth, Jiaquan Xu, and Melissa A. Jim. 2014. "Period life tables for the non-Hispanic American Indian and Alaska Native population, 2007–2009." *American Journal of Public Health*, 104: S312–S319.

Armelagos, George J., and Kristin N. Harper. 2010. "Emerging infectious diseases, urbanization, and globalization in the time of global warming." Pp. 291–311 in *New Blackwell Companion to Medical Sociology*, edited by William C. Cockerham. Chichester, UK: Wiley-Blackwell.

Armstrong, Elizabeth M. 2000. "Lessons in control: Prenatal education in the hospital." *Social Problems*, 47: 583–605.

Arno, Peter S., Karen Bonuck, and Robert Padgug. 1995. "The economic impact of high-tech home care." Pp. 220–234 in *Bringing the Hospital Home: Ethical and Social Implications of High-tech Home Care*, edited by John D. Arras. Baltimore: Johns Hopkins University Press.

Arras, John D., and Nancy Neveloff Dubler. 1995. "Ethical and social implications of high-tech home care." Pp. 1–34 in *Bringing the Hospital Home: Ethical and Social Implications of High-tech Home Care*, edited by John D. Arras. Baltimore: Johns Hopkins University Press.

Association of American Medical Colleges. 2010a. "Diversity of U.S. medical school students by parental education." *Analysis in Brief*, 9(10). https://www.aamc.org/download/142770/data/aibvol9_no10.pdf, accessed November 2010.

______. 2014a. *Applicants and Matriculants Data*. http://www.aamc.org/data/facts/, accessed November 2014.

______. 2014b. Debt Fact Card. https://www.aamc.org/download/152968/data/debtfactcard.pdf, accessed November 2014.

Association of Asian Pacific Community Health Organizations. 2008. *AAPI Limited English Proficiency*. http://www.aapcho.org/wp/wp-content/uploads/2012/02/AAPCHO_FactSheet-AAPI_LEP_2005.pdf, accessed April 2008.

Avis, N. E., and S. M. McKinlay. 1991. "A longitudinal analysis of women's attitudes toward the menopause: Results from the Massachusetts women's health study." *Maturitas* 13: 65–79.

Avison, William R., and Stephanie S. Thomas. 2010. "Stress." Pp. 242–267 in *New Blackwell Companion to Medical Sociology*, edited by William C. Cockerham. Chichester, UK: Wiley-Blackwell.

Azevedo, Kathryn, and Hilda Ochoa Bogue. 2001. "Health and occupational risks of Latinos living in rural America." Pp. 359–380 in *Health Issues in the Latino Community*, edited by Marilyn Aguirre-Molina, Carlos W. Molina, and Ruth Enid Zambrana. San Francisco: Jossey-Bass.

Baer, Hans A. 2010. "Complementary and alternative medicine: Processes of legitimization, professionalization, and cooption." Pp. 373–390 in *New Blackwell Companion to Medical Sociology*, edited by William C. Cockerham. Chichester, UK: Wiley-Blackwell.

Barker, Kristin. 2005. *The Fibromyalgia Story: Biomedical Authority and Women's Worlds of Pain*. Philadelphia: Temple University Press.

______. 2008. "Electronic support groups, patient-consumers, and medicalization: The case of contested illness." *Journal of Health and Social Behavior*, 49: 20–36.

Barker-Benfield, Graham J. 1976. *The Horrors of the Half-Known Life: Male Attitudes Toward Women and Sexuality in Nineteenth Century America*. New York: Harper & Row.

Barnes, Patrica M., Patricia F. Adams, and Eve Powell-Griner. 2010. "Health characteristics of the American Indian or Alaska Native adult population: United States, 2004–2008." *National Health Statistics Reports*, Number 20.

Barnes, Patricia M., Barbara Bloom, and Richard L. Nahin. 2008. "Complementary and alternative medicine use among adults and children: United States, 2007." *National Health Statistics Reports*, Number 12.

Barry, Michael J. 2009. "Screening for prostate cancer—the controversy that refuses to die." *New England Journal of Medicine*, 360: 1351–1354.

Barstow, Anne Llewellyn. 1994. *Witchcraze: A New History of the European Witch Hunts*. San Francisco: Pandora.

Basch, Paul F. 1999. *Textbook of International Health*, 2nd ed., New York: Oxford University Press.

Basnett, Ian. 2001. "Healthcare professionals and their attitudes toward and decisions affecting disabled people." Pp. 450–467 in *Handbook of Disability Studies*, edited by Gary L. Albrecht, Katherine D. Seelman, and Michael Bury. Thousand Oaks, CA: Sage Publications.

Bates, David W., Suchi Saria, Lucila Ohno-Machado, Anand Shah, and Gabriel Escobar. 2014. "Big data in health care: Using analytics to identify and manage high-risk and high-cost patients." *Health Affairs*, 33: 1123–1131.

Beck, Ulrich. 1992. *Risk Society: Towards a New Modernity*. London: Sage.

______. 2006. "Living in the world risk society." *Economy and Society*, 35: 329–345.

Becker, Marshall H. (ed.). 1974. *The Health Belief Model and Personal Health Behavior*. San Francisco: Society for Public Health Education.

______. 1993. "A medical sociologist looks at health promotion." *Journal of Health and Social Behavior*, 34: 1–6.

Beecher, Henry K. 1966. "Ethics and clinical research." *New England Journal of Medicine*, 274: 1354–1360.

Beeson, Paul B. 1980. "Changes in medical therapy during the past half century." *Medicine*, 59: 79–99.

Beider, Perry, and Stuart Hagen. January 2004. *Limiting Tort Liability for Medical Malpractice*. Washington, DC: Congressional Budget Office Economic & Budget Issue Brief.

Benyshek, Daniel C., John F. Martin, and Carol S. Johnston. 2001. "A reconsideration of the origins of the Type 2 diabetes epidemic among Native Americans and the implications for intervention policy." *Medical Anthropology*, 20: 25–64.

Berger, Joseph. 2014. "The D.O. is in now." *New York Times Education Life*, August 3: 14–15.

Berlinger, Nancy. 2008. "Conscience clauses, health care providers, and parents." Pp. 35–40 in *From Birth to Death and Bench to Clinic: The Hastings Center Bioethics Briefing Book for Journalists, Policymakers, and Campaigns*, edited by Mary Crowley. Garrison, NY: The Hastings Center.

Bettie, Julie. 2003. *Women without Class: Girls, Race, and Identity*. Berkeley: University of California Press.

Bilefsky, Dan. 2010. "Seven charged in Kosovo organ-trafficking ring." *New York Times*, November 15: A4.

Bird, Chloe E., and Patricia R. Rieker. 2008. *Gender and Health: The Effects of Constrained Choices and Social Policies*. New York: Cambridge University Press.

Blackless, Melanie, Anthony Charuvastra, Amanda Derryck, Anne Fausto-Sterling, Karl Lauzanne, and Ellen Lee. 2000. "How sexually dimorphic are we? Review and synthesis." *American Journal of Human Biology*, 12: 151–166.

Blair, Stephen N., and Timothy S. Church. 2004. "The fitness, obesity, and health equation: Is physical activity the common denominator?" *Journal of the American Medical Association*, 292: 1232–1234.

Blue Cross Blue Shield Association. 2014. "About Blue Cross Blue Shield Association." http://www.bcbs.com/about-the-association, accessed September 2014.

Bodenheimer, Thomas. 1999. "The American health care system—physicians and the changing medical marketplace." *New England Journal of Medicine*, 340(7): 584–588.

______. 2000. "Uneasy alliance: Clinical investigators and the pharmaceutical industry." *New England Journal of Medicine*, 342: 1539–1543.

______. 2005a. "High and rising health care costs. Part 1: Seeking an explanation." *Annals of Internal Medicine*, 142: 847–854.

______. 2005b. "High and rising health care costs. Part 2: Technologic innovation." *Annals of Internal Medicine*, 142: 932–937.

______. 2005c. "High and rising health care costs. Part 3: The role of health care providers. *Annals of Internal Medicine*, 142: 996–1002.

Bosk, Charles L. 2003. *Forgive and Remember: Managing Medical Failure*, 2nd ed. Chicago: University of Chicago Press.

______. 2010. "Bioethics, raw and cooked: Extraordinary conflict and everyday practice." *Journal of Health and Social Behavior*, 51: S133–S146.

Boyer, Carol A., and Karen E. Lutfey. 2010. "Examining critical health policy issues within and beyond the clinical encounter: Patient–provider relationships and help-seeking behaviors." *Journal of Health and Social Behavior*, 51: S80–S93.

Bradby, Hannah, and James Y. Nazroo. 2010. "Health, ethnicity, and race." Pp. 113–130 in *New Blackwell Companion to Medical Sociology*, edited by William C. Cockerham. Chichester, UK: Wiley-Blackwell.

Brandt, Allan M., and Paul Rozin. 1997. *Morality and Health*. New York: Routledge.

Brannon, Robert L. 1996. "Restructuring hospital nursing: Reversing the trend toward a professional workforce." *International Journal of Health Services*, 26: 643–654.

Brawley, Otis Webb. 2011. *How We Do Harm*. New York: St. Martin's.

British Medical Journal. 2014. "What conclusions has *Clinical Evidence* drawn about what works, what doesn't based on randomised controlled trial evidence?" http://clinicalevidence.bmj.com/x/set/static/cms/efficacy-categorisations.html, accessed September 2014.

Brooks, Nancy A., and Ronald R. Matson. 1987. "Managing multiple sclerosis." *Research in the Sociology of Health Care*, 6: 73–106.

Brosnan, Caragh, Alan P. Cribb, Steven Wainwright, and Clare Williams. 2013. "Neuroscientists' everyday experiences of ethics: The interplay of regulatory, professional, personal and tangible ethical spheres." *Sociology of Health and Illness*, 35: 1133–1148.

Brown, Patricia Leigh. 2009. "A doctor for disease, a shaman for the soul." *New York Times*, September 20: A20.

Brown, Phil. 1990. "The name game: Toward a sociology of diagnosis." *Journal of Mind and Behavior*, 11: 385–406.

Brown, Phil, Steve Kroll-Smith, and Valerie J. Gunter. 2000. "Knowledge, citizens, and organizations: An overview of environments, diseases, and social conflict." Pp. 9–28 in *Illness and the Environment: A Reader in Contested Medicine*, edited by Steve Kroll-Smith, Phil Brown, and Valerie J. Gunter. New York: New York University Press.

Brown, Phil, Brian Mayer, Stephen Zavestoski, Theo Luebke, Joshua Mandelbaum, and Sabrina McCormick. 2003. "The health politics of asthma: Environmental justice and collective illness experience in the United States." *Social Science & Medicine*, 57: 453–464.

Brown, Phil, Stephen Zavestoski, Brian Mayer, Sabrina McCormick, and

Pamela S. Webster. 2002. "Policy issues in environmental health disputes." *Annals of the American Academy of Political and Social Sciences*, 584: 175–202.

Brown, Phil, Stephen Zavestoski, Sabrina McCormick, Brian Myer, Rachel Morello-Frosch, and Rebecca Gasior Altman. 2004. "Embodied health movements: New approaches to social movements in health." *Sociology of Health and Illness*, 26: 50–80.

Brulle, Robert J., and David N. Pellow. 2006. "Environmental justice: Human health and environmental inequalities." *Annual Review of Public Health*, 27: 103–24.

Brumberg, Joan Jacobs. 1997. *The Body Project: An Intimate History of American Girls*. New York: Random House.

Bullard, Robert D., Rueben C. Warren, and Glenn S. Johnson. 2001. "The quest for environmental justice." Pp. 471–488 in *Health Issues in the Black Community*, 2nd ed., edited by Ronald L. Braithwaite and Sandra E. Taylor, San Francisco: Jossey-Bass.

Bunker, John P., Howard S. Frazier, and Frederick Mosteller. 1994. "Improving health: Measuring effects of medical care." *Milbank Quarterly*, 72: 225–258.

Bureau of Labor Statistics. 2010. *Consumer Expenditures—2009*. Washington, DC: US Government Printing Office.

______. 2014. *Occupational Outlook Handbook, 2010/2011*. Washington, DC: US Government Printing Office.

Burton, Elizabeth C., and Kim A. Collins. 2014. "Autopsy rate and physician attitudes toward autopsy." http://emedicine.medscape.com/article/1705948-overview, accessed October 2014.

Byrne, Rhonda. 2006. *The Secret*. New York: Atria Books.

Byock, Ira. 2013. *The Best Care Possible: A Physician's Quest to Transform Care Through the End of Life*. New York: Avery.

Calderisi, Robert. 2006. *The Trouble with Africa: Why Foreign Aid Isn't Working*. New York: Palgrave Macmillan.

Caldwell, John C. 1993. "Health transition: The cultural, social, and behavioral determinants of health in the Third World." *Social Science & Medicine*, 36: 125–135.

Callahan, Daniel. 1998. *False Hopes: Why America's Quest for Perfect Health Is a Recipe for Failure*. New York: Simon & Schuster.

Campbell, James B., Jason W. Busse, and H. Stephen Injeyan. 2000. "Chiropractors and vaccination: A historical perspective." *Pediatrics*, 105, http://pediatrics.aappublications.org/cgi/content/full/105/4/e43, accessed March 2008.

Campos, Paul. 2004. *The Obesity Myth: Why America's Obsession with Weight Is Hazardous to Your Health*. New York: Gotham Books.

Canadian Institute for Health Information. 2010. "Most patients receiving care within recommended wait times for priority areas." http://www.cihi.ca/CIHI-ext-portal/internet/en/Document/health+system+performance/access+and+wait+times/RELEASE_24MAR10, accessed November 2010.

Cancian, Francesca M., and Stacey J. Oliker. 2000. *Caring and Gender*. Thousand Oaks, CA: Pine Forge.

Cane, Leslie, and Carol Peckham. 2014. Medscape Physician Compensation Report 2014. http://www.medscape.com/features/slideshow/compensation/2014/public/overview#2, accessed September 2014.

Carlat, Daniel. 2010. *Unhinged: The Trouble With Psychiatry—A Doctor's*

Revelations About a Profession in Crisis. New York: Free Press.

Carmel, Simon. 2006. "Boundaries obscured and boundaries reinforced: Incorporation as a strategy of occupational enhancement for intensive care." *Sociology of Health and Illness,* 28: 154–177.

Casper, Monica J., and Daniel R. Morrison. 2010. "Medical sociology and technology: Critical engagements." *Journal of Health and Social Behavior,* 51: S120–S132.

Cejka Search. 2010. *Physician Compensation Data.* http://www.cejkasearch.com/compensation/amga_physician_compensation_survey.htm, accessed November 2010.

Center for Responsive Politics. 2010. *Gun Control vs. Gun Rights.* http://www.opensecrets.org/news/issues/guns, accessed October 2010.

______. 2011. *Beer, Wine & Liquor.* http://www.opensecrets.org/industries/indus.php?ind=N02, accessed March 2011.

Center for the Evaluative Clinical Sciences, Dartmouth Medical School. 1996. *The Dartmouth Atlas of Health Care.* Chicago: American Hospital Association.

Centers for Disease Control and Prevention. 1995. *MMWR Weekly: Poverty and Infant Mortality—United States 1988.* http://www.cdc.gov/mmwr/preview/mmwrhtml/00039818.htm, accessed November 2010.

______. 2006. "Youth exposure to alcohol advertising on radio—United States." *Morbidity and Mortality Weekly Report,* 55: 937–940.

______. 2012. "Overweight and Obesity: Home." http://www.cdc.gov/obesity/adult/causes/index.html, accessed October 2014.

Centers for Medicare and Medicaid Services. 2014. *National Health Expenditure Projections 2012–2022.* http://www.cms.gov/Research-Statistics-Data-and-Systems/Statistics-Trends-and-Reports/NationalHealthExpendData/Downloads/Proj2012.pdf, accessed July 2014.

Champaneria, Manish C., and Sara Axtell. 2004. "Cultural competence training in U.S. medical schools." *Journal of the American Medical Association,* 291: 2142.

Charmaz, Kathy. 1991. *Good Days, Bad Days: The Self in Chronic Illness and Time.* New Brunswick, NJ: Rutgers University Press.

Charmaz, Kathy, and Dana Rosenfeld. 2010. "Chronic illness." Pp. 312–334 in *New Blackwell Companion to Medical Sociology,* edited by William C. Cockerham. Chichester, UK: Wiley-Blackwell.

Chen, Ian, James Kurz, Mark Pasanen, Charles Faselis, Mukta Panda, Lisa Staton, Jane O'Rorke, et al. 2005. "Racial differences in opioid use for chronic nonmalignant pain." *Journal of General Internal Medicine,* 20: 593–598.

Chen, Lena M., Wildon R. Farwell, and Ashish K. Jha. 2009. "Primary care visit duration and quality: Does good care take longer?" *Archives of Internal Medicine,* 169: 1866–1872.

Chen, Meei-Shia. 2001. "The great reversal: Transformation of health care in the People's Republic of China." Pp. 456–482 in *The Blackwell Companion to Medical Sociology,* edited by William C. Cockerham. Malden, MA: Blackwell.

Cherkin, Daniel C., Richard A. Deyo, Michele Battié, Janet Street, and William Barlow. 1998. "A comparison of physical therapy, chiropractic manipulation, and provision of an educational booklet for the treatment of patients with low back pain." *New England Journal of Medicine,* 339: 1021–1029.

Cherry, Mark J. 2005. *Kidney for Sale by Owner: Human Organs, Transplantation, and the Market*. Washington, DC: Georgetown University Press.

Chivers, Sally, and Nicole Markotic. 2005. "Film." *Encyclopedia of Disability*. SAGE Publications. http://www.sage-ereference.com/disability/Article_n338.html, accessed November 2010.

Christakis, Dmitri A., and Christopher Feudtner. 1997. "Temporary matters: The ethical consequences of transient social relationships in medical training." *Journal of American Medical Association*, 278: 739–743.

Clarke, Adele E., Janet K. Shim, Laura Mamo, Jennifer Ruth Fosket and Jennifer R. Fishman. 2003. "Biomedicalization: Technoscientific Transformations of Health, Illness, and U.S. Biomedicine." *American Sociological Review*, 68: 161–194.

Clarke, Adele E., Laura Mamo, Jennifer Ruth Fosket, Jennifer R. Fishman, and Janet K. Shim (eds.). 2010. *Biomedicalization: Technoscience, Health, and Illness in the U.S.* Durham, NC: Duke University Press.

Cockerham, William. 2005. "Health lifestyle theory and the convergence of agency and structure." *Journal of Health and Social Behavior*, 46: 51–67.

Cohen, Susan, and Christine Cosgrove. 2009. *Normal at Any Cost: Tall Girls Short Boys, and the Medical Industry's Quest to Manipulate Height*. New York: Penguin.

Collins, Michael J. 2005. *Hot Lights, Cold Steel: Life, Death and Sleepless Nights in a Surgeon's First Years*. New York: St. Martin's Press.

Collins, Sara R., Petra W. Rasmussen, and Michelle M. Doty. 2014a. *Gaining Ground: Americans' Health Insurance Coverage and Access to Care After the Affordable Care Act's First Open Enrollment Period*. Commonwealth Fund Pub 1760, Vol. 16.

Collins, Sara R., Petra W. Rasmussen, Michelle M. Doty, and Sophie Beutel. 2014b. *Too high a price: Out-of-pocket health care costs in the United States*. Commonwealth Fund Pub 1784, Vol. 29.

Commonwealth Fund. 2013. *International Profiles of Health Care Systems*. Washington, DC.

Commonwealth Fund. 2014. From Chapter 8, Table 8.1, p. 14.

Connor, Stephen R., Bruce Pyenson, Kathryn Fitch, Carol Spence, and Kosuke Iwasaki. 2007. "Comparing hospice and nonhospice patient survival among patients who die within a three-year window." *Journal of Pain and Symptom Management*, 33: 238–246.

Conrad, Peter. 1985. "The meaning of medications: Another look at compliance." *Social Science & Medicine*, 20: 29–37.

______. 1987. "The experience of illness: Recent and new directions." *Research in the Sociology of Health Care*, 6: 1–32.

______. 2005. "The shifting engines of medicalization." *Journal of Health and Social Behavior*, 46: 3–14.

______. 2007. *The Medicalization of Society: On the Transformation of Human Conditions into Treatable Disorders*. Baltimore: Johns Hopkins University Press.

Conrad, Peter, and Cheryl Stults. 2010. "The Internet and the experience of illness." Pp. 179–191 in *Handbook of Medical Sociology*, edited by Chloe E. Bird, Peter Conrad, Allen M. Fremont, and Stefan Timmermans. Nashville, TN: Vanderbilt University Press.

Contexts. 2004. "Assisted suicide." *Contexts*, 3(3): 58.

Copelton, Denise A., and Giuseppina Valle. 2009. "You don't need a prescription to go gluten-free: The scientific self-diagnosis of celiac disease." *Social Science & Medicine*, 69(4): 623–631.

Corbin, Juliet M., and Anselm Strauss. 1987. "Accompaniments of chronic illness: Changes in body, self, biography, and biographical time." *Research in the Sociology of Health Care*, 6: 249–282.

Coulehan, John, and Peter C. Williams. 2001. "Vanquishing virtue: The impact of medical education." *Academic Medicine*, 76: 598–605.

Council for Responsible Genetics. 2001. *Genetic Discrimination*. http://www.councilforresponsiblegenetics.org/ViewPage.aspx?pageId=85, accessed June 2008.

Council on Ethical and Judicial Affairs, American Medical Association. 1991. "Gender disparities in clinical decision making." *Journal of the American Medical Association*, 266: 559–562.

Critser, Greg. 2003. *Fat Land: How Americans Became the Fattest People in the World*. New York: Houghton Mifflin.

Crosby, Alfred J. 1986. *Ecological Imperialism: The Biological Expansion of Europe, 900–1900*. New York: Cambridge University Press.

Cutler, David, and Grant Miller. 2005. "The role of public health improvements in health advances: The twentieth-century United States." *Demography*, 42: 1–22.

Daniels, Cynthia R. 1993. *At Women's Expense: State Power and the Politics of Fetal Rights*. Cambridge, MA: Harvard University Press.

Daniels, Norman, and Marc Roberts. 2008. "Health care reform." Pp. 83–88 in *From Birth to Death and Bench to Clinic: The Hastings Center Bioethics Briefing Book for Journalists, Policymakers, and Campaigns*, edited by Mary Crowley. Garrison, NY: The Hastings Center.

Davis, Devra. 2007. *The Secret History of the War on Cancer*. New York: Basic.

Davis, Karen, Kristof Stremikis, David Squires, and Cathy Schoen. 2014. *Mirror, mirror on the wall: How the performance of the U.S. Health care system compares internationally*. Washington, DC: Commonwealth Fund.

DeBellonia, Renato Rocco, Steven Marcus, Richard Shih, John Kashani, Joseph G. Rella, and Bruce Ruck. 2008. "Curanderismo: Consequences of folk medicine." *Pediatric Emergency Care*, 24: 228–229.

Degenhardt, Louisa, Jessica Singleton, Bianca Calabria, Jennifer McLaren, Thomas Kerr, Shruti Mehta, Gregory Kirk, and Wayne D. Hall. 2011. "Mortality among cocaine users: A systematic review of cohort studies." *Drug and Alcohol Dependence*, 113: 88–95.

Dettwyler, Katherine A. 1995. "Beauty and the breast." Pp. 167–213 in *Breastfeeding: Biocultural Perspectives*, edited by Patricia Stuart-Macadam and Katherine A. Dettwyler. New York: Aldine De Gruyter.

Diamond, Timothy. 1992. *Making Gray Gold: Narratives of Nursing Home Care*. Chicago: University of Chicago Press.

Dickman, Sam, David Himmelstein, Danny McCormick, and Steffie Woolhandler. 2014. "Opting out of Medicaid expansion: The health and financial impacts ." *Health Affairs Blog*. http://healthaffairs.org/blog/2014/01/30/opting-out-of-medicaid-expansion-the-health-and-financial-impacts, accessed July 2014.

Diller, Lawrence H. 1998. *Running on Ritalin: A Physician Reflects on Children, Society, and Performance in a Pill*. New York: Bantam.

Dobash, Russell P., and Rebecca Emerson Dobash. 1998. *Rethinking Violence*

Against Women. Newbury Park, CA: Sage.

Doctors Without Borders. 2014. *Everyday Emergency: Silent Suffering in Democratic Republic of Congo*. http://www.doctorswithoutborders.org/news-stories/special-report/everyday-emergency-silent-suffering-democratic-republic-congo, accessed November 2014.

Dodson, Lisa, and Rebekah M. Zincavage. 2007. "It's like a family: Caring labor, exploitation, and race in nursing homes." *Gender & Society*, 21: 905–928.

Dove, Edward S. 2013. "Back to blood: The sociopolitics and law of compulsory DNA testing of refugees," *University of Massachusetts Law Review*, 8: 466–530.

Dreze, Jean, and Amartya Sen. 1989. *Hunger and Public Action*. Oxford, UK: Clarendon.

Duncan, R. Paul, Michael E. Morris, and Linda A. McCarey. 2009. "Canada." Pp. 59–82 in *Comparative Health Systems: Global Perspectives for the 21st Century*, edited by James A. Johnson and Carleen H. Stoskopf. Sudbury, MA: Jones & Bartlett.

Dunn, Kyla. 2002. "Cloning Trevor." *Atlantic Monthly*, 289(6): 31–53.

Dworkin, Shari, and Faye Linda Wachs. 2009. *Body Panic: Gender, Health, and the Selling of Fitness*. New York: New York University Press.

Earley, Pete. 2007. *Crazy: A Father's Search Through America's Mental Health Madness*. New York: Penguin.

Easterly, William. 2006. *The White Man's Burden: Why the West's Efforts to Aid the Rest Have Done So Much Ill and So Little Good*. New York: Penguin.

Eaton, William W., and Carles Muntaner. 1999. "Social stratification and mental disorder." Pp. 259–285 in *A Handbook for the Study of Mental Health: Social Contexts, Theories, and Systems*, edited by Allan V. Horwitz and Teresa L. Scheid. Cambridge, UK: Cambridge University Press.

Eckholm, Erik. 2013. "Case explores rights of fetus versus mother." *New York Times*, October 23: A1+.

Edgar, Dahl, S. Gupta Ruchi, Beutel Manfred, Stoebel-Richter Yve, Brosig Burkhard, Tinneberg Hans-Rudolf, and Jain Tarun. 2006. "Preconception sex selection demand and preferences in the United States." *Fertility and Sterility*, 85: 468–473.

Elliott, Carl. 2004. "Pharma goes to the laundry: Public relations and the subject of medical education." *Hastings Center Review*, 34: 18–23.

______. 2008. "Guinea-pigging." *Atlantic Monthly*, January 7: 36–41.

Ensel, Walter M., and Nan Lin. 1991. "The life stress paradigm and psychological distress." *Journal of Health and Social Behavior*, 32: 321–341.

Epstein, Arnold M., John Z. Ayanian, Joseph H. Keogh, Susan J. Noonan, Nancy Armistead, Paul D. Cleary, Joel S. Weissman, Jo Ann David-Kasdan, Diane Carlson, Jerry Fuller, Douglas Marsh, and Rena M. Conti. 2000. "Racial disparities in access to renal transplantation: Clinically appropriate or due to under-use or overuse?" *New England Journal of Medicine*, 343: 1537–1544.

Epstein. 1996. *Impure Science: AIDS, Activism, and the Politics of Knowledge*. Berkeley: University of California Press.

Epstein, Steven. 2007. *Inclusion: The Politics of Difference in Medical Research*. Chicago: University of Chicago Press.

Evans, Dylan. 2003. *Placebo: The Belief Effect*. New York: HarperCollins.

Exley, Catherine. 2009. "Bridging a gap: The (lack of a) sociology of oral health and healthcare." *Sociology of Health & Illness*, 31: 1093–1108.

Ezzati, Majid, Ari B. Friedman, Sandeep C. Kulkarni, and Christopher J. L. Murray. 2008. "The reversal of fortunes: Trends in county mortality and cross-county mortality disparities in the United States." *PLoS Medicine*, 5(4): 1–12.

Fadiman, Anne. 1997. *The Spirit Catches You and You Fall Down: A Hmong Child, Her American Doctors, and the Collision of Two Cultures*. New York: Farrar, Straus and Giroux.

Farmer, Paul. 1999. *Infections and Inequalities: The Modern Plagues*. Berkeley: University of California Press.

Farmer, Paul, Jim Yong Kim, and Arthur Kleinman. 2013. *Reimagining Global Health*. Berkeley: University of California Press.

Farmer, Paul E., Bruce Nizeye, Sara Stulac, and Salmaan Keshavjee. 2006. "Structural violence and clinical medicine." *PLoS Med*, 3: e449.

Favvaza, Titus. 2013. "Seeking and utilizing a curandero in the United States: A literature review." *Journal of Holistic Nursing*, 32: 189–201.

Federal Interagency Forum on Child and Family Statistics. 1999. *America's Children 1999*. Washington, DC: US Government Printing Office.

Feng, Zhanlian, Mary L. Fennell, Denise Tyler, Melissa A. Clark, and Vincent Mor. 2010. "Shifts in racial/ethnic composition of nursing home residents in the United States, 2000–2007." Conference Series on Aging in the Americas. http://www.utexas.edu/lbj/caa/2010/presentations/FengZ1.pdf, accessed December 2010.

Fennell, Mary L., Zhanlian Feng, Melissa A. Clark, and Vincent Mor. 2010. "Elderly Hispanics more likely to reside in poor-quality nursing homes." *Health Affairs*, 29: 65–73.

Ferraro, Kenneth F., and Teyana P. Shippee. 2009. "Aging and cumulative inequality: How does inequality get under the skin?" *Gerontologist*, 49: 333–43.

Feshbach, Morris. 1999. "Dead souls." *Atlantic Monthly*, 283(1): 26–27.

Feshbach, Morris, and Alfred Friendly. 1992. *Ecocide in the USSR: Health and Nature Under Siege*. New York: Basic.

Fisher, Elliott S., David E. Wennberg, Therese A. Stukel, Daniel J. Gottlieb, F. L. Lucas, and Etoile L. Pinder. 2003. "Implications of regional variation in Medicare spending." *Annals of Internal Medicine*, 138: 273–287.

Fisher, Jill A. 2009. *Medical Research for Hire: The Political Economy of Pharmaceutical Clinical Trials*. New Brunswick, NJ: Rutgers University Press.

Fisher, Jill A., and Lorna M. Ronald. 2008. "Direct to consumer responsibility: Medical neoliberalism in pharmaceutical advertising and drug development." *Advances in Medical Sociology*, 10: 29–51.

Flegal, Katherine M., Barry I. Graubard, David F. Williamson, and Mitchell H. Gail. 2005. "Excess deaths associated with underweight, overweight, and obesity." *Journal of the American Medical Association*, 293: 1861–1867.

Katherine M. Flegal, Brian K. Kit, Heather Orpana, and Barry I. Graubard. 2013. *Association of All-Cause Mortality With Overweight and Obesity Using Standard Body Mass Index Categories: A Systematic Review and Meta-analysis. Journal of the American Medical Association*, 309: 71–82.

Flynn, D. P. 2008. "Pharmacist conscience clauses and access to oral contraceptives." *Journal of Medical Ethics*, 34: 517–520.

Foreman, Judy. 2014. *A Nation in Pain: Healing Our Biggest Health Problem*. New York: Oxford University Press.

Fortson, Jane J. 2008. "The gradient in sub-Saharan Africa: Socioeconomic status and HIV/AIDS." *Demography*, 45: 303–322.

Fox, Renee C. 2000. "Medical uncertainty revisited." Pp. 409–425 in *The Handbook of Social Studies in Health and Medicine*, edited by Gary L. Albrecht, Ray Fitzpatrick, and Susan C. Scrimshaw. Thousand Oaks, CA: Sage.

Fox, Renee C., and Judith Swazey. 1974. *The Courage to Fail*. Chicago: University of Chicago Press.

Fox, Susannah. 2012. *The Social Life of Health Information*. http://www.pewinternet.org/2011/05/12/the-social-life-of-health-information-2011, accessed November, 2014.

Fox, Susannah, and Maeve Duggan. 2013. *Health Online 2013*. Pew Internet and American Life Project. http://www.pewinternet.org/Reports/2013/Health-online.aspx, accessed November 2014.

Frances, Allen. 2012. "DSM5 in distress." *Psychology Today*. http://www.psychologytoday.com/blog/dsm5-in-distress, accessed September 2013.

Frei, Patrizia, Aslak H Poulsen, Christoffer Johansen, Jørgen H. Olsen, Marianne Steding-Jessen, and Joachim Schüz. 2011. "Use of mobile phones and risk of brain tumours: Update of Danish cohort study." *British Medical Journal*, 343: 1–9.

Freidson, Eliot. 1970. *Profession of Medicine: A Study of the Sociology of Applied Knowledge*. New York: Dodd, Mead.

______. 1994. *Professionalism Reborn*. Chicago: University of Chicago Press.

French, S. A., M. Story, D. Neumark-Sztainer, J. A. Fulkerson, and P. Hannan. 2001. "Fast food restaurant use among adolescents: Associations with nutrient intake, food choices and behavioral and psychosocial variables." *International Journal of Obesity*, 25: 1823–1833.

Frenk, Julio, Eduardo González-Pier, Octavio Gómez-Dantés, Miguel A. Lezana, and Felicia Marie Knaul. 2006. "Health system reform in Mexico: Comprehensive reform to improve health system performance in Mexico." *Lancet*, 368: 1524–1534.

Furman, Lydia. 2009. "ADHD: What do we really know?" Pp. 21–57 in *Rethinking ADHD: International Perspectives*, edited by Sami Timimi and Jonathan Leo. London: Palgrave/Macmillan.

Garfield, Craig F., Paul J. Chung, and Paul J. Rathouz. 2003. "Alcohol advertising in magazines and adolescent readership." *Journal of the American Medical Association*, 289: 2424–2429.

Garfield, Rachel, Anthony Damico, Jessica Stephens, and Saman Rouhani. 2014. *The Coverage Gap: Uninsured Poor Adults in States that Do Not Expand Medicaid—An Update*. http://files.kff.org/attachment/the-coverage-gap-uninsured-poor-adults-in-states-that-do-not-expand-medicaid-issue-brief, accessed January 2015.

Gawande, Atul. 2014. *Being Mortal: Medicine and What Matters in the End*. New York: Metropolitan.

Geiger, H. Jack, and Robert M. Cook-Deegan. 1993. "The role of physicians in conflicts and humanitarian crises: Case studies from the field missions of Physicians for Human Rights, 1988 to 1993." *Journal of the American Medical Association*, 270: 616–620.

Gerberding, Julie L., and James S. Marks. 2004. "Making America fit and trim." *American Journal of Public Health*, 94: 1478–1479.

Gevitz, Norman. 1988. "Osteopathic medicine: From deviance to difference." Pp. 124–156 in *Other Healers: Unorthodox Medicine in America*, edited by Norman Gevitz. Baltimore: Johns Hopkins University Press.

Gibbs, W. Wayt. 2005. "Obesity: An overblown epidemic?" *Scientific American*, 292(6): 70–77.

Givel, Michael, and Stanton A. Glantz. 2004. "The 'global settlement' with the tobacco industry: 6 years later." *American Journal of Public Health*, 94: 218–229.

Glasser, Ronald J. 2005. "A war of disabilities: Iraq's hidden costs are coming home." *Harper's Magazine*, 311(1862): 59–62.

Glazer, Nona Y. 1993. *Women's Paid and Unpaid Labor: The Work Transfer in Health Care and Retailing*. Philadelphia: Temple University Press.

Glenton, Claire. 2003. "Chronic back pain sufferers—striving for the sick role." *Social Science & Medicine*, 57: 2243–2252.

Godlee, Fiona, Jane Smith, and Harvey Marcovitch. 2011. "Wakefield's article linking MMR vaccine and autism was fraudulent." *British Medical Journal*, 342: c7452.

Goffman, Erving. 1961. *Asylums*. Garden City, NY: Doubleday.

Goldacre, Ben. 2010. *Bad Science: Quacks, Hacks, and Big Pharma Flacks*. New York: Faber & Faber.

Goleman, Daniel. 1995. "Making room on the couch for culture." *New York Times*, December 5: C1+.

Goodnough, Abby, and Robert Pear. 2014. "Unable to meet the deductible or the doctor." *New York Times*, October 17: A1+.

Goosby, Bridget J. 2013. "Early life course pathways of adult depression and chronic pain." *Journal of Health and Social Behavior*, 54: 75–91.

Goozner, Merrill. 2008. "A report from the Russian front in the global fight against drug-resistant tuberculosis." http://www.scientificamerican.com/article/siberia-drug-resistant-tuberculosis, accessed November 2010.

Gordon, Suzanne. 2005. *Nursing Against the Odds: How Health Care Cost Cutting, Media Stereotypes, and Medical Hubris Undermine Nurses and Patient Care*. Ithaca, NY: Cornell University Press.

Gostin, Lawrence O., Chai Feldblum, and David W. Webber. 1999. "Disability discrimination in America." *Journal of the American Medical Association*, 281: 745–752.

Gottfried, Robert S. 1983. *The Black Death*. New York: Free Press.

Goyal, Madhav, Ravindra L. Mehta, Lawrence J. Schneiderman, and Ashwini R. Sehgal. 2002. "Economic and health consequences of selling a kidney in India." *Journal of the American Medical Association*, 288: 1589–1593.

Green, Carolyn J., Arminée Kazanjian, and Diane Helmer. 2004. "Informing, advising, or persuading? An assessment of bone mineral density testing information from consumer health websites." *International Journal of Technology Assessment in Health Care*, 20: 156–166.

Greenhouse, Steven. 2001. "Fear and poverty sicken many migrant workers in U.S." *New York Times*, May 13: A14.

______. 2006. "Hotel rooms get plusher, adding to maids' injuries." *New York Times*, April 21: A20.

Grob, Gerald N. 1997. "Deinstitutionalization: The illusion of policy." *Journal of Policy History*, 9: 48–73.

Grob, Gerald N., and Allan V. Horwitz. 2009. *Diagnosis, Therapy, and Evidence: Conundrums in Modern American Medicine*. New Brunswick, NJ: Rutgers University Press.

Gross, Cary P., Benjamin D. Smith, Elizabeth Wolf, and Martin Andersen. 2008. "Racial disparities in cancer therapy: Did the gap narrow between 1992 and 2002?" *Cancer*, 112: 900–908.

Guallar, Eliseo, Saverio Stranges, Cynthia Mulrow, Lawrence J. Appel, and

Edgar R. Miller, III. 2013. "Enough is enough: Stop wasting money on vitamin and mineral supplements." *Annals of Internal Medicine*, 159: 850–851.

Gwyther, Marni E., and Melinda Jenkins. 1998. "Migrant farmworker children: Health status, barriers to care, and nursing innovations in health care delivery." *Journal of Pediatric Health Care*, 12: 60–66.

Hadler, Nortin M. 2008. *Worried Sick*. Chapel Hill, NC: University of North Carolina Press.

Hafferty, Frederic W. 1991. *Into the Valley: Death and the Socialization of Medical Students*. New Haven, CT: Yale University Press.

Hafferty, Frederic W., and Ronald Franks. 1994. "The hidden curriculum, ethics teaching, and the structure of medical education." *Academic Medicine*, 69: 861–871.

Hahn, Harlan. 1985. "Toward a politics of disability definitions, disciplines, and policies." *Social Science Journal*, 22 (October): 87–105.

Harrington, Charlene, Steffie Woolhandler, Joseph Mullan, Helen Carrillo, and David U. Himmelstein. 2001. "Does investor ownership of nursing homes compromise the quality of care?" *American Journal of Public Health*, 91: 1452–1455.

Harris, Gardiner. 2009. "In Hawaii's health system, lessons for lawmakers." *New York Times*, October 17: A1+.

______. 2011. "When the nurse wants to be called 'doctor.'" *New York Times*, October 1: A1+.

______. 2014. "Malnutrition in well-fed children is linked to poor sanitation." *New York Times*, July 15: A1+.

Harris Interactive. 2010. *Kessler Foundation/NOD 2010 Survey of Americans with Disabilities*. http://www.2010disabilitysurveys.org/pdfs/surveyresults.pdf, accessed November 2010.

Harris Poll. 2010a. "Firefighters, scientists, and doctors seen as most prestigious occupations." Number 86.

______. 2010b. "Virtually no change in annual Harris Poll confidence index from last year." Number 33.

Hartley, Heather. 2006. "The 'pinking' of viagra culture: Drug industry efforts to create and repackage sex drugs for women." *Sexualities*, 9: 363–378.

Hayden, Dolores. 2003. *Building Suburbia: Green Fields and Urban Growth, 1820–2000*. New York: Pantheon.

HealthGrades. 2004. *Patient Safety in American Hospitals*. Lakewood, CO: Author.

Heath, Christian, Paul Luff, and Marcus Sanchez Svensson. 2003. "Technology and medical practice." *Sociology of Health and Illness*, 25: 75–96.

Helzer, John L., Lee N. Robin, Mitchell Taibleson, Robert A. Woodruff, Theodore Reich, and Eric D. Wish. 1977. "Reliability of psychiatric diagnosis: A methodological review." *Archives of General Psychiatry*, 34: 129–133.

Hendlin, Herbert, Chris Rutenfrans, and Zbigniew Zylicz. 1997. "Physicianassisted suicide and euthanasia in the Netherlands: Lessons from the Dutch." *Journal of the American Medical Association*, 277: 1720–1722.

Heritage, John, and Douglas W. Maynard (eds.). 2006. *Communication in Medical Care: Interaction Between Primary Care Physicians and Patients*. Cambridge, UK: Cambridge University Press.

Heron, Melanie. 2013. "Deaths: Leading causes for 2010." *National Vital Statistics Report*, 62(6).

Higgins, Paul C. 1992. *Making Disability: Exploring the Social Transformation of Human Variation*. Springfield, IL: Charles C. Thomas.

Hilts, Philip J. 1999. "In tests on people, who watches the watchers?" *New York Times*, May 25: D1+.

Himmelstein, David U., Miraya Jun, Reinhard Busse, Karine Chevreul, Alexander Geissler, Patrick Jeurissen, Sarah Thomson, Marie-Amelie Vinet, and Steffie Woolhandler. 2014. "A comparison of hospital administrative costs in eight nations: U.S. costs exceed all others by far." *Health Affairs*, 33: 1586–1594.

Hinze, Susan W. 2004. "'Am I being oversensitive?' Women's experience of sexual harassment during medical training." *Health*, 8: 101–127.

Hochschild, Arlie. 1983. *The Managed Heart: The Commercialization of Human Feelings*. Berkeley: University of California Press.

Hoff, Timothy. 2010. *Practice under Pressure: Primary Care Physicians and Their Medicine in Twenty-First Century America*. New Brunswick, NJ: Rutgers University Press.

Hoffman, Diane E., and Anita J. Tarzian. 2001. "The girl who cried pain: A bias against women in the treatment of pain." *Journal of Law, Medicine, and Ethics*, 29: 13–27.

Hogan, Margaret C., J. Foreman Kyle, Naghavi Mohsen, Stephanie Y. Ahn, Wang Mengru, Susanna M. Makela, Alan D. Lopez, Rafael Lozano, and Christopher J. L. Murray. 2010. "Maternal mortality for 181 countries, 1980–2008: A systematic analysis of progress towards millennium development goal 5." *The Lancet*, 375: 1609–1623.

Holt-Giménez, Eric, and Loren Peabody. 2008. "From food rebellions to food sovereignty: Urgent call to fix a broken food system." http://international.uiowa.edu/files/international.uiowa.edu/files/file_uploads/bgrspring2008-FoodRebellionstoFoodSovereignty.pdf, accessed June 2008.

Hoover, Eric. 2014. "Going Pro." *New York Times*, Education Life: August 3–6.

Horn, Joshua S. 1969. *"Away with all Pests . . .": An English Surgeon in People's China*. London: Paul Hamlyn.

Horne, Jed. 2006. *Breach of Faith: Hurricane Katrina and the Near Death of a Great American City*. New York: Random House.

Horwitz, Allan V. 1982. *Social Control of Mental Illness*. New York: Academic.

______. 2002. *Creating Mental Illness*. Chicago: University of Chicago Press.

______. 2007. *Loss of Sadness: How Psychiatry Transformed Normal Sorrow into Depressive Disorder*. New York: Oxford University Press.

House, James S. 2002. "Understanding social factors and inequalities in health: 20th century progress and 21st century prospects." *Journal of Health and Social Behavior*, 43: 125–142.

Human Rights Watch. 2005. *Blood, Sweat, and Fear: Workers' Rights in U.S. Meat and Poultry Plants*. New York: Author.

______. 2009. *Mental Illness, Human Rights, and U.S. Prisons*. http://www.hrw.org/en/news/2009/09/22/mental-illness-human-rights-and-usprisons, accessed April 2011.

Hunt, Charles W. 1989. "Migrant labor and sexually transmitted disease: AIDS in Africa." *Journal of Health and Social Behavior*, 30: 353–373.

______. 1996. "Social vs. biological: Theories on the transmission of AIDS in Africa." *Social Science & Medicine*, 42: 1283–1296.

Indianz.com. 2012. "Native Sun News: Study shows high Indian infant death rate." http://www.indianz.com/News/2012/004543.asp, accessed September 2014.

Institute of Medicine. 2002. *Care Without Coverage: Too Little Too Late*. Washington, DC: National Academy Press.

_____. 2014. *Relieving Pain in America: A Blueprint for Transforming Prevention, Care, Education, and Research*. Washington, DC: National Academies Press.

Intergovernmental Panel on Climate Change. 2007. *Climate Change 2007: The Physical Science Basis*. New York: Cambridge University Press.

Inungu, Joseph. 2010. "Democratic Republic of Congo." Pp. 287–300 in *Comparative Health Systems: Global Perspectives for the 21st Century*, edited by James A. Johnson and Carleen H. Stoskopf. Sudbury, MA: Jones & Bartlett.

Jacobs, Lawrence R., and Theda Skocpol. 2010. *Health Care Reform and American Politics: What Everyone Needs to Know*. New York: Oxford University Press.

Jacobson, Matthew Frye. 1998. *Whiteness of a Different Color: European Immigrants and the Alchemy of Race*. Cambridge, MA: Harvard University Press.

James, Doris J., and Lauren E. Glaze. 2006. *Mental Health Problems of Prison and Jail Inmates*. NCJ 213600. Washington, DC: US Department of Justice, Bureau of Justice Statistics.

James, John T. 2013. "New, evidence-based estimate of patient harms associated with hospital care." *Journal of Patient Safety*, 9: 122–128.

Johnson, Kenneth C., and Betty-Anne Daviss. 2005. "Outcomes of planned home births with certified professional midwives: Large prospective study in North America." *British Medical Journal*, 330: 1416.

Jones, James. 1993. *Bad Blood: The Tuskegee Syphilis Experiment*, rev. ed. New York: Free Press.

Judd, Deborah, Kathleen Sitzman, and G. Megan Davis. 2009. *A History of American Nursing: Trends and Eras*. Sudbury, MA: Jones & Bartlett.

Kaiser Commission on Medicaid and the Uninsured. 2010. *The Uninsured and the Difference Health Insurance Makes. Fact Sheet 1420–12*. Washington, DC: Kaiser Family Foundation.

_____. 2014. *Key Facts about the Uninsured Population*. Washington, DC: Kaiser Family Foundation.

Kaiser Family Foundation. 2011. *Summary of New Health Reform Law*. http://www.kff.org/healthreform/upload/8061.pdf, accessed January 2011.

Kaminer, Ariel. 2012. "On the defensive." *New York Times*, March 30: MM15.Kangas, Julie L., and James D. Calvert. 2014. "Ethical issues in mental health background checks for firearm ownership." Professional Psychology 45: 76–83.

Karnieli-Miller, Orit, and Zvi Eisikovits. 2009. "Physician as partner or salesman? Shared decision-making in realtime encounters." *Social Science & Medicine*, 69: 1–8.

Keen, John D. 2010. "Promoting screening mammography: Insight or uptake?" *Journal of American Board of Family Medicine*, 23: 775–782.

Kellerman, Arthur L., Frederick P. Rivara, Norman B. Rushforth, Joyce G. Banton, Donald T. Reay, Jerry T. Francisco, Ana B. Locci, Janice Prodzinski, Bela B. Hackman, and Grant Somes. 1993. "Gun ownership as a risk factor for homicide in the home." *New England Journal of Medicine*, 329(15): 1084–1091.

Kellogg, J. H. 1880. *Plain Facts for Young and Old*. Burlington, IA: Segner and Condit.

Kelly, John. 2005. *The Great Mortality: An Intimate History of the Black Death, the Most Devastating Plague of All Time*. New York: HarperCollins.

Kelly, Susan E., Patricia A. Marshall, Lee M. Sanders, Thomas A. Raffin, and Barbara A. Koenig. 1997. "Understanding the practice of ethics consultation: Results of an ethnographic multi-site study." *Journal of Clinical Ethics*, 8: 136–149.

Kessler, Ronald C., Patricia Berglund, Olga Demler, Robert Jim, and Ellen E. Walters. 2005a. "Lifetime prevalence and age-of-onset distributions of *DSM IV* disorders in the National Comorbidity Survey Replication." *Archives of General Psychiatry*, 62: 593–602.

Kessler, Ronald C., Olga Demler, Richard G. Frank, Mark Olfson, Harold Alan Pincus, Ellen E. Walters, Philip Wang, Kenneth B. Wells, and Alan M. Zashavsky. 2005b. "Prevalence and treatment of mental disorders, 1990–2003." *New England Journal of Medicine*, 352: 2515–2523.

Kessler, Ronald C., Katherine A. McGonagle, Shanyang Zhao, Christopher B. Nelson, Michael Hughes, Suzann Eshleman, Hans-Ulrich Wittchen, and Kenneth S. Kendler. 1994. "Lifetime and 12-month prevalence of *DSM-III-R* psychiatric disorders in the United States. Results from the National Comorbidity Survey." *Archives of General Psychiatry*, 51: 8–19.

Kessler, Ronald C., and Harold W. Neighbors. 1986. "A new perspective on the relationships among race, social class, and psychological distress." *Journal of Health and Social Behavior*, 27: 107–115.

Kessler, Suzanne J. 1998. *Lessons from the Intersexed*. New Brunswick, NJ: Rutgers University Press.

Kiesler, Charles A., and Amy E. Sibulkin. 1987. *Mental Hospitalization: Myths and Facts About a National Crisis*. Newbury Park, CA: Sage.

Kiple, Kenneth F. 1993. *Cambridge World History of Human Disease*. New York: Cambridge University Press.

Kirk, Stuart A. 1992. *The Selling of DSM: The Rhetoric of Science in Psychiatry*. New York: Aldine de Gruyter.

Kirsch, Irving. 2011. *The Emperor's New Drugs: Exploding the Antidepressant Myth*. Basic Books: New York.

Kirschen, Matthew P., Amy Tsou, Sarah B. Nelson, James Russell, and Daniel Larriviere. 2014. "Legal and ethical implications in the evaluation and management of sports-related concussion." *Neurology*, 83: 352–358.

Kitchener, Martin, and Charlene Harrington. 2004. "The U.S. long-term care field: A dialectic analysis of institution dynamics." *Journal of Health and Social Behavior*, 45: 87–101.

Kangas, Julie L.; Calvert, James D. 2014. "Ethical issues in mental health background checks for firearm ownership." *Professional Psychology*, 45: 76–83.

Kaplan, C. P., A. Napoles-Springer, S. L. Stewart, and E. J. Perez-Stable. 2001. "Smoking acquisition among adolescents and young Latinas: The role of socioenvironmental and personal factors." *Addictive Behavior*, 26: 531–550.

Knaul, Felicia Marie, Eduardo González-Pier, Octavio Gómez-Dantés, David García-Junco, Héctor Arreola-Ornelas, Mariana Barraza-Lloréns, Rosa Sandoval, Francisco Caballero, Mauricio Hernández-Avila, Mercedes Juan, David Kershenobich, Gustavo Nigenda, Enrique Ruelas, Jaime Sepúlveda, Roberto Tapia, Guillermo Soberón, Salomón Chertorivski, and Julio Frenk. 2012. "The quest for universal health coverage: Achieving social protection for all in Mexico." *Lancet*, 380: 1259–1279.

Kohler, Pamela K., Lisa E. Manhart, and William E. Lafferty. 2007. "Abstinence-only and comprehensive sex education and the initiation of sexual activity and teen pregnancy," *Journal of Adolescent Health*, 42: 344–351.

Kohn, Linda T., Janet M. Corrigan, and Molla S. Donaldson. 1999. *To Err is*

Human: Building a Safer Health System. Washington, DC: National Academy Press.

Kolata, Gina. 2004. "Stem cells: Promise, in search of results." *New York Times*, August 24: D1+.

______. 2005. "PSA test no longer gives clear answers." *New York Times*, June 20: E1+.

Kolata, Gina, and Andrew Pollack. 2008. Costly cancer drug offers hope, but also a dilemma. *New York Times*, July 6: A1+.

Kolko, Gabriel. 1999. "Ravaging the poor: The International Monetary Fund indicted by its own data." *International Journal of Health Services*, 29: 51–57.

Kozhimannil, Katy Backes, Michael R. Law, and Beth A. Virnig. 2013. "Cesarean delivery rates vary tenfold among us hospitals; reducing variation may address quality and cost issues." *Health Affairs*, 32: 527–535.

Kozol, Jonathan. 2005. *The Shame of the Nation: The Restoration of Apartheid Schooling in America.* New York: Crown.

Kramer, Peter. 1993. *Listening to Prozac.* New York: Viking.

Krieger, Nancy, David H. Rehkopf, Jarvis T. Chen, Pamela D. Waterman, Enrico Marcelli, and Malinda Kennedy. 2008. "The fall and rise of U.S. inequities in premature mortality: 1960–2002." *PLoS Medicine*, 5(2): e46.

Kristof, Nicholas D., and Sheryl WuDunn. 2010. *Half the Sky: Turning Oppression into Opportunity for Women Worldwide.* New York: Vintage.

Krug, E. G., K. E. Powell, and L. L. Dahlberg. 1998. "Firearm-related deaths in the United States and 35 other high- and upper-middle income countries." *International Journal of Epidemiology*, 27: 214–221.

Kutner, Nancy G. 1987. "Social worlds and identity in end-stage renal disease (ESRD)." *Research in the Sociology of Health Care*, 6: 33–71.

Kwate, Naa, Meghan Jernigan, and Tammy Lee. 2007. "Prevalence, proximity and predictors of alcohol ads in central Harlem." *Alcohol and Alcoholism*, 42: 635–640.

LaFraniere, Sharon. 2010. "Chinese hospitals are battlegrounds of discontent." *New York Times*, August11: A1.

Lahelma, Eero. 2010. "Health and social stratification." Pp. 71–96 in *New Blackwell Companion to Medical Sociology*, edited by William C. Cockerham. Chichester, UK: Wiley-Blackwell.

Laing, Ronald D. 1967. *The Politics of Experience.* New York: Ballantine.

Lakhani, Sarah Morando, and Stefan Timmermans. 2014. "Biopolitical citizenship in the immigration adjudication process." *Social Problems*, 61: 360–379.

Landecker, Hannah, and Aaron Panofsky. 2013. "From social structure to gene regulation, and back: A critical introduction to environmental epigenetics for sociology." *Annual Review of Sociology*, 39: 333–357.

Lane, Christopher. 2007. *Shyness: How Normal Behavior Became a Sickness.* New Haven: Yale University Press.

Lappé, Frances Moore, Joseph Collins, and Peter Rosset. 1998. *World Hunger: Twelve Myths*, 2nd ed. New York: Grove.

Lawn, Joy E., Simon Cousens, and Jelka Zupan. 2005. "4 million neonatal deaths: When? Where? Why?" *Lancet*, 365: 891–900.

Leape, Lucian L. 1994. "Error in medicine." *Journal of the American Medical Association*, 272: 1851–1857.

Leape, Lucian L., and Donald M. Berwick. 2005. "Five years after *To Err is*

Human: What have we learned?" *Journal of the American Medical Association*, 293: 2384–2390.

Leavitt, Judith Walzer. 1983. "Science enters the birthing room: Obstetrics in America since the eighteenth century." *Journal of American History*, 70: 281–304.

______. 1986. *Brought to Bed: Childbearing in America, 1750–1950*. New York: Oxford University Press.

Leavitt, Judith Walzer, and Ronald L. Numbers. 1985. *Sickness and Health in America*. Madison: University of Wisconsin Press.

Leith, Katherine H., Astrid Knott, Alexander Mayer, and Jorg Westermann. 2009. "Germany." Pp. 147–166 in *Comparative Health Systems: Global Perspectives for the 21st Century*, edited by James A. Johnson and Carleen H. Stoskopf. Sudbury, MA: Jones & Bartlett.

Lemmens, Trudo. 2004. "Piercing the veil of corporate secrecy about clinical trials." *Hastings Center Review*, 34: 14–18.

Lemmens, Trudo, and Benjamin Freedman. 2000. "Ethics review for sale? Conflict of interest and commercial research review boards." *Milbank Quarterly*, 78: 547–584.

Lennon, Mary Claire, and Laura Limonic. 2010. "Work and unemployment as stressors." Pp. 213–225 in *Handbook for the Sociology of Mental Health*, 2nd ed., edited by Teresa L. Scheid and Tony N. Brown. New York: Cambridge University Press.

Leo, Jonathan. 2004. "The biology of mental illness." *Society*, 41: 45–53.

Leonhardt, David. 2010. "Opposition to health law is steeped in tradition." *New York Times, December* 14: A1+.

Lewis, Caroline T. 1993. "Midwife-attended births." Pp. 247–250 in *Encyclopedia of Childbearing: Critical Perspectives*, edited by Barbara Katz Rothman. Phoenix, AZ: Oryx.

Lifton, Robert J. 1986. *The Nazi Doctors: Medical Killing and the Psychology of Genocide*. New York: Basic.

Light, Donald W. 2010. "Health care professions, markets, and countervailing powers." Pp. 270–249 in *Handbook of Medical Sociology*, edited by Chloe E. Bird, Peter Conrad, Allen M. Fremont, and Stefan Timmermans. Nashville, TN: Vanderbilt University Press.

Lingard, L., K. Garwood, C. F. Schryer, and M. M. Spafford. 2003. "A certain art of uncertainty: Case presentation and the development of professional identity." *Social Science & Medicine*, 56: 603–616.

Link, Bruce G., Mary Claire Lennon, and Bruce P. Dohrenwend. 1993. "Socioeconomic status and depression." *American Journal of Sociology*, 98: 1351–1387.

Link, Bruce G., and Jo C. Phelan. 2010. "Social conditions as fundamental causes of health inequalities." Pp. 3–17 in *Handbook of Medical Sociology*, edited by Chloe E. Bird, Peter Conrad, Allen M. Fremont, and Stefan Timmermans. Nashville, TN: Vanderbilt University Press.

Link, Bruce G., Jo C. Phelan, Michaeline Bresnahan, Ann Stueve, and Bernice A. Pescosolido. 1999. "Public conceptions of mental illness: Labels, causes, dangerousness, and social distance." *American Journal of Public Health*, 89: 1328–1333.

Link, Bruce G., Elmer L. Struening, Michael Rahav, Jo C. Phelan, and Larry Nuttbrock. 1997. "On stigma and its consequences: Evidence from a longitudinal study of men with dual diagnoses of mental illness and substance abuse." *Journal of Health and Social Behavior*, 38: 177–190.

Liska, Ken. 1997. *Drugs and the Human Body*. Upper Saddle River, NJ: Prentice Hall.

Loe, Meika. 2004. *The Rise of Viagra: How the Little Blue Pill Changed Sex in America*. New York: New York University Press.

LongtermCare.gov. 2014. "Costs of care." http://longtermcare.gov/costs-how-to-pay/costs-of-care/, accessed September 2014.

Lonsdale, Susan. 1990. *Women and Disability*. Basingstoke, UK: Macmillan.

Lopes, John E. Jr., M. Nicholas Coppola, and Lisa Riste. 2009. "United Kingdom." Pp. 83–110 in *Comparative Health Systems: Global Perspectives for the 21st Century*, edited by James A. Johnson and Carleen H. Stoskopf. Sudbury, MA: Jones & Bartlett.

Loviglio, Joann. 2005. "Alternative goes mainstream." *Washington Times*, June 10, http://washtimes.com/culture/20050609-114805-1956r.htm, accessed July 2005.

Ludmerer, Kenneth M. 1985. *Learning to Heal: The Development of American Medical Education*. New York: Basic.

Luhrmann, T. M. 2000. *Of Two Minds: The Growing Disorder in American Psychiatry*. New York: Knopf.

Luke, Douglas, Emily Esmundo, and Yael Bloom. 2000. "Smoke signs: Patterns of tobacco billboard advertising in a metropolitan region." *Tobacco Control*, 9: 16–23.

Luker, Kristin. 1984. *Abortion and the Politics of Motherhood*. Berkeley: University of California Press.

Lundberg, George D. 2001. *Severed Trust: Why American Medicine Hasn't Been Fixed*. New York: Basic.

Lupton, Deborah. 2013a. Quantifying the body: Monitoring, performing and configuring health in the age of mHealth technologies. *Critical Public Health*, 23(4), 393–403.

______. 2013b. "The digitally engaged patient: Self-monitoring and self-care in the digital health era." *Social Theory & Health*, 11: 256–270.

Lynch, Michael. 1983. "Accommodation practices: Vernacular treatments of madness." *Social Problems*, 31: 152–164.

MacDorman, Marian F., and Gopal K. Singh. 1999. "Midwifery care, social and medical risk factors and birth outcomes in the USA." *Journal of Epidemiology and Community Health*, May: 310–317.

Mah, Timothy, and Daniel Halperin. 2010. "Concurrent sexual partnerships and the HIV epidemics in Africa: Evidence to move forward." *AIDS and Behavior*, 14: 11–16.

Malka, Susan Gelland. 2007. *Daring to Care*. Urbana, IL: University of Illinois Press.

Markoff, John. 2002. "Technology's toxic trash is sent to poor nations." *New York Times*, February 25: C1+.

Marmot, Michael G. 2002. "The influence of income on health." *Health Affairs*, 21: 31–46.

______. 2004. *The Status Syndrome: How Your Social Standing Directly Affects Your Health and Life Expectancy*. London: Bloomsbury.

Marteau, Theresa, and Martin Richards (eds.). 1996. *The Troubled Helix. Social and Psychological Implications of the New Human Genetics*. Cambridge: Cambridge University Press.

Martin, Brook I., Anna N.A. Tosteson, Jon D. Lurie, and Sohail K. Mirza. 2014. *Variation in the Care of Surgical Conditions: Spinal Stenosis*. Dartmouth, NH: Dartmouth Institute of Health Policy & Clinical Practice.

Martin, Joyce A., Brady E. Hamilton, Paul D. Sutton, Stephanie J. Ventura, Fay Menacker, Sharon Kirmeyer, and Martha L. Munson. 2007. "Births:

Final data for 2005." *National Vital Statistics Reports,* 56*(6):* 1–104.

Martin, Joyce A., Michelle J. K. Osterman, and Paul D. Sutton. 2010. "Are preterm births on the decline in the United States? Recent data from the National Vital Statistics System." *NCHS Data Brief*, Number 39.

Mathews, T. J., and Marian F. MacDorman. 2010. "Infant mortality statistics from the 2006 period linked birth/infant death data set." *National Vital Statistics Reports*, 58(17).

McGinnis, J. Michael, Pamela Williams-Russo, and James R. Knickman. 2002. "The case for more active policy attention to health promotion." *Health Affairs*, 21: 78–93.

McGinty, Emma E., Shannon Frattaroli, Paul S. Appelbaum, Richard J. Bonnie, Anna Grilley, Joshua Horwitz, Jeffrey W. Swanson, and Daniel W. Webster. 2014. "Using research evidence to reframe the policy debate around mental illness and guns: Process and recommendations." *American Journal of Public Health*, 104: e22–e26.

McKeown, Thomas. 1979. *The Role of Medicine: Dream, Mirage, or Nemesis?* Princeton, NJ: Princeton University Press.

McKinlay, John B. 1994. "A case for refocusing upstream: The political economy of illness." Pp. 509–530 in *The Sociology of Health and Illness*, edited by Peter Conrad and Rachelle Kern. New York: St. Martin's Press.

McKinlay John B., and Lisa D. Marceau. 2002. "The end of the golden age of doctoring." *International Journal of Health Services*, 32: 379–416.

McKinlay, John B., and Sonja J. McKinlay. 1977. "The questionable effect of medical measures on the decline of mortality in the United States in the twentieth century." *Milbank Memorial Fund Quarterly*, 55: 405–428.

McNeil, Donald G. 2002. "With folk medicine on rise, health group is monitoring." *New York Times*, May 17: A9.

Mechanic, David. 1989. *Mental Health and Social Policy*, 3rd ed. Englewood Cliffs, NJ: Prentice Hall.

______. 1995. "Sociological dimensions of illness behavior." *Social Problems*, 41: 1207–1216.

______. 1997. "Managed mental health care." *Society*, 35(1): 44–52.

______. 1999. *Mental Health and Social Policy: The Emergence of Managed Care*, 4th ed. Boston: Allyn & Bacon.

______. 2004. "The rise and fall of managed care." *Journal of Health and Social Behavior*, 45: 76–86.

______. 2006. *The Truth about Health Care.* New Brunswick, NJ: Rutgers University Press.

Mechanic, David, and David A. Rochefort. 1990. "Deinstitutionalization: An appraisal of reform." *Annual Review of Sociology*, 16: 301–327.

Mendel, Peter, and W. Richard Scott. 2010. "The dynamics of institutional disarray: 'Muddling through' profound change in U.S. Health Care." Pp. 249–269 in *Handbook of Medical Sociology*, edited by Chloe E. Bird, Peter Conrad, Allen M. Fremont, and Stefan Timmermans. Nashville, TN: Vanderbilt University Press.

Messer, Ellen. 1997. "Intra-household allocation of food and health care: Current findings and understandings—introduction." *Social Science & Medicine*, 44: 1675–1684.

Meyer, Pamela A., Timothy Pivetz, Timothy A. Dignam, David M. Homa, Jaime Schoonover, and Debra Brody. 2003. "Surveillance for elevated blood lead levels among children—United States, 1997–2001."

Morbidity and Mortality Weekly Report, 52(No. SS-10): 1–21.

Meyerowitz, Beth E., Janice G. Williams, and Jocelyne Gessner. 1987. "Perceptions of controllability and attitudes toward cancer and cancer patients." *Journal of Applied Social Psychology*, 17: 471–492.

Michon, Heather K. 2008. "Congo, Democratic Republic of the." *Encyclopedia of Global Health*. http://www.sage-reference.com/globalhealth/Article_n307.html, accessed October 2010.

Midwives Association of North America. 2013. "Big Push Campaign." http://www.mana.org/healthcare-policy/big-push-campaign, accessed July 2014.

Miller, Katherine. 1998. "The evolution of professional identity: The case of osteopathic medicine." *Social Science & Medicine*, 47: 1739–1748.

Miller, Robert H., and Harold T. Luft. 1997. "Does managed care lead to better or worse quality of care?" *Health Affairs*, 16(5): 7–25.

Miller, Thomas P. 2010. "Health reform: Only a cease-fire in a political hundred years' war." *Health Affairs*, 29: 1101–1105.

Mills, C. Wright. 1959. *The Sociological Imagination*. New York: Grove.

Mirowsky, John, and Catherine E. Ross. 1989. "Psychiatric diagnosis as reified measurement." *Journal of Health and Social Behavior*, 30: 11–25.

Mishler, Elliot G. 1990. "The struggle between the voice of medicine and the voice of the lifeworld." Pp. 295–307 in *The Sociology of Health and Illness: Critical Perspectives*, 3rd edition, edited by Peter Conrad and Rochelle Kern. New York: St. Martin's Press.

Mokdad, Ali H., James S. Marks, Donna F. Stroup, and Julie L. Gerberding. 2004. "Actual causes of death in the United States, 2000." *Journal of the American Medical Association*, 291: 1238–1245.

Molinari, Victor, David Chiriboga, Laurence G. Branch, Soyeon Cho, Kristen Turner, Jing Guo, and Kathryn Hyer. 2010. "Provision of psychopharmacological services in nursing homes." *The Journals of Gerontology Series B: Psychological Sciences and Social Sciences*, 65B: 57–60.

Moloney, Mairead Eastin, Thomas R. Konrad, and Catherine R. Zimmer. 2011. "The medicalization of sleeplessness: A public health concern." *American Journal of Public Health*, 101: 1429–1433.

Mootz, Robert D., Daniel C. Cherkin, Carson E. Odegard, David M. Eisenberg, James P. Barassi, and Richard A. Deyo. 2005. "Characteristics of chiropractic practitioners, patients, and encounters in Massachusetts and Arizona." *Journal of Manipulative and Physiological Therapeutics*, 28: 645–653.

Morbidity and Mortality Weekly Report. 2010. "Vital signs: Health insurance coverage and health care utilization—United States, 2006–2009 and January–March 2010: Early Release." http://www.cdc.gov/mmwr/preview/mmwrhtml/mm59e1109a1.htm?s_cid=mm59e1109a1_w, accessed November 2010.

Moses, Tally. 2009. "Self-labeling and its effects among adolescents diagnosed with mental disorders." *Social Science & Medicine*, 68: 570–578.

Mosher, James F. 1995. "The merchants, not the customers: Resisting the alcohol and tobacco industries' strategy to blame young people for illegal alcohol and tobacco sales." *Journal of Public Health Policy*, 16: 412–432.

Muennig, Peter A., and Sherry A. Glied. 2010. "What changes in survival rates tell us about U.S. health care." *Health Affairs*, 29: 2105–2113.

Mullan, Fitzhugh. 2002. *Big Doctoring in America: Profiles in Primary Care*. Berkeley: University of California Press.

Mundinger, Mary O., Robert L. Kane, Elizabeth R. Lenz, Annette M. Totten, Wei-Yann Tsai, Paul D. Cleary, William T. Friedewald, Albert L. Siu, and Michael L. Shelanski. 2000. "Primary care outcomes in patients treated by nurse practitioners or physicians: A randomized trial." *Journal of the American Medical Association*, 283: 59–68.

Muntaner, Carles, William W. Eaton, Richard Miech, and Patricia O'Campo. 2004. "Socioeconomic position and major mental disorders." *Epidemiologic Reviews*, 26: 53–62.

National Alliance for Caregiving and AARP.2009. *Caregiving in the U.S.* Bethesda, MD: National Alliance for Caregiving.

National Cancer Institute. 2010. *Laetrile/Amygdalin (PDQ®)*. http://www.cancer.gov/cancertopics/pdq/cam/laetrile/patient/allpages-Section_20/, accessed November 2010.

National Center for Health Statistics. 2014a. *Health, United States, 2013*. Hyattsville, MD: US Public Health Service.

______. 2014b. *Leading Causes of Death, 1900–2011*. http://www.cdc.gov/nchs/data/dvs/LCWK9_2011.pdf, accessed October 2014.

National Center for Injury Prevention and Control. 2009. *Understanding Intimate Partner Violence*. http://www.cdc.gov/violenceprevention/pdf/IPV_factsheet-a.pdf, accessed October 2009.

National Hospice and Palliative Care Organization. 2014. *NHPC's Facts and Figures: Hospice Care in America*. Alexandria, VA: Author.

Navarro, Mireya. 2004. "When gender isn't a given." *New York Times*, September 19: I1+.

Naylor, C. David. 1995. "Grey zones of clinical practice: Some limits to evidence-based medicine." *Lancet*, 345: 840–842.

Nelson, Alan R., Brian D. Smedley, and Adrienne Y. Stith. 2002. *Unequal Treatment: Confronting Racial and Ethnic Disparities in Health Care*. Washington, DC: Institute of Medicine, National Academy Press.

Nestlé, Marion. 2002. *Food Politics: How the Food Industry Influences Nutrition and Health*. Berkeley: University of California Press.

Neubauer, Deane. 1997. "Hawaii: The health state revisited." Pp. 163–188 in *Health Policy Reform in America: Innovations from the States*, edited by Howard M. Leichter. Armonk, NY: M. E. Sharpe.

New York Times. 1999. "Registered nurses in short supply at hospitals nationwide." March 23: A14.

______. 2010. *Haiti Earthquake of 2010*. http://www.nytimes.com/info/haiti-earthquake-2010, accessed November 2010.

Norrish, Barbara R., and Thomas G. Rundall. 2001. "Hospital restructuring and the work of registered nurses." *Milbank Quarterly*, 79: 55–79.

North, Fiona M., S. Leonard Syme, Amanda Feeney, Martin Shipley, and Michael Marmot. 1996. "Psychosocial work environment and sickness absence among British civil servants: The Whitehall II Study." *American Journal of Public Health*, 86: 332–340.

Nossiter, Adam. 2008. "Rural Alabama County cracks down on pregnant drug users." *New York Times*, March 15: 9.

______. 2014. "Ebola is taking a second toll, on economies." *New York Times*, September 5: 1+.

Oberlander, Jonathan. 2010. "Long time coming: Why health reform finally passed." *Health Affairs*, 29: 1112–1116.

OECD. 2014. *OECD Health Statistics*. http://www.oecd.org/els/health-systems/health-data.htm, accessed November 2014.

Office of Minority Health. 2014. "American Indian/Alaska native profile." http://minorityhealth.hhs.gov/templates/browse.aspx?lvl=2&lvlID=52, accessed December 2014.

Offit, Paul A., and Sarah Erush. 2013. "Skip the Supplements." *New York Times*, December 14: SR7.

Ofri, Danielle. 2013. *What Doctors Feel*. Boston: Beacon.

Oldstone, Michael B. A. 2010. *Viruses, Plagues and History: Past, Present and Future*. New York: Oxford University Press.

Olivier, Degomme, and Guha-Sapir Debarati. 2010. "Patterns of mortality rates in Darfur conflict." *Lancet*, 375: 294–300.

Omer, Saad B., Daniel A. Salmon, Walter A. Orenstein, M. Patricia deHart, and Neal Halsey. 2009. "Vaccine refusal, mandatory immunization, and the risks of vaccine-preventable diseases." *New England Journal of Medicine*, 360: 1981–1988.

Omran, Abdel R. 1971. "The epidemiological transition." *Milbank Memorial Fund Quarterly*, 49: 509–538.

Orlander, Jay D., and Graeme Fincke. 2003. "Morbidity and mortality conferences: A survey of academic internal medicine departments." *Journal of General Internal Medicine*, 18: 656–661.

Orr, Robert D., and Eliot Moon. 1993. "Effectiveness of an ethics consultation service." *Journal of Family Practice*, 36: 49–53.

Oswald, Andrew J., and Nattavudh Powdthavee. 2008. "Does happiness adapt? A longitudinal study of disability with implications for economists and judges." *Journal of Public Economics*, 92: 1061–1077.

Oudshoorn, Nelly. 2011. *Telecare Technologies and the Transformation of Healthcare*. Houndmills, UK: Palgrave Macmillan.

Palloni, Alberto, and Elizabeth Arias. 2004. "Paradox lost: Explaining the Hispanic adult mortality advantage." *Demography*, 41: 385–415.

Pan American Health Organization. 2012. "Mexico." http://www.paho.org/saludenlasamericas/index.php?option=com_docman&task=doc_view&gid=137&Itemid=, accessed November 2014.

Parsons, Talcott. 1951. *The Social System*. New York: Free Press.

Payer, Lynn. 1996. *Medicine and Culture*, rev. ed. New York: Holt.

Peabody, John W. 1996. "Economic reform and health sector policy: Lessons from structural adjustment programs." *Social Science & Medicine*, 43: 823–835.

Pear, Robert. 2008. "Violations reported at 94% of nursing homes." *New York Times*, September 29: A20.

______. 2010. "Pay practices in health care are investigated." *New York Times*, August 9: A15.

Peregoy, Jennifer A., Tainya C. Clarke, Lindsey I. Jones, Barbara J. Stussman, and Richard L. Nahin. 2014. "Regional variation in use of complementary health approaches by U.S. adults." *NCHS data brief*, no 146. Hyattsville, MD: National Center for Health Statistics.

Perrone, Bobette, H. Henrietta Stockel, and Victoria Krueger. 1989. *Medicine Women, Curanderas, and Women Doctors*. Norman: University of Oklahoma Press.

Persson, Diane I., and Sharon K. Ostwald. 2009. "Younger residents in nursing homes." *Journal of Gerontological Nursing*, 35: 22–31.

Pescosolido, Bernice A. 1992. "Beyond rational choice: The social dynamics of how people seek help." *American Journal of Sociology*, 97: 1096–1138.

Pescosolido, Bernice A. 2013. "The public stigma of mental illness: What do we think; what do we know; what can we prove?" *Journal of Health and Social Behavior*, 54: 1–21.

Pescosolido, Bernice A., Carol Brooks Gardner, and Keri M. Lubell. 1998. "How people get into mental health services: Stories of choice, coercion and 'muddling through' from 'first-timers.'" *Social Science & Medicine*, 46: 275–286.

Phelan, Jo C., Bruce G. Link, Ana Diez-Roux, Ichiro Kawachi, and Bruce Levin. 2004. "'Fundamental causes' of social inequality in mortality: A test of the theory." *Journal of Health and Social Behavior*, 45: 265–285.

Phillips, Marilynn J. 1990. "Damaged goods: Oral narratives of the experience of disability in American culture." *Social Science & Medicine*, 30: 849–857.

Pierluissi, Edgar, Melissa A. Fischer, Andre R. Campbell, and C. Seth Landefeld. 2003. "Discussion of medical errors in morbidity and mortality conferences." *Journal of the American Medical Association*, 290: 2838–2842.

Pierret, Janine. 2003. "The illness experience: state of knowledge and perspectives for research." *Sociology of Health & Illness*, 25: 4–22.

Plotnick, Robert D. 1992. "The effects of attitudes on teenage premarital pregnancy and its resolution." *American Sociological Review*, 57: 800–811.

Polakovic, Gary. 2002. "Asia's wind-born pollution a hazardous export to U.S." *Los Angeles Times*, April 26: A1+.

Pollack, Andrew. 2005. "Marketing a disease, and also a drug to treat it." *New York Times*, May 9: C1+.

Population Reference Bureau. 2014. *Datafinder*. http://www.prb.org/DataFinder.aspx, accessed November 2014.

Powdthavee, Nattavudh. 2009. "What happens to people before and after disability? Focusing effects, lead effects, and adaptation in different areas of life." *Social Science & Medicine*, 69: 1834–1844.

Prasad, Vinay, Andrae Vandross, Caitlin Toomey, Michael Cheung, Jason Rho, Steven Quinn, Satish Jacob Chacko, Durga Borkar, Victor Gall, Senthil Selvaraj, Nancy Ho, and Adam Cifu. 2013. "A decade of reversal: An analysis of 146 contradicted medical practices." *Mayo Clinic Proceedings*, 88: 790–798.

Preves, Sharon E. 2003. *Intersex and Identity: The Contested Self*. New Brunswick, NJ: Rutgers University Press.

Quadagno, Jill. 2005. *One Nation Uninsured: Why the U.S. Has No National Health Insurance*. New York: Oxford University Press.

Quammen, David. 2013. *Spillover: Animal Infections and the Next Human Pandemic*. New York: Norton.

Quinn, Sandra Crouse. 2004. "Protecting human subjects: The role of community action boards." *American Journal of Public Health*, 94: 918–922.

Rao, Birju, and Ida Hellander. 2014. "The Widening U.S. Health Care Crisis Three Years After the Passage of 'Obamacare.'" *International Journal of Health Services*, 44: 215–232.

Read, Jen'nan Ghazal, and Bridget K. Gorman. 2010. "Gender and health inequality." *Annual Review of Sociology*, 36: 371–386.

Reading, Richard. 1997. "Social disadvantage and infection in childhood." *Sociology of Health and Illness*, 19: 395–414.

Redlich, Fredrick C. 1978. "Medical ethics under National Socialism." Pp. 1015–1019 in *Encyclopedia of Bioethics*, edited by Warren T. Reich. New York: Free Press.

Reid, T. C. 2009. *The Healing of America: A Global Quest for Better, Cheaper, and Fairer Health Care*. New York: Penguin.

Reinhard, Susan, and Allan V. Horwitz. 1996. "Caregiver burden: Differentiating the content and consequences of family caregiving." *Journal of Marriage and the Family*, 57: 741–750.

Reinhardt, Uwe E., Peter S. Hussey, and Gerard F. Anderson. 2004. "U.S. health care spending in an international context." *Health Affairs*, 23: 10–25.

Reverby, Susan. 1987. *Ordered to Care: The Dilemma of American Nursing*. New York: Cambridge University Press.

Revkin, Andrew C. 2005. "In disaster's wake: The future of calamity." *New York Times*, January 2: D4.

Richards, Peter. 1977. *The Medieval Leper and His Northern Heirs*. Totowa, NJ: Rowman & Littlefield.

Rieker, Patricia P., Chloe E. Bird, and Martha E. Lang. 2010. "Understanding gender and health: Old patterns, new trends, and future directions." Pp. 52–74 in *Handbook of Medical Sociology*, edited by Chloe E. Bird, Peter Conrad, Allen M. Fremont, and Stefan Timmermans. Nashville, TN: Vanderbilt University Press.

Riley, James C. 2007. *Low Income, Social Growth, and Good Health: A History of Twelve Countries*. Berkeley: University of California Press.

Riska, Elianne. 2010. "Health professions and occupations." Pp. 337–354 in *New Blackwell Companion to Medical Sociology*, edited by William C. Cockerham. Chichester, UK: Wiley-Blackwell.

Risse, Guenter B. 1988. "Epidemics and history: Ecological perspectives and social responses." Pp. 33–66 in *AIDS: The Burdens of History*, edited by Elizabeth Fee and Daniel M. Fox. Berkeley: University of California Press.

Roberts, Les, Riyadh Lafta, Richard Garfield, Jamal Khudhairi, and Gilbert Burnham. 2004. "Mortality before and after the 2003 invasion of Iraq: Cluster sample survey." *Lancet*, 364: 1857–1864.

Roberts, Nicholas J., Joshua T. Vogelstein, Giovanni Parmigiani, Kenneth W. Kinzler, Bert Vogelstein, and Victor E. Velculescu. 2012. "The Predictive Capacity of Personal Genome Sequencing." *Science Translational Medicine*. http://stm.sciencemag.org/content/early/2012/04/02/scitranslmed.3003380, accessed June 2014.

Rodriguez, Jason. 2014. *Labors of Love: Nursing Homes and the Structures of Care Work*. New York: New York University Press.

Roeder, Beatrice A. 1988. *Chicano Folk Medicine from Los Angeles, California*, Vol. 34 in Folklore and Mythology Series. Los Angeles: University of California Press.

Rohter, Larry. 2004. "Tracking the sale of a kidney on a path of poverty and hope." *New York Times*, May 23: A1+.

Rooks, Judith. 1997. *Midwifery and Childbirth in America*. Philadelphia: Temple University Press.

Rosenberg, Charles E. 1987. *The Care of Strangers: The Rise of America's Hospital System*. New York: Basic.

Rosenfeld, Dana, and Christopher A. Faircloth (eds.). 2005. *Medicalized Masculinities*. Philadelphia: Temple University Press.

Rosenfield, Sarah. 1997. "Labeling mental illness: The effects of received services and perceived stigma on life satisfaction." *American Sociological Review*, 62: 660–672.

Rosenhan, David L. 1973. "On being sane in insane places." *Science*, 179: 250–258.

Rosenstein, Alan H. 2002. "Nurse-physician relationships: Impact on nurse satisfaction and retention." *American Journal of Nursing*, 102: 26–34.

Rosenstock, Irwin M. 1966. "Why people use health services." *Milbank Memorial Fund Quarterly*, 44: 94–127.

Rosenthal, Elizabeth. 2014a. "As health care shifts, U.S. doctors switch to salaried jobs. *New York Times*, February 13: A1+.

_____. 2014b. "After Surgery, Surprise $117,000 Medical Bill From Doctor He Didn't Know." *New York Times*, September 21: A1+.

Roth, Rachel. 2003. *Making Women Pay: The Hidden Costs of Fetal Rights*. Ithaca, NY: Cornell University Press.

Rothman, Barbara Katz. 1986. *The Tentative Pregnancy: Prenatal Diagnosis and the Future of Motherhood*. New York: Penguin.

Rothman, David J. 1971. *The Discovery of the Asylum*. Boston: Little, Brown.

______. 1991. *Strangers at the Bedside: A History of How Law and Bioethics Transformed Medical Decision-Making*. New York: Basic.

______. 1997. *Beginnings Count: The Technological Imperative in American Health Care*. New York: Oxford University Press.

Rothman, Sheila M., and David J. Rothman. 2003. *The Pursuit of Perfection*. New York: Pantheon.

Roush, Sandra W., Trudy V. Murphy, and the Vaccine-Preventable Disease Table Working Group. 2007. "Historical comparisons of morbidity and mortality for vaccine-preventable diseases in the United States." *Journal of the American Medical Association*, 298: 2155–2163.

Ruggieri, Paul A. 2014. *The Cost of Cutting: A Surgeon Reveals the Truth Behind a Multibillion-Dollar Industry*. New York: Berkley.

Ryan, Ellen B., Selina Bajorek, Amanda Beaman, and Ann P. Anas. 2005. "'I just want you to know that "them" is me': Intergroup perspectives on communication and disability." Pp. 117–137 in *Intergroup communication: Multiple perspectives*, edited by Jake Harwood and Howard Giles. New York: Peter Lang Publishing Group.

Ryan, William. 1976. *Blaming the Victim*, rev. ed. New York: Pantheon.

Sack, Kevin. 2008a. "Doctors say 'I'm sorry' before 'See you in court.'" *New York Times*, May 18: A1+.

______. 2008b. "Illegal farm workers resort to health care in the shadows." *New York Times*, May 10: 1+.

Sacks, Peter. 2007. *Tearing Down the Gates: Confronting the Class Divide in American Education*. Berkeley: University of California Press.

Saha, Somnath, Darren D. Coffman, and Ariel K. Smits. 2010. "Giving teeth to comparative-effectiveness research—the Oregon experience." *New England Journal of Medicine*, 362(7): e18.

Sakr, M., J. Angus, J. Perrin, C. Nixon, J. Nicholl, and J. Wardrope. 1999. "Care of minor injuries by emergency NPs or junior MDs: A randomized controlled trial." *Lancet*, 354: 1321–1326.

Sandall, Jane, Hora Soltani, Simon Gates, Andrew Shennan, and Declan Devane. 2013. "Midwife-led continuity models versus other models of care for childbearing women." *Cochrane Database of Systematic Reviews*, Issue 8. http://onlinelibrary.wiley.com/doi/10.1002/14651858.CD004667.pub3/abstract, accessed July 2014.

Sandhaus, Sonia. 1998. "Migrant health: A harvest of poverty." *American Journal of Nursing*, 98: 52–53.

Sapolsky, Robert M. 2004. "Social status and health in humans and other animals." *Annual Review of Anthropology*, 33: 393–418.

Saul, Stephanie. 2009. "21st century babies: Grievous choice on risky path to parenthood." *New York Times*, October 12: A1+.

Scheff, Thomas J. 1984. *Being Mentally Ill: A Sociological Theory*, rev. ed. Chicago: Aldine.

Scheid, Teresa L. 2001. "Rethinking professional prerogative: Managed mental health care providers." Pp. 153–171 in *Rethinking the Sociology of Mental Health*, edited by Joan Busfield. Oxford, UK: Blackwell.

Schmidt, Charles W. 2000. "Bordering on environmental disaster." *Environmental Health Perspectives*, 108: A308–A313.

Schneider, Andrew, and David McCumber. 2004. *An Air That Kills: How the Asbestos Poisoning of Libby, Montana, Uncovered a National Scandal.* New York: Putnam's Sons.

Schneirov, Matthew, and Jonathan David Geczik. 1996. "A diagnosis for our times: Alternative health's submerged networks and the transformation of identities." *Sociological Quarterly*, 37: 627–644.

Schnittker, Jason. 2007. "Working more and feeling better: Women's health, employment, and family life, 1974–2004." *American Sociological Review*, 72: 221–238.

Schnittker, Jason, and Jane D. McLeod. 2005. "The social psychology of health disparities." *Annual Review of Sociology*, 31: 75–103.

Schoen, Cathy, Sara R. Collins, Jennifer L. Kriss, and Michelle M. Doty. 2008. "How many are underinsured? Trends among U.S. adults, 2003 and 2007." *Health Affairs* Web Exclusive, June 10, 2008. http://content.healthaffairs.org/cgi/reprint/hlthaff.27.4.w298v1?ijkey=rhRn2Tr4HAKZ.&keytype=ref&siteid=healthaff, accessed June 2008.

Schoen, Cathy, Robin Osborn, David Squires, Michelle M. Doty, Roz Pierson, and Sandra Applebaum. 2010. "How health insurance design affects access to care and costs, by income, in eleven countries." *Health Affairs*, 29: 2323–2334.

Schroeder, Steven A. 2007. "We can do better: Improving the health of the American people." *New England Journal of Medicine*, 357: 1221–1228.

Schulein, T. M. 2004. "A chronology of dental education in the United States." *Journal of the History of Dentistry*, 52: 97–108.

Schwartz, Sharon, and Ilan H. Meyer. 2010. "Mental health disparities research: The impact of within and between group analyses on tests of social stress hypotheses." *Social Science & Medicine*, 70: 1111–1118.

Schwarz, Alan. 2014. "Thousands of toddlers are medicated for A.D.H.D., report finds, raising worries," *New York Times*, May 16: A1+.

Schwarz, Alan, and Sarah Cohen. 2013. "A.D.H.D. seen in 11% of U.S. children as diagnoses rise," *New York Times*, March 31: A1+.

Scott, S., L. Prior, F. Wood, and J. Gray. 2005. "Repositioning the patient: The implications of being 'at risk.'" *Social Science & Medicine*, 60: 1869–1879.

Scull, Andrew. 1977. *Decarceration, Community Treatment and the Deviant: A Radical View.* Englewood Cliffs, NJ: Prentice Hall.

______. 1989. *Social Order/Mental Disorder: Anglo-American Psychiatry in Historical Perspective.* Berkeley: University of California Press.

Seale, Clive. 2010. "Death, dying, and the right to die." Pp. 210–228 in *Handbook of Medical Sociology*, edited by Chloe E. Bird, Peter Conrad, Allen M. Fremont, and Stefan Timmermans. Nashville, TN: Vanderbilt University Press.

Sedgh, Gilda, Stanley Henshaw, Susheela Singh, Elisabeth Åhman, and Iqbal H. Shah. 2007. "Induced abortion: Estimated rates and trends worldwide." *The Lancet*, 370: 1338–1345.

Sen, Amartya. 1999. *Development as Freedom*. New York: Knopf.

Sered, Susan Starr, and Rushika Fernandopulle. 2005. *Uninsured in America: Life and Death in the Land of Opportunity*. Berkeley: University of California Press.

Shabecoff, Philip, and Alice Shabecoff. 2010. *Poisoned for Profit: How Toxins Are Making Our Children Chronically Ill*. White River Junction, VT: Chelsea Green Publishing.

Shah, Sonia. 2010. *The Fever: How Malaria Has Ruled Humankind for 500,000 Years*. New York: Farrar, Straus, and Giroux.

Shanahan, Michael J., Shawn Bauldry, and Jason Freeman. 2010. "Beyond Mendel's ghost." *Contexts*, 9(4): 34–39.

Shapiro, Joseph P. 1993. *No Pity: People with Disabilities Forging a New Civil Rights Movement*. New York: Random House.

Shilling, Chris, and Phillip A. Mellor. 2001. *The Sociological Ambition: Elementary Forms of Social and Moral Life*. London: Sage.

Shim, Janet K. 2010. "Cultural health capital." *Journal of Health and Social Behavior*, 51(1): 1–15.

Shostak, Sara, Peter Conrad, and Allan V. Horwitz. 2008. "Sequencing and its consequences: Path dependence and the relationships between genetics and medicalization." *American Journal of Sociology*, 114: S287–S316.

Siegrist, Johannes. 2010. "Stress in the workplace." Pp. 268–288 in *New Blackwell Companion to Medical Sociology*, edited by William C. Cockerham. Chichester, UK: Wiley-Blackwell.

Sigsbee, Bruce. 1997. "Medicare's resource-based relative value scale, a *de facto* national fee schedule: Its implications and uses for neurologists." *Neurology*, 49: 315–320.

Silverman, B. L., B. E. Metzger, N. H. Cho, and C. A. Loeb. 1995. "Impaired glucose tolerance in adolescent offspring of diabetic mothers." *Diabetes Care*, 18: 611–617.

Simmons, Janie, Paul Farmer, and Brooke G. Schoepf. 1996. "A global perspective." Pp. 39–90 in *Women, Poverty, and AIDS: Sex, Drugs, and Structural Violence*, edited by Paul Farmer, Margaret Connors, and Janie Simmons. Monroe, ME: Common Courage.

Simonds, Wendy, Barbara Katz Rothman, and Bari Meltzer Norman. 2007. *Laboring On: Birth in Transition in the United States*. London: Routledge.

Singer, Lynn T., Robert Arendt, Sonia Minnes, Kathleen Farkas, Ann Salvator, H. Lester Kirchner, and Robert Kliegman. 2002. "Cognitive and motor outcomes of cocaineexposed infants." *Journal of the American Medical Association*, 287: 1952–1960.

Skocpol, Theda. 1996. *Boomerang: Clinton's Health Security Effort and the Turn Against Government in U.S. Politics*. New York: Norton.

Smith, David Barton, Zhanlian Feng, Mary L. Fennell, Jacqueline S. Zinn, and Vincent Mor. 2007. "Separate and unequal: Racial segregation and disparities in quality across U.S. nursing homes." *Health Affairs*, 26: 1448–1458.

Smith, David P., and Benjamin Bradshaw. 2006. "Rethinking the Hispanic paradox: Death rates and life expectancy for U.S. Non-Hispanic white and Hispanic populations." *American Journal of Public Health*, 96: 1686–1692.

Smith, Kirsten P., and Nicholas A. Christakis. 2008. "Social networks and health." *Annual Review of Sociology*, 34: 405–429.

Smyke, Patricia. 1991. *Women & Health*. London: Zed.

Snyder, Karrie Ann, and Adam Isaiah Green. 2008. "Revisiting the glass escalator: The case of gender segregation in a female dominated occupation." *Social Problems*, 55: 271–299.

Song, Lijun. 2011. "Social capital and psychological distress." *Journal of Health and Social Behavior*, 52: 478–492.

Song, Lijun, Joonmo Son, and Nan Lin. 2010. "Social Capital and Health." Pp. 184–210 in *New Blackwell Companion to Medical Sociology*, edited by William C. Cockerham. Chichester, UK: Wiley-Blackwell.

Spitzer, Robert L., Janet B. W. Williams, and Andrew E. Skodol. 1980. "*DSM-III*: The major achievements and an overview." *American Journal of Psychiatry*, 137: 151–164.

Starr, Paul. 1982. *The Social Transformation of American Medicine*. New York: Basic.

______. 1994. *The Logic of Health Care Reform: Why and How the President's Plan Will Work*. New York: Penguin.

Steinberg, Jacques. 1999. "Expanded school drug tests face a challenge." *New York Times*, August 18: A14.

Steingart, Richard M. 1991. "Sex differences in the management of coronary artery disease." *New England Journal of Medicine*, 325: 226–230.

Steinhauer, Jennifer. 2008. "Public health risk seen as parents reject vaccines." *New York Times*, March 21: 1A+.

Stevens, Fred. 2010. "The convergence and divergence of modern health care systems." Pp. 434–454 in *New Blackwell Companion to Medical Sociology*, edited by William C. Cockerham. Chichester, UK: Wiley-Blackwell.

Stevens, Rosemary. 1989. *In Sickness and in Wealth: American Hospitals in the Twentieth Century*. New York: Basic.

Stine, Gerald J. 2005. *AIDS Update 2005*. San Francisco: Benjamin Cummings.

Stockl, Andrea. 2007. "Complex syndromes, ambivalent diagnosis, and existential uncertainty: The case of systemic lupus erythematosus (SLE)." *Social Science & Medicine*, 65: 1549–1559.

Stratton, Kathleen, Christopher B. Wilson, and Marie C. McCormick (eds.). 2002. *Immunization Safety Review: Multiple Immunizations and Immune Dysfunction*. Washington, DC: National Academy Press.

Straus, Robert. 1957. "The nature and status of medical sociology." *American Sociological Review*, 22: 200–204.

Street, Richard. 1991. "Information-giving in medical consultations: The influence of patients' communicative styles and personal characteristics." *Social Science & Medicine*, 32: 541–548.

Sullivan, Deborah A. 2001. *Cosmetic Surgery: The Cutting Edge of Commercial Medicine in America*. Brunswick, NJ: Rutgers University Press.

Sullivan, Deborah A., and Rose Weitz. 1988. *Labor Pains: Modern Midwives and Home Birth*. New Haven, CT: Yale University Press.

Sutton, John R. 1991. "The political economy of madness: The expansion of the asylum in progressive America." *American Sociological Review*, 56: 665–678.

Swedish Institute. 1997. *Social Insurance in Sweden*. Stockholm: Author. http://www.si.se, accessed 2007.

______. 1999. *The Care of the Elderly in Sweden*. Stockholm: Author. http://www.si.se, accessed 2007.

Swoboda, Debra A. 2008. "Negotiating the diagnostic uncertainty of contested illnesses: Physician practices and paradigms." *Health*, 12: 453–478.

Tanielian, Terri, and Lisa H. Jaycox (eds.). 2008. *Invisible Wounds of War: Psychological and Cognitive Injuries, Their Consequences, and Services to Assist Recovery*. Santa Monica, CA: Rand Corporation.

Tate, Susan, and David Goldstein. 2004. "Will tomorrow's medicines work for everyone?" *Nature Genetics*, 36: S34–S42.

Taylor, Donald H. Jr., Jan Ostermann, Courtney H. Van Houtven, James A. Tulsky, and Karen Steinhauser. 2007. "What length of hospice use maximizes reduction in medical expenditures near death in the U.S. Medicare program?" *Social Science & Medicine*, 65: 1466–1478.

Taylor, Dorceta E. 2014. *Toxic Communities: Environmental Racism, Industrial Pollution, and Residential Mobility*. New York: New York University Press.

Teal, Cayla R., and Richard L. Street. 2009. "Critical elements of culturally competent communication in the medical encounter: A review and model." *Social Science & Medicine*, 68: 533–543.

Temel, Jennifer S., Joseph A. Greer, Alona Muzikansky, Emily R. Gallagher, Sonal Admane, Vicki A. Jackson, Constance M. Dahlin, Craig D. Blinderman, Juliet Jacobsen, William F. Pirl, J. Andrew Billings, and Thomas J. Lynch. 2010. "Early palliative care for patients with metastatic non-small-cell lung cancer." *New England Journal of Medicine*, 363: 733–742.

Tesh, Sylvia. 1988. *Hidden Arguments: Political Ideology and Disease Prevention Policy*. New Brunswick, NJ: Rutgers University Press.

Tessler, Richard, and Gail Gamache. 1994. "Continuity of care, residence, and family burden in Ohio." *Milbank Quarterly*, 72: 149–169.

Thernstrom, Melanie. 2010. *The Pain Chronicles: Cures, Myths, Mysteries, Prayers, Diaries, Brain Scans, Healing, and the Science of Suffering*. New York: Farrar, Straus and Giroux.

Thoits, Peggy A. 1985. "Self-labeling processes in mental illness: The role of emotional deviance." *American Journal of Sociology*, 91: 221–249.

______. 2010. "Stress and health: Major findings and policy implications." *Journal of Health and Social Behavior*, 51: S41–S53.

Timmermans, Stefan. 1999. *Sudden Death and the Myth of CPR*. Philadelphia: Temple University Press.

Timmermans, Stefan, and Marc Berg. 2003a. *The Gold Standard: The Challenge of Evidence-Based Medicine and Standardization in Health Care*. Philadelphia: Temple University Press.

______. 2003b. "The practice of medical technology." *Sociology of Health and Illness*, 25: 97–114.

______. 2005. "Death brokering: Constructing culturally appropriate deaths." *Sociology of Health and Illness*, 27: 993–1013.

Timmermans, Stefan, and Mara Buchbinder. 2010. "Patients-in-waiting: Living between sickness and health in the genomics era." *Journal of Health and Social Behavior*, 51: 408–423.

Timmermans, Stefan, and Hyeyoung Oh. 2010. "The continued social transformation of the medical profession." *Journal of Health and Social Behavior*, 51: S94–S106.

Tindle, Hilary A., Roger B. Davis, Russell S. Phillips, and David M. Eisenberg.

2005. "Trends in use of complementary and alternative medicine by U.S. adults: 1997–2002." *Alternative Therapies*, 11: 42–49.

Tjaden, Patricia, and Nancy Thoennes. 2000. "Extent, nature, and consequences of intimate partner violence." *National Institute of Justice Research Report.*

Tomashek, Kay M., Cheng Qin, Jason Hsia, Solomon Iyasu, Wanda D. Barfield, and Lisa M. Flowers. 2006. "Infant mortality trends and differences between American Indian/Alaska Native Infants and White infants in the United States, 1989–1991 and 1998–2000." *American Journal of Public Health*, 96: 2222–2227.

Toole, Michael J., and Ronald J. Waldman. 1993. "Refugees and displaced persons: War, hunger, and public health." *Journal of the American Medical Association*, 270: 600–605.

Toscano, Vicki. 2005. "Misguided retribution: Criminalization of pregnant women who take drugs." *Social Legal Studies*, 14: 359–386.

Trent, James W. "Sterilization." *Encyclopedia of Disability*. 2005. SAGE Publications. http://www.sage-ereference.com/disability/Article_n746.html, accessed November 2010.

Trials of War Criminals before the Nuremberg Military Tribunals under Control Council Law No. 10. 1949. Vol. 2, Pp. 181–182. Washington, DC: US Government Printing Office.

Tu, Jack V., Chris L. Pashos, David Naylor, Erluo Chen, Sharon-Lise Normand, Joseph P. Newhouse, and Barbara J. McNeil. 1997. "Use of cardiac procedures and outcomes in elderly patients with myocardial infarction in the United States and Canada." *New England Journal of Medicine*, 336: 1500–1505.

Turner, Erick H., Annette M. Matthews, Eftihia Linardatos, Robert A. Tell, and Robert Rosenthal. 2008. "Selective publication of antidepressant trials and its influence on apparent efficacy." *New England Journal of Medicine*, 358: 252–260.

Turner, R. Jay, and William R. Avison. 2003. "Status variations in stress exposure: Implications for the interpretation of research on race, socioeconomic status, and gender." *Journal of Health and Social Behavior*, 44: 488–505.

Turner, R. Jay, and Robyn Lewis Brown. 2010. "Social support and mental health." Pp. 200–212 in *Handbook for the Sociology of Mental Health,* 2nd ed., edited by Teresa L. Scheid and Tony N. Brown. New York: Cambridge University Press.

Twardowski, Linda, Twila M. C. Innis, Carleton C. Cappuccino, James M. Donald, and Jason Rhodes. 2014. "Professional responsibilities for treatment of patients with Ebola: Can a healthcare provider refuse to treat a patient with Ebola? *Public Health Briefing*, Rhode Island Department of Health. http://health.ri.gov/publications/briefs/20141007ProfessionalResponsibilitiesForTreatmentOfPatientsWithEbola.pdf, accessed November 2014.

UNAIDS/WHO. 2010. *AIDS Epidemic Update*. Geneva, Switzerland: Author.

Underwood, Felix J. 1926. "Development of midwifery in Mississippi." *Southern Medical Journal*, 19: 683–685.

UNICEF. 2005. "Press release: Poor feeding for children under two leads to nearly one-fifth of child deaths." http://www.unicef.org/media/media_27814.html, accessed March 2008.

______. 2010. *Fact Sheet: Child Soldiers*. http://www.unicef.org/emerg/files/childsoldiers.pdf, accessed October 2010.

______. 2013. *Female Genital Mutilation/Cutting: A Statistical Overview*. New York: United Nations.

______. 2014. *Committing to Child Survival: A Promise Renewed*. New York: United Nations.

US Bureau of the Census. 1975. *Historical Statistics of the United States, Colonial Times to 1970*. Washington, DC: US Government Printing Office.

______. 2013. *Statistical Abstract of the United States*. Washington, DC: US Government Printing Office.

US Congressional Budget Office. 2014. *Effects of the Affordable Care Act on Health Insurance Coverage—Baseline Projections*. http://www.cbo.gov/sites/default/files/cbofiles/attachments/43900-2014-04-ACAtables2.pdf, accessed December 2014.

US Department of Health and Human Services. 2010. *2009 OPTN/SRTR Annual Report*. http://www.srtr.org/annual_reports/archives/2009/2009_Annual_Report/, accessed November 2010.

US Environmental Protection Agency. 2008. *America's Children and the Environment*. http://www.epa.gov/envirohealth/children/index.htm, accessed March 2008.

Valenstein, Elliot S. 1986. *Great and Desperate Cures*. New York: Basic.

Van Olphen-Fehr, Juliana. 1998. *Diary of a Midwife: The Power of Positive Childbearing*. Westport, CT: Bergin & Garvey.

Vanderminden, Jennifer, and Sharyn J. Potter. 2010. "Challenges to the doctor–patient relationship in the twenty-first century." Pp. 355–372 in *New Blackwell Companion to Medical Sociology*, edited by William C. Cockerham. Chichester, UK: Wiley-Blackwell.

Vastag, Brian. 2001. "Pay attention: Ritalin acts much like cocaine." *Journal of the American Medical Association*, 286: 905–906.

Vega, William A., Bohdan Kolody, Sergio Aguilar-Gaxiola, Ethel Alderete, Ralph Catalano, and Jorge Caraveo-Anduaga. 1998. "Lifetime prevalence of DSM-III-R psychiatric disorders among urban and rural Mexican Americans in California." *Archives of General Psychiatry*, 55: 771–778.

Vesely, Rebecca. 2014. "The great migration." *Hospitals and Health Networks*. http://www.hhnmag.com/display/HHN-news-article.dhtml?dcrPath=/templatedata/HF_Common/NewsArticle/data/HHN/Magazine/2014/Mar/cover-story-great-migration, accessed November 2014.

Vincent, Norah. 2009. *Voluntary Madness: My Year Lost and Found in the Loony Bin*. New York: Viking.

Vuckovic, Nancy, and Mark Nichter. 1997. "Changing patterns of pharmaceutical practice in the United States." *Social Science & Medicine*, 44: 1285–1302.

Wade, Nicholas. 2010. "A decade later, genetic map yields few new cures." *New York Times*, June 12: A1+.

Waitzkin, Howard. 1991. *The Politics of Medical Encounters*. New Haven, CT: Yale University Press.

______. 1993. *The Second Sickness: Contradictions of Capitalist Health Care*. Rev. ed. New York: Free Press.

Waldman, H. Barry, Mary Rose Truhlar, and Steven P. Perlman. 2005. "Medicare Dentistry: The Next Logical Step." *Public Health Reports*, 120: 6–10.

Wang, Houli, Tengda Xu, and Jin Xu. 2007. "Factors contributing to high costs and inequality in China's health care system." *Journal of the American Medical Association*, 298: 1928–1930.

Wardwell, Walter I. 1979. "Limited and marginal practitioners." Pp. 230–250 in *Handbook of Medical Sociology*, edited by Howard E. Freeman, Sol Levine, and Leo G. Reeder. Englewood Cliffs, NJ: Prentice Hall.

______. 1988. "Chiropractors: Evolution to acceptance." Pp. 157–191 in *Other Healers: Unorthodox Medicine in America*, edited by Norman Gevitz. Baltimore: Johns Hopkins University Press.

Weil, Andrew, and Winifred Rosen. 1998. *From Chocolate to Morphine*, rev. ed. New York: Houghton Mifflin.

Weissert, William G. 1991. "A new policy agenda for home care." *Health Affairs*, 10: 67–77.

Weitz, Rose. 1991. *Life with AIDS*. New Brunswick, NJ: Rutgers University Press.

______. 1999. "Watching Brian die: The rhetoric and reality of informed consent." *Journal for the Social Study of Health, Illness and Medicine*, 3: 209–227.

Wennberg, John E. 2010. *Tracking Medicine: A Researcher's Quest to Understand Health Care*. New York: Oxford University Press.

Werner, Anne, and Kirsti Malterud. 2003. "'It is hard work behaving as a credible patient': Encounters between women with chronic pain and their doctors." *Social Science & Medicine*, 57: 1409–1419.

Wertz, Richard, and Dorothy Wertz. 1989. *Lying-In*. New Haven, CT: Yale University Press.

Whitaker, Robert. 2011. *Anatomy of an Epidemic: Magic Bullets, Psychiatric Drugs, and the Astonishing Rise of Mental Illness in America*. New York: Crown.

Whitehouse, Beth. 2010. *The Match: "Savior Siblings" and One Family's Battle to Heal Their Daughter*. Boston: Beacon Press.

Whitehouse, Peter J., Eric Juengst, Maxwell Mehlman, and Thomas H. Murray. 1997. "Enhancing cognition in the intellectually intact." *Hastings Center Report*, 27(3): 14–22.

WHO/UNICEF. 2014. Progress on drinking water and sanitation. Geneva, Switzerland: World Health Organization, Joint Monitoring Programme for Water Supply and Sanitation.

Wicclair, Mark R., and Douglas B. White. 2014. "Surgeons, intensivists, and discretion to refuse requested treatments." *Hastings Center Report*, 44(5): 33–42.

Wilkinson, Richard G. 1996. *Unhealthy Societies: The Afflictions of Inequality*. London: Routledge.

______. 2005. *The Impact of Inequality*. London: New Press.

Williams, David R., and Pamela Braboy Jackson. 2005. "Social sources of racial disparities in health." *Health Affairs*, 24: 325–335.

Williams, D. R., S. A. Mohammed, J. Leavell, and C. Collins. 2010. "Race, socioeconomic status, and health: Complexities, ongoing challenges, and research opportunities." *Annals of the New York Academy of Sciences*, 1186: 69–101.

Williams, David R., and Michelle Sternthal. 2010. "Understanding racial-ethnic disparities in health: Sociological contributions. *Journal of Health and Social Behavior*, 51: S15–S27.

Wilson, Duff. 2010a. "Child's ordeal shows risks of psychosis drugs for young." *New York Times*, September 1: A1+.

______. 2010b. "Side effects may include lawsuits." *New York Times*, October 3: BY 1+.

Wilt, Timothy J., Tatyana Shamliyan, Brent Taylor, Roderick MacDonald, James Tacklind, Indulis Rutks, Kenneth Koeneman, Chin-Soo Cho, and Robert L. Kane. 2008.

"Comparative effectiveness of therapies for clinically localized prostate cancer." *Comparative Effectiveness Review*, No. 13. Rockville, MD: Agency for Healthcare Research and Quality.

Wingood, Gina M., and Ralph J. DiClemente. 1997. "The effects of an abusive primary partner on the condom use and sexual negotiation practices of African-American women." *American Journal of Public Health*, 87: 1016–1018.

Wirth, Louis. 1985. "The problem of minority groups." Pp. 309–315 in *Theories of Society: Foundations of Modern Sociological Theory*, edited by T. Parsons, E. Shils, K. D. Naegele, and J. R. Pitts. New York: Free Press.

Wolpe, Paul Root. 1985. "Acupuncture and the American physician." *Social Problems*, 32: 409–424.

Women Physicians' Congress, American Medical Association. 2010. *AMA Statistics History*. http://www.ama-assn.org/ama/pub/about-ama/our-people/member-groups-sections/women-physicians-congress/statistics-history/table-4-womenresidents-specialty-2005.shtml, accessed November 2010.

World Bank.1998. *Assessing Aid: What Works, What Doesn't, and Why*. New York: Oxford University Press.

World Health Organization. 2003. *Fact Sheet No. 134: Traditional Medicine*. Geneva, Switzerland: Author.

______. 2009a. *WHO Report on the Global AIDS Epidemic*. Geneva, Switzerland: Author.

______. 2009b. *Fact Sheet No. 330: Diarrhoeal disease*. Geneva, Switzerland: Author.

______. 2010. *World Health Statistics 2010*. http://www.who.int/whosis/whostat/2010/en/, accessed October 2010.

______. 2014a. *Fact Sheet No. 339:* Tobacco. Geneva, Switzerland: Author.

______. 2014b. *Top Ten Causes of Death*. http://www.who.int/mediacentre/factsheets/fs310/en/index1.html, accessed October 2014.

______. 2014c. *Fact Sheet No. 104: Tuberculosis*. Geneva, Switzerland: Author.

World Health Organization, Global Alert and Response, 2014. 2014. "Ebola Response." http://www.who.int/csr/disease/ebola/en, accessed September 10, 2014.

World Health Organization Regional Office for Europe. 2006. *Highlights on Health, Russian Federation—2005*. http://www.euro.who.int/eprise/main/WHO/Progs/CHHRUS/cismortality/20051202_2, accessed June 2008.

World Health Organization, Reproductive Health and Research Department. 2004. *Low Birthweight: National, Regional, and Global Estimates*. Geneva, Switzerland: Author.

______. 2013. *Global and Regional Estimates of Violence against Women*. Geneva, Switzerland: Author.

Yach, Derek, Corinna Hawkes, C. Linn Gould, and Karen J. Hofman. 2004. "The global burden of chronic diseases: Overcoming impediments to prevention and control." *Journal of the American Medical Association*, 291: 2616–2622.

Yardley, Jim. 2005. "A deadly fever, once defeated, lurks in a Chinese lake." *New York Times*, February 22: A2.

Zezima, Katie, and Abby Goodnough. 2010. "Drug testing poses quandary for employers." *New York Times*, October 25: A1+.

Zhu, Wei Xing, Li Lu, and Therese Hesketh. 2009. "China's excess males, sex selective abortion, and one child policy: Analysis of data from 2005 national intercensus survey." *British Medical Journal*, 338: b1211.

Zimmerman, Mary K. 1993. "Caregiving in the welfare state: Mothers' informal health care work in Finland." *Research in the Sociology of Health Care*, 10: 193–211.

Ziporyn, Terra D. 1992. *Nameless Diseases*. New Brunswick, NJ: Rutgers University Press.

Zupan, Mark, and Tim Swanson. 2006. *Gimp*. New York: HarperCollins.

Zussman, Robert. 1992. *Intensive Care: Medical Ethics and the Medical Profession*. Chicago: University of Chicago Press.

______. 1997. "Sociological perspectives on medical ethics and decision-making." *Annual Review of Sociology*, 23: 171–189.

Zwarun, Lara. 2006. "Ten years and one master settlement agreement later: The nature and frequency of alcohol and tobacco promotion in televised sports, 2000 through 2002." *American Journal of Public Health*, 96: 1492–1497.

Index

Page numbers in *italic* indicate figures or tables. Page numbers in **bold** indicate glossary terms.